Mathematical Data Science with Applications in Business, Industry, and Medicine

Mathematical Data Science with Applications in Business, Industry, and Medicine

Guest Editors

**Arne Johannssen
Nataliya Chukhrova**

Basel • Beijing • Wuhan • Barcelona • Belgrade • Novi Sad • Cluj • Manchester

Guest Editors

Arne Johannssen
Harz University of Applied Sciences
Wernigerode
Germany

Nataliya Chukhrova
University of Southern Denmark
Odense
Denmark

Editorial Office
MDPI AG
Grosspeteranlage 5
4052 Basel, Switzerland

This is a reprint of the Special Issue, published open access by the journal *Mathematics* (ISSN 2227-7390), freely accessible at: https://www.mdpi.com/journal/mathematics/special_issues/OGV8RSFGNP.

For citation purposes, cite each article independently as indicated on the article page online and as indicated below:

Lastname, A.A.; Lastname, B.B. Article Title. *Journal Name* **Year**, *Volume Number*, Page Range.

ISBN 978-3-7258-2741-1 (Hbk)
ISBN 978-3-7258-2742-8 (PDF)
https://doi.org/10.3390/books978-3-7258-2742-8

© 2024 by the authors. Articles in this book are Open Access and distributed under the Creative Commons Attribution (CC BY) license. The book as a whole is distributed by MDPI under the terms and conditions of the Creative Commons Attribution-NonCommercial-NoDerivs (CC BY-NC-ND) license (https://creativecommons.org/licenses/by-nc-nd/4.0/).

Contents

About the Editors vii

Arne Johannssen, Nataliya Chukhrova
Mathematical Data Science with Applications in Business, Industry and Medicine
Reprinted from: *Mathematics* **2024**, *12*, 2756, https://doi.org/10.3390/math12172756 1

Hanan Haj Ahmad, Ehab Almetwally, and Dina A. Ramadan
Investigating the Relationship between Processor and Memory Reliability in Data Science: A Bivariate Model Approach
Reprinted from: *Mathematics* **2023**, *11*, 2142, https://doi.org/10.3390/math11092142 4

Gholamreza Hesamian, Arne Johannssen and Nataliya Chukhrova
A Three-Stage Nonparametric Kernel-Based Time Series Model Based on Fuzzy Data
Reprinted from: *Mathematics* **2023**, *11*, 2800, https://doi.org/10.3390/math11132800 27

Vasileios Alevizakos
Process Capability and Performance Indices for Discrete Data
Reprinted from: *Mathematics* **2023**, *11*, 3457, https://doi.org/10.3390/math11163457 44

Hamed Sabahno and Seyed Taghi Akhavan Niaki
New Machine-Learning Control Charts for Simultaneous Monitoring of Multivariate Normal Process Parameters with Detection and Identification
Reprinted from: *Mathematics* **2023**, *11*, 3566, https://doi.org/10.3390/ math11163566 66

Unarine Netshiozwi, Ali Yeganeh, Sandile Charles Shongwe and Ahmad Hakimi
Data-Driven Surveillance of Internet Usage Using a Polynomial Profile Monitoring Scheme
Reprinted from: *Mathematics* **2023**, *11*, 3650, https://doi.org/10.3390/math11173650 97

Eftychia Mamzeridou and Athanasios C. Rakitzis
A Combined Runs Rules Scheme for Monitoring General Inflated Poisson Processes
Reprinted from: *Mathematics* **2023**, *11*, 4671, https://doi.org/10.3390/math11224671 120

Ioannis S. Triantafyllou
Archimedean Copulas-Based Estimation under One-Parameter Distributions in Coherent Systems
Reprinted from: *Mathematics* **2024**, *12*, 334, https://doi.org/10.3390/math12020334 136

Walena Anesu Marambakuyana and Sandile Charles Shongwe
Composite and Mixture Distributions for Heavy-Tailed Data—An Application to Insurance Claims
Reprinted from: *Mathematics* **2024**, *12*, 335, https://doi.org/10.3390/math12020335 150

David Enck, Mario Beruvides, Víctor G. Tercero-Gómez and Alvaro E. Cordero-Franco
Addressing Concerns about Single Path Analysis in Business Cycle Turning Points: The Case of Learning Vector Quantization
Reprinted from: *Mathematics* **2024**, *12*, 678, https://doi.org/10.3390/math12050678 175

Aylin Pakzad, Ali Yeganeh, Rassoul Noorossana and Sandile Charles Shongwe
Process Capability Index for Simple Linear Profile in the Presence of Within- and Between-Profile Autocorrelation
Reprinted from: *Mathematics* **2024**, *12*, 2549, https://doi.org/10.3390/math12162549 190

About the Editors

Arne Johannssen

Arne Johannssen holds a Master's degree in Economics and a PhD in Statistics from the University of Hamburg, Germany. He is a Professor of Statistics and Data Analytics at Harz University of Applied Sciences, Germany. His research interests are Artificial Intelligence, Computational Statistics, Data Science, Soft Computing, and Statistical Learning with applications in Business and Economics. He authored 50 publications in renowned international journals, and especially in the top-tier journals of the disciplines "Artificial Intelligence" and "Statistics & Probability". His research is funded by the German Research Foundation (DFG), among others.

Nataliya Chukhrova

Nataliya Chukhrova is an Assistant Professor in Applied AI and Data Science at the Maersk Mc-Kinney Moller Institute, Faculty of Engineering, University of Southern Denmark (SDU). Her interests, in both research and teaching, are computational statistics, statistical inference, statistical learning, and artificial intelligence with applications in medicine and industry. Her research is funded by the DFG and the Federal Ministry of Education and Research (BMBF), among others. Nataliya has given numerous lectures on mathematics and statistics and has been awarded the Excellence in Higher Education Award.

Editorial

Mathematical Data Science with Applications in Business, Industry, and Medicine

Arne Johannssen [1,*] and Nataliya Chukhrova [2]

[1] Faculty of Business Studies, Harz University of Applied Sciences, 38855 Wernigerode, Germany
[2] Faculty of Engineering, University of Southern Denmark, 5230 Odense, Denmark; nach@mmmi.sdu.dk
* Correspondence: ajohannssen@hs-harz.de

Citation: Johannssen, A.; Chukhrova, N. Mathematical Data Science with Applications in Business, Industry and Medicine. *Mathematics* **2024**, *12*, 2756. https://doi.org/10.3390/math12172756

Received: 21 August 2024
Accepted: 26 August 2024
Published: 5 September 2024

Copyright: © 2024 by the authors. Licensee MDPI, Basel, Switzerland. This article is an open access article distributed under the terms and conditions of the Creative Commons Attribution (CC BY) license (https://creativecommons.org/licenses/by/4.0/).

Mathematical data science is a field that combines mathematical techniques with data science methods to extract insights and knowledge from data. It includes working with data at all stages of the data life cycle, from collection and storage to cleansing and processing, analysis and visualization of data, and communication of the results and findings. Data scientists use a variety of tools and techniques to analyze data, including mathematical concepts and models, artificial intelligence techniques, machine learning algorithms, statistical analysis, and data visualization. Data science can be used to make predictions, identify patterns, and draw conclusions from data, and it is applied in a variety of areas, including business, industry, and medicine. It is a rapidly evolving field, and data scientists are expected to stay up to date with new tools, techniques, and technologies.

We are pleased to present this Special Issue of *Mathematics*, titled "Mathematical Data Science with Applications in Business, Industry, and Medicine". This special issue brings together ten insightful papers that highlight innovative mathematical data science techniques in solving complex problems across various domains, including business, industry, and medicine. These publications are listed below, ordered by date of publication:

1. Haj Ahmad, H.; Almetwally, E.M.; Ramadan, D.A. Investigating the Relationship between Processor and Memory Reliability in Data Science: A Bivariate Model Approach. *Mathematics* **2023**, *11*, 2142. https://doi.org/10.3390/math11092142.
2. Hesamian, G.; Johannssen, A.; Chukhrova, N. A Three-Stage Nonparametric Kernel-Based Time Series Model Based on Fuzzy Data. *Mathematics* **2023**, *11*, 2800. https://doi.org/10.3390/math11132800.
3. Alevizakos, V. Process Capability and Performance Indices for Discrete Data. *Mathematics* **2023**, *11*, 3457. https://doi.org/10.3390/math11163457.
4. Sabahno, H.; Niaki, S.T.A. New Machine-Learning Control Charts for Simultaneous Monitoring of Multivariate Normal Process Parameters with Detection and Identification. *Mathematics* **2023**, *11*, 3566. https://doi.org/10.3390/math11163566.
5. Netshiozwi, U.; Yeganeh, A.; Shongwe, S.C.; Hakimi, A. Data-Driven Surveillance of Internet Usage Using a Polynomial Profile Monitoring Scheme. *Mathematics* **2023**, *11*, 3650. https://doi.org/10.3390/math11173650.
6. Mamzeridou, E.; Rakitzis, A.C. A Combined Runs Rules Scheme for Monitoring General Inflated Poisson Processes. *Mathematics* **2023**, *11*, 4671. https://doi.org/10.3390/math11224671.
7. Triantafyllou, I.S. Archimedean Copulas-Based Estimation under One-Parameter Distributions in Coherent Systems. *Mathematics* **2024**, *12*, 334. https://doi.org/10.3390/math12020334.
8. Marambakuyana, W.A.; Shongwe, S.C. Composite and Mixture Distributions for Heavy-Tailed Data—An Application to Insurance Claims. *Mathematics* **2024**, *12*, 335. https://doi.org/10.3390/math12020335.
9. Enck, D.; Beruvides, M.; Tercero-Gómez, V.G.; Cordero-Franco, A.E. Addressing Concerns about Single Path Analysis in Business Cycle Turning Points: The Case of

Learning Vector Quantization. *Mathematics* **2024**, *12*, 678. https://doi.org/10.3390/math12050678.
10. Pakzad, A.; Yeganeh, A.; Noorossana, R.; Shongwe, S.C. Process Capability Index for Simple Linear Profile in the Presence of Within- and Between-Profile Autocorrelation. *Mathematics* **2024**, *12*, 2549. https://doi.org/10.3390/math12162549.

The remainder of this editorial contains a summary of the contributions to this special issue.

Haj Ahmad et al. (1.) introduce a bivariate model using a copula function to model the failure times of processors and memories in computers. Properties of the model are analyzed, and inferential statistics for the parameters are performed under the assumption of a Type-II censored sampling scheme. The parameters are estimated by both maximum likelihood and Bayesian estimation methods, and the estimations are compared. The model's efficiency was validated with real data, demonstrating its excellent fit compared to other bivariate models.

Hesamian et al. (2.) present a nonlinear time series model tailored for data reported as *LR* fuzzy numbers, using a three-stage nonparametric kernel-based estimation procedure. The Nadaraya–Watson estimator is employed in each stage to estimate the center and spreads of the fuzzy smooth function, with a hybrid algorithm determining the optimal bandwidths and autoregressive order. The model's effectiveness is validated through a simulation study and real-life applications, demonstrating its improved performance over other fuzzy time series models.

The paper from Alevizakos (3.) focuses on adapting classical process capability indices (PCIs) and performance indices for discrete data following Poisson, binomial, or negative binomial distributions using various transformation techniques. The methodology simplifies the computation of these indices, allowing for a straightforward assessment of process capability and performance. The paper includes a simulation study comparing these indices with existing ones for discrete data and provides three examples to demonstrate the practical application of the transformation techniques.

Sabahno and Niaki (4.) propose three control charts based on artificial neural networks, support vector machines, and random forests for monitoring process parameters in multivariate normal processes. These tools are designed to classify processes as in-control or out-of-control variables and are evaluated under different input scenarios, training methods, and process control objectives, including detection and identification of out-of-control variables. The developed control charts demonstrate very good performance compared to traditional memory-less statistical control charts, with an illustrative example provided for application in a healthcare process.

Netshiozwi et al. (5.) study a data-driven monitoring framework based on profile monitoring to assess internet usage patterns in a telecom company. By defining a polynomial model between the hours of each day and internet usage, the framework aims to detect unnatural patterns, assess the impact of policies like discounts, and investigate social behavior variations in usage. The study compares charting statistics, showing that the MEWMA scheme outperforms Hoteling T^2 chart in detecting small shifts, offering quicker detection capabilities.

Mamzeridou and Rakitzis (6.) propose a Shewhart-type control chart with multiple runs rules to monitor inflated processes, particularly when the process mean level is very low, making it difficult to detect decreases in the mean. The chart is supplemented with two runs rules: one for detecting decreases in the process mean and another for enhancing sensitivity to small and moderate increases in the mean. Using the Markov chain method, the study evaluates the performance of these schemes, showing their effectiveness in detecting shifts in process parameters, with practical applications demonstrated through two examples based on healthcare data.

Triantafyllou (7.) deals with a signature-based framework for estimating the mean lifetime and variance of a continuous distribution in a coherent system with exchangeable components. The dependency among components is modeled using Archimedean multivariate copulas, specifically the Frank and Joe copulas. Numerical experiments are conducted to illustrate the methodology across all possible coherent systems with three components.

Marambakuyana and Shongwe (8.) analyze two-component non–Gaussian composite and mixture models for insurance claims data, emphasizing their flexibility in curve fitting. A total of 256 composite and 256 mixture models, derived from 16 popular parametric distributions, are evaluated. The study applies these models to two real insurance datasets, using model selection criteria and risk metrics to identify the top 20 models in each category. Results indicate that composite models offer superior risk estimates compared to mixture models for both datasets.

Enck et al. (9.) highlight the limitations of single path analysis (SPA) in economic analysis, particularly in identifying business cycle (BC) turning points using machine learning. SPA fails to account for temporal dependence in BCs, which can lead to inadequate evaluation and calibration of algorithms. Using learning vector quantization (LVQ) as a case study, the study employs a multivariate Monte Carlo simulation to incorporate change points, autocorrelations, and cross-correlations, offering a more robust understanding of LVQ's uncertainties. The results reveal the shortcomings of SPA and underscore the importance of considering temporal dependence to improve the robustness of data-driven approaches in economic and financial analysis.

The paper from Pakzad et al. (10.) addresses the issue of autocorrelation in simple linear profiles (SLPs) when assessing process capability, which is often overlooked in traditional PCIs. The study introduces novel methods to evaluate SLP capability under within-profile, between-profile, and simultaneous autocorrelation, using an AR(1) model. A new functional index is proposed and modified to account for these autocorrelation effects, with simulation results showing improved performance in terms of bias and mean square error. Bootstrap confidence intervals are provided, and an illustrative example from the chemical industry demonstrates the method's practical applicability.

In conclusion, the papers presented in this Special Issue highlight the crucial role of mathematical data science in tackling complex problems across various fields such as business, industry, and medicine. As the landscape of data science continues to evolve, the integration of advanced mathematical techniques with emerging technologies will be pivotal in driving innovation. Future research should focus on enhancing the scalability and adaptability of these methods to address increasingly diverse and large-scale datasets. We anticipate that the insights and methodologies discussed in this Special Issue will inspire further exploration and contribute to the ongoing advancement of mathematical data science, ultimately leading to more robust solutions and transformative applications across different domains.

Conflicts of Interest: The authors declare no conflicts of interest.

Disclaimer/Publisher's Note: The statements, opinions and data contained in all publications are solely those of the individual author(s) and contributor(s) and not of MDPI and/or the editor(s). MDPI and/or the editor(s) disclaim responsibility for any injury to people or property resulting from any ideas, methods, instructions or products referred to in the content.

Article

Investigating the Relationship between Processor and Memory Reliability in Data Science: A Bivariate Model Approach

Hanan Haj Ahmad [1,*], Ehab M. Almetwally [2] and Dina A. Ramadan [3]

[1] Department of Basic Science, Preparatory Year Deanship, King Faisal University, Hofuf 31982, Al Ahsa, Saudi Arabia
[2] Faculty of Business Administration, Delta University for Science and Technology, Gamasa 11152, Egypt
[3] Department of Mathematics, Faculty of Science, Mansoura University, Mansoura 35516, Egypt
* Correspondence: hhajahmed@kfu.edu.sa

Abstract: Modeling the failure times of processors and memories in computers is crucial for ensuring the reliability and robustness of data science workflows. By understanding the failure characteristics of the hardware components, data scientists can develop strategies to mitigate the impact of failures on their computations, and design systems that are more fault-tolerant and resilient. In particular, failure time modeling allows data scientists to predict the likelihood and frequency of hardware failures, which can help inform decisions about system design and resource allocation. In this paper, we aimed to model the failure times of processors and memories of computers; this was performed by formulating a new type of bivariate model using the copula function. The modified extended exponential distribution is the suggested lifetime of the experimental units. It was shown that the new bivariate model has many important properties, which are presented in this work. The inferential statistics for the distribution parameters were obtained under the assumption of a Type-II censored sampling scheme. Therefore, point and interval estimation were observed using the maximum likelihood and the Bayesian estimation methods. Additionally, bootstrap confidence intervals were calculated. Numerical analysis via the Markov Chain Monte Carlo method was performed. Finally, a real data example of processors and memories failure time was examined and the efficiency of the new bivariate distribution of fitting the data sample was observed by comparing it with other bivariate models.

Keywords: data science; computer processor; bivariate models; copula; modified extended exponential; Markov Chain Monte Carlo; maximum likelihood estimation; Bayesian estimation; bootstrap; simulation

MSC: 62F10; 62F15; 62N05; 62N02

Citation: Haj Ahmad, H.; Almetwally, E.M.; Ramadan, D.A. Investigating the Relationship between Processor and Memory Reliability in Data Science: A Bivariate Model Approach. *Mathematics* 2023, 11, 2142. https://doi.org/10.3390/math11092142

Academic Editors: Arne Johannssen and Nataliya Chukhrova

Received: 19 March 2023
Revised: 21 April 2023
Accepted: 30 April 2023
Published: 3 May 2023

Copyright: © 2023 by the authors. Licensee MDPI, Basel, Switzerland. This article is an open access article distributed under the terms and conditions of the Creative Commons Attribution (CC BY) license (https://creativecommons.org/licenses/by/4.0/).

1. Introduction

The guide to designing and elaborating a reliable system is to know the kind of failures and errors that may happen during operation. One way to do this is to study the failures occurring in the observed production systems. System planners and data center developers can use the ideas obtained from such a study to improve the efficiency of the system and increase the resistance to failure for future systems by determining some issues and developing stronger techniques, operational policies, and application designs. Modeling failure time can also improve the process of error correction techniques and other mitigation strategies. For instance, if a particular type of memory failure occurs frequently, data scientists may develop algorithms that can detect and correct errors in real time or implement hardware redundancy to ensure continuity of processing. Overall, modeling failure time is an essential aspect of ensuring the reliability of data science workflows. By accounting for the potential failures of hardware components, data scientists can develop more robust and resilient systems that are better equipped to handle the challenges of large-scale data processing and analysis.

One way to measure chip reliability and its predicted lifetime is by using statistical analysis tools, which will estimate the lifetime of a processor for a large batch of manufactured chips.

Bivariate models are used to model the relationship between two variables. They have a wide range of applications in different fields such as finance, economics, technology, engineering, medicine, and many more. They allow the separation of the marginal distributions of the variables from the dependence structure between them. This makes it a flexible method for modeling complex distributions, especially in multivariate contexts. Many authors have proposed and studied bivariate models and shown their broad applications in many fields of science. In the last decade, many studies have discussed the idea of generating new bivariate models based on a multiform copula function. A copula function is an effective tool for modeling the dependency structure of variables, regardless of their marginals. They have a rich class of dependence models, such as Gaussian, t, Clayton, and Gumbel copulas, which can be used to model an immense range of dependence structures. Copulas are defined over the unit square. By using a copula function, it is possible to generate random numbers with a specified dependence structure, see Nelsen [1].

Flores [2] discussed various bivariate Weibull models based on different copula functions, such as the Farlie–Gumbel–Morgenstern, Clayton, Gumbel–Hougaard, Ali–Mikhail–Haq, and Gumbel–Barnett copulas. A new bivariate Gaussian–Weibull model was proposed by Verrill et al. [3]. Later on, many authors focused their statistical analysis on bivariate models; see [4] who developed a bivariate generalized Rayleigh distribution depending on the Clayton copula function, [5] who introduced Bivariate power Lomax distribution and [6] created a new bivariate Fréchet distribution using the Farlie–Gumbel–Morgenstern and the Ali–Mikhail–Haq copulas. A bivariate Weibull distribution was proposed by [7] based on the FGM copula function. In [8], the authors studied some families of bivariate Kumaraswamy distribution using different copulas. Others used the Marshall–Olkin bivariate copulas; one may refer to ([9–18]). More recent work on bivariate models with copula functions with applications can be found in [19,20], among others.

Along with the method for estimating parameters, it is also worth noting that the selection depends on the specific application and the assumptions, robustness, and availability of the data. It is necessary to evaluate the performance of each method by using the appropriate statistical measures and possibly comparing it with other methods to confirm the results.

Yet, there are still many important issues in some practical cases, where the classical bivariate models do not support a suitable fit to the data from real-life experiments. Accordingly, there is an extensive need to develop new flexible bivariate distributions. One of the main motivations for using a bivariate distribution with copulas is its ability to capture the non-linear and asymmetric dependencies between variables. Copulas provide a rich class of dependence models, which can be used to model many cases of dependence structures. In addition, copulas facilitate the computation of various probability measures, such as the probability of a joint event, the probability of one event conditional on another, and the probability of exceeding certain thresholds. Another important motivation is the ability to use copulas for risk assessment and portfolio optimization. The flexibility of copulas in modeling the dependence structure makes it a suitable tool for financial and insurance applications, where the non-linear and asymmetric dependencies between variables are often present.

Sklar [21] found the probability density function (pdf) and the cumulative distribution function (CDF) for the two-dimension copula function. As a result, if X and Y are two random variables for the distribution functions $F(x)$ and $F(y)$, respectively, the joint CDF and pdf for bivariate copula are given as

$$F(x,y) = C(F(x;\Lambda_1), F(y;\Lambda_2)), \qquad (1)$$

and
$$f(x,y) = f(x; \Lambda_1) f(y; \Lambda_2) c(F(x; \Lambda_1), F(y; \Lambda_2)). \tag{2}$$

Different kinds of copula functions have been defined based on the above equations, such as the Farlie–Gumbel–Morgenstern copula. This copula is a well-known parametric family of the copula; this family was first introduced by Gumbel [22]. The joint CDF and pdf of the Farlie–Gumbel–Morgenstern copula are given respectively as

$$C(\tau, \nu) = \tau \nu \{1 + \lambda \left[(1 - \tau)(1 - \nu)\right]\}, \tag{3}$$

$$c(\tau, \nu) = 1 + \lambda(1 - 2\tau)(1 - 2\nu), \tag{4}$$

where $\tau = F(x; \Lambda_1)$, $\nu = F(y; \Lambda_2)$, and Λ_1 and Λ_2 are vectors of the parameters for the variables X and Y, respectively. λ represents the copula parameter and it has a value in the range $[0, 1]$.

In this work, we formulated a bivariate model based on a new extension of the modified exponential lifetime, see El-Damcese and Ramadan [23]. This new extension of the exponential distribution is called the modified extended exponential (MExE) distribution. This new model has three parameters and offers great flexibility to the density function; it also has an increasing failure rate and increasing reliability function. Many of its properties and inferential analysis were discussed in [24–26], where in the latter, the estimation under progressively hybrid Type-II and progressive hybrid Type-I censored samples were considered and applied to fit some mechanical data. The attractive properties of the MExE that have been discussed in the literature encourage us to use it as a bivariate model as we expect to fit many real data from different fields of science.

The suggested bivariate distribution is based on the Farlie–Gumbel–Morgenstern copula function, which is applied to the modified extended exponential distribution, and it is denoted by bivariate modified extended exponential (BMExE) distribution. Statistical properties for this distribution are discussed. Point and interval estimation for the parameters was performed using the maximum likelihood estimation and the Bayesian estimation methods, where a numerical technique such as the Metropolis–Hastings algorithm was utilized for evaluating estimators. Three types of confidence intervals to estimate the unknown parameters were considered, namely, asymptomatic, Bayesian credible, and bootstrap intervals. Some measures of goodness-of-fit were used to fit the bivariate model such as the Akaike information criterion (AIC), the Kolmogorov–Smirnov test, the corrected Akaike information criterion (CAIC), the Bayesian information criterion (BIC), and the Hannan–Quinn information criterion (HQIC). The real data examples show the suitability of the suggested distribution compared with some competitive traditional distributions.

The remaining parts of this paper appear as follows: the BMExE distribution is defined in Section 2. Statistical characteristics of the BMExE distribution are presented in Section 3, while the maximum likelihood estimation and the Bayesian estimation were performed in Section 4. In Section 5, confidence intervals are discussed. Section 6 shows the suitability of the new model, which was illustrated by a simulation study. In Section 7, the applications of computers' processors and memories of real data units are discussed. Finally, the conclusion and some remarks for the BMExE distribution are presented in Section 8.

2. The Bivariate Modified Extended Exponential Model

The modified extended exponential distribution was first studied by El-Damcese and Ramadan [23], in which the pdf and CDF are given, respectively, by:

$$f(x; \theta, \sigma, \delta) = \theta(\delta + 2\sigma x)(1 + \delta x + \sigma x^2)^{\theta-1} e^{[1-(1+\delta x+\sigma x^2)^\theta]}, \quad x > 0, \theta, \sigma, \delta > 0 \tag{5}$$

and

$$F(x; \theta, \sigma, \delta) = 1 - e^{[1-(1+\delta x+\sigma x^2)^\theta]}, \quad x > 0, \theta, \sigma, \delta > 0. \tag{6}$$

The reliability and hazard functions of MEED are given by

$$S(x;\theta,\sigma,\delta) = e^{[1-(1+\delta x+\sigma x^2)^\theta]}, \quad x \geq 0, \theta, \sigma, \delta > 0 \tag{7}$$

and

$$h(x;\theta,\sigma,\delta) = \theta(\delta + 2\sigma x)(1+\delta x + \sigma x^2)^{\theta-1}, \quad x \geq 0, \theta, \sigma, \delta > 0, \tag{8}$$

respectively.

According to Equations (1) and (2), and using the copula defined in Equations (3) and (4) with X and Y are two random variables following the MExE function as in Equations (5) and (6); the joint pdf and CDF of the bivariate MExE distribution with the Farlie–Gumbel–Morgenstern copula function are given as follows:

$$f(x,y) = \theta_1\theta_2(\delta_1 + 2\sigma_1 x)(\delta_2 + 2\sigma_2 y)(1+\delta_1 x + \sigma_1 x^2)^{\theta_1-1}(1+\delta_2 y + \sigma_2 y^2)^{\theta_2-1}e^{[u(x)+v(y)]} \\ \left[1 + \lambda\left(2e^{u(x)} - 1\right)\left(2e^{v(y)} - 1\right)\right], \tag{9}$$

and

$$F(x,y;\theta,\sigma,\delta) = \left(1-e^{u(x)}\right)\left(1-e^{v(y)}\right)\left[1+\lambda e^{u(x)}e^{v(y)}\right], \quad x > 0, \theta_1, \theta_2, \sigma_1, \sigma_2, \delta_1, \delta_2 > 0,$$

where $u(x) = 1 - e^{[1-(1+\delta_1 x+\sigma_1 x^2)^{\theta_1}]}$ and $v(y) = 1 - e^{[1-(1+\delta_2 y+\sigma_2 y^2)^{\theta_2}]}$.

3. Properties of the New Model

In this section, various important statistical properties were studied and are presented for the new bivariate model.

3.1. Marginal and Conditional Distributions

Let X and Y be two random variables following the MExE distribution, then the marginal density functions are given by

$$f_1(x;\theta_1,\sigma_1,\delta_1) = \theta_1(\delta_1 + 2\sigma_1 x)(1+\delta_1 x + \sigma_1 x^2)^{\theta_1-1}e^{[1-(1+\delta_1 x+\sigma_1 x^2)^{\theta_1}]}, \quad x > 0, \theta_1, \sigma_1, \delta_1 > 0 \tag{10}$$

and

$$f_2(y;\theta_2,\sigma_2,\delta_2) = \theta_2(\delta_2 + 2\sigma_2 y)(1+\delta_2 x + \sigma_1 x^2)^{\theta_1-1}e^{[1-(1+\delta_2 y+\sigma_2 y^2)^{\theta_2}]}, \quad y > 0, \theta_2, \sigma_2, \delta_2 > 0, \tag{11}$$

respectively. For the conditional pdf of X given Y, it can be written as

$$f(x \mid y) = \theta_1(\delta_1 + 2\sigma_1 x)(1+\delta_1 x + \sigma_1 x^2)^{\theta_1-1}(1-u(x))(1+\lambda - 2\lambda v(y)) + (2\lambda u(x)(2v(y)-1)),$$

where $u(x) = 1 - e^{[1-(1+\delta_1 x+\sigma_1 x^2)^{\theta_1}]}$ and $v(y) = 1 - e^{[1-(1+\delta_2 y+\sigma_2 y^2)^{\theta_2}]}$.

Additionally, we can define the conditional CDF of X given Y as:

$$F(x \mid y) = u(x)[(1+\lambda - 2\lambda v(y)) + \lambda u(x)(2v(y)-1)].$$

On the other hand, the conditional pdf of Y given X is given as:

$$f(y \mid x) = \theta_2(\delta_2 + 2\sigma_2 y)(1+\delta_2 x + \sigma_1 x^2)^{\theta_1-1}(1-v(y))[(1+\lambda - 2\lambda u(x)) + (2\lambda v(y)(2u(x)-1))].$$

The conditional CDF of Y given X is obtained as:

$$F(y \mid x) = v(y)[(1+\lambda - 2\lambda u(x)) + \lambda v(y)(2u(x)-1)]. \tag{12}$$

A bivariate sample can be generated from the MExE distribution by using the conditional way as summarized below:

- Generate independent variables W and Z from uniform (0,1) distribution.

- Let $X = Q_{MExE}(W) = \frac{1}{2\sigma_1} \sqrt{4\sigma_1(1 - \ln(1-W))^{\frac{1}{\theta_1}} + (\delta_1^2 - 4\sigma_1)} - \delta_1$.
- Let $F(y|x) = Z$ in Equation (12) to find y by a numerical method such as Newton–Raphson.
- To obtain (x_i, y_i), $i = 1, ..., n$, repeat the above steps n times.

3.2. Product Moments

If the random variables X and Y are distributed as a BMExE distribution, then its mth and l^{th} moments about zero can be obtained as follows:

$$\mu'_{m,l} = E(x^m y^l) = \int_0^\infty \int_0^\infty x^m y^l f(x,y) dx dy.$$

Then,

$$\mu'_{m,l} = \sum_{r=i=j=0}^{\infty} \sum_{n=s=0}^{\infty} \sum_{k=d=0}^{\infty} \Omega_{m,l}\, \Phi_{m,l}\, \lambda^r(i+1)^{\frac{-s}{\theta_1}-1} (j+1)^{\frac{-d}{\theta_2}-1} \phi_l(x) \Gamma\left(\frac{d}{\theta_1}+1, j+1\right) \Psi_m(y) \Gamma\left(\frac{s}{\theta_2}+1, i+1\right), \quad (13)$$

where $\theta_1, \sigma_1, \delta_1, \theta_2, \sigma_2, \delta_2 > 0$, $\Omega_{m,l} = \binom{1}{r}\binom{r}{i}\binom{r}{j}\binom{l}{n}\binom{\frac{r-n}{2}}{s}\binom{m}{k}\binom{\frac{m-k}{2}}{d}$,

$\Phi_{m,l} = (-1)^{n+k} 2^{2s+2d-l-m+i+j} e^{i+j+2}$, $\phi_l(x) = \delta_1^n \sigma_1^{s-l} (\delta_1^2 - 2\sigma_1)^{\frac{l-n-2s}{2}}$ and

$\Psi_m(y) = \delta_2^k \sigma_2^{d-m} (\delta_2^2 - 2\sigma_2)^{\frac{m-k-2d}{2}}$.

The above result can be obtained using some integration techniques to obtain the needed result.

3.3. Moment Generating Function

Let X and Y be random variables with the pdf presented in Equation (9). Then, the moment generating function ($\mathcal{MGF}$) of X, Y is given by

$$M_{x,y}(t_1, t_2) = \int_0^\infty \int_0^\infty e^{t_1 x} e^{t_2 y} f(x,y) dx\, dy,$$

By using the exponential series expansion $e^t = \sum_{r=0}^\infty \frac{t^r}{r!}$, we observe

$$M_{x,y}(t_1, t_2) = \sum_{m=0}^\infty \sum_{l=0}^\infty \frac{t_1^m t_2^l}{m!\, l!} \int_0^\infty \int_0^\infty x^m y^l f(x,y) dx\, dy$$

$$= \sum_{m=0}^\infty \sum_{l=0}^\infty \frac{t_1^m t_2^l}{m!\, l!} \mu'_{m,l}.$$

Using the product moment defined in the previous section in Equation (13), the $\mathcal{MGF}$ of X, Y is written as

$$M_{x,y}(t_1, t_2) = \sum_{m=l=0}^\infty \sum_{r=i=j=0}^\infty \sum_{n=s=0}^\infty \sum_{k=d=0}^\infty \frac{t_1^m t_2^l}{m!\, l!} \Omega_{m,l}\, \Phi_{m,l}\, \lambda^r(i+1)^{\frac{-s}{\theta_1}-1} (j+1)^{\frac{-d}{\theta_2}-1} \phi_l(x)$$

$$\Gamma\left(\frac{d}{\theta_1}+1, j+1\right) \Psi_m(y) \Gamma\left(\frac{s}{\theta_2}+1, i+1\right); \theta_1, \sigma_1, \delta_1, \theta_2, \sigma_2, \delta_2 > 0,$$

where $\Omega_{m,l} = \binom{1}{r}\binom{r}{i}\binom{r}{j}\binom{l}{n}\binom{\frac{r-n}{2}}{s}\binom{m}{k}\binom{\frac{m-k}{2}}{d}$, $\Phi_{m,l} = (-1)^{n+k} 2^{2s+2d-l-m+i+j} e^{i+j+2}$,

$\phi_l(x) = \delta_1^n \sigma_1^{s-l} (\delta_1^2 - 2\sigma_1)^{\frac{l-n-2s}{2}}$ and $\Psi_m(y) = \delta_2^k \sigma_2^{d-m} (\delta_2^2 - 2\sigma_2)^{\frac{m-k-2d}{2}}$.

3.4. Reliability Function

Osmetti and Chiodini [27] demonstrated the idea of finding the reliability of a joint survival function and they showed that it is more convenient to work with X and Y as random variables with survival functions $1 - F(x)$ and $1 - F(y)$ as follows:

The reliability functions of the marginal distributions are defined as

$$S(x;\theta_1,\sigma_1,\delta_1) = 1 - F(x) = e^{[1-(1+\delta_1 x+\sigma_1 x^2)^{\theta_1}]}; \; \theta_1,\sigma_1,\delta_1 > 0$$

and

$$S(y;\theta_2,\sigma_2,\delta_2) = 1 - F(y) = e^{[1-(1+\delta_2 y+\sigma_2 y^2)^{\theta_2}]}; \; \theta_2,\sigma_2,\delta_2 > 0.$$

The joint survival function for copula is expressed as

$$S(x,y) = C(S(x),S(y)).$$

Hence, the reliability function of the BMExE distribution is

$$S(x,y) = [1-u(x)][1-v(y)][1+\lambda u(x)v(y)],$$

where $u(x) = 1 - e^{[1-(1+\delta_1 x+\sigma_1 x^2)^{\theta_1}]}$ and $v(y) = 1 - e^{[1-(1+\delta_2 y+\sigma_2 y^2)^{\theta_2}]}$.

Basu [28] was the first who defined the bivariate failure rate function which is given as

$$h(x,y) = \frac{f(x,y)}{S(x,y)}.$$

Hence, the hazard rate function of the BMExE distribution is

$$h(x,y) = \theta_1 \theta_2 (\delta_1 + 2\sigma_1 x)(\delta_2 + 2\sigma_2 y)(1+\delta_1 x + \sigma_1 x^2)^{\theta_1 - 1}(1+\delta_2 y + \sigma_2 y^2)^{\theta_2 - 1}$$
$$[1-u(x)][1-v(y)][1+\lambda(1-2u(x))(1-2v(y))] \, [1+\lambda u(x)v(y)]^{-1}; \theta_1,\sigma_1,\delta_1,\theta_2,\sigma_2,\delta_2 > 0.$$

4. Inference under Complete and Censored Samples

In this section, the classical and non-classical estimation problems were considered; for classical, we used the maximum likelihood estimation and discussed the point and interval estimation for the unknown parameters of the BMExE distribution. Bayesian estimation is the most powerful non-classical estimation method; hence, point and credible intervals were obtained for the unknown parameters. These estimation issues were handled for both complete samples and Type-II censored samples.

4.1. The Maximum Likelihood Estimation

In this subsection, we explored the maximum likelihood estimation (MLE) for the unknown parameters $\omega = (\theta_1,\sigma_1,\delta_1,\theta_2,\sigma_2,\delta_2,\lambda)$, subject to complete and Type-II censored samples. Suppose that $(x_{1:n},y_{1:n}), (x_{2:n},y_{2:n}), \ldots, (x_{n:n},y_{n:n})$ are the n observed values from the BMExE distribution. The likelihood function for a bivariate model was discussed by Kim et al. [29]. Therefore, the likelihood function for ω under a complete sample is defined as follows:

$$L(\omega) = \prod_{i=1}^n f_{X,Y}(x_i,y_i)$$
$$= \prod_{i=1}^n f_X(x_i) f_Y(y_i) c(F(x_i),F_Y(y_i).\omega).$$

Then,

$$L(\omega) = \prod_{i=1}^n \theta_1 \theta_2 (\delta_1 + 2\sigma_1 x_i)(\delta_2 + 2\sigma_2 y_i)(1+\delta_1 x_i + \sigma_1 x_i^2)^{\theta_1 - 1}(1+\delta_2 y_i + \sigma_2 y_i^2)^{\theta_2 - 1}$$
$$\prod_{i=1}^n [1-u(x_i)] \, [1-v(y_i)]$$
$$\prod_{i=1}^n [1+\lambda(1-2u(x_i))(1-2v(y_i))], \tag{14}$$

where $u(x_i) = 1 - e^{[1-(1+\delta_1 x_i + \sigma_1 x_i^2)^{\theta_1}]}$ and $v(y_i) = 1 - e^{[1-(1+\delta_2 y_i + \sigma_2 y_i^2)^{\theta_2}]}$.

The log-likelihood function $\ell(\omega)$ can be expressed as follows:

$$\begin{aligned}\ell(\omega) =\ & n\ln\theta_1 + n\ln\theta_2 + \sum_{i=1}^{n}\ln(\delta_1 + 2\sigma_1 x_i) + \sum_{i=1}^{n}\ln(\delta_2 + 2\sigma_2 y_i) + \\ & (\theta_1 - 1)\sum_{i=1}^{n}\ln(1 + \delta_1 x_i + \sigma_1 x_i^2) + (\theta_2 - 1)\sum_{i=1}^{n}\ln(1 + \delta_2 y_i + \sigma_2 y_i^2) + \sum_{i=1}^{n}\left[1 - (1+\delta_1 x_i + \sigma_1 x_i^2)^{\theta_1}\right] + \\ & \sum_{i=1}^{n}\left[1 - (1+\delta_2 y_i + \sigma_2 y_i^2)^{\theta_2}\right] + \sum_{i=1}^{n}\ln[1 + \lambda(1 - 2u(x_i))(1 - 2v(y_i))]. \end{aligned} \quad (15)$$

Differentiating Equation (15) partially with respect to the model's parameters, we have:

$$\begin{aligned}\frac{\partial \ell}{\partial \delta_1} =\ & \sum_{i=1}^{n} \frac{1}{(\delta_1 + 2\sigma_1 x_i)} + (\theta_1 - 1)\sum_{i=1}^{n}\frac{x_i}{(1+\delta_1 x_i + \sigma_1 x_i^2)} - \theta_1 \sum_{i=1}^{n} x_i (1+\delta_1 x_i + \sigma_1 x_i^2)^{\theta_1 - 1} \\ & - \sum_{i=1}^{n}\frac{2\lambda\theta_1 x_i (1+\delta_1 x_i + \sigma_1 x_i^2)^{\theta_1 - 1}(1 - u(x_i))(1 - 2v(y_i))}{[1 + \lambda(1 - 2u(x_i))(1 - 2v(y_i))]},\end{aligned}$$

$$\begin{aligned}\frac{\partial \ell}{\partial \delta_2} =\ & \sum_{i=1}^{n} \frac{1}{(\delta_2 + 2\sigma_2 y_i)} + (\theta_2 - 1)\sum_{i=1}^{n}\frac{y_i}{(1+\delta_2 y_i + \sigma_2 y_i^2)} - \theta_2 \sum_{i=1}^{n} y_i (1+\delta_2 y_i + \sigma_2 y_i^2)^{\theta_2 - 1} \\ & - \sum_{i=1}^{n}\frac{2\lambda\theta_2\, y_i (1+\delta_2 y_i + \sigma_2 y_i^2)^{\theta_2 - 1}(1 - v(y_i))(1 - 2u(x_i))}{[1 + \lambda(1 - 2u(x_i))(1 - 2v(y_i))]},\end{aligned}$$

$$\begin{aligned}\frac{\partial \ell}{\partial \sigma_1} =\ & \sum_{i=1}^{n} \frac{2x_i}{(\delta_1 + 2\sigma_1 x_i)} + (\theta_1 - 1)\sum_{i=1}^{n}\frac{x_i^2}{(1+\delta_1 x_i + \sigma_1 x_i^2)} - \theta_1 \sum_{i=1}^{n} x_i^2 (1+\delta_1 x_i + \sigma_1 x_i^2)^{\theta_1 - 1} \\ & - \sum_{i=1}^{n}\frac{2\lambda\theta_1 x_i^2 (1+\delta_1 x_i + \sigma_1 x_i^2)^{\theta_1 - 1}(1 - 2v(y_i))(1 - u(x_i))}{[1 + \lambda(1 - 2u(x_i))(1 - 2v(y_i))]},\end{aligned}$$

$$\begin{aligned}\frac{\partial \ell}{\partial \sigma_2} =\ & \sum_{i=1}^{n} \frac{2y_i}{(\delta_2 + 2\sigma_2 y_i)} + (\theta_2 - 1)\sum_{i=1}^{n}\frac{y_i^2}{(1+\delta_2 y_i + \sigma_2 y_i^2)} - \theta_2 \sum_{i=1}^{n} y_i^2 (1+\delta_2 y_i + \sigma_2 y_i^2)^{\theta_2 - 1} \\ & - \sum_{i=1}^{r}\frac{2\lambda\theta_2\, y_i^2 (1+\delta_2 y_i + \sigma_2 y_i^2)^{\theta_2 - 1}(1 - v(y_i))(1 - 2u(x_i))}{[1 + \lambda(1 - 2u(x_i))(1 - 2v(y_i))]},\end{aligned}$$

$$\begin{aligned}\frac{\partial \ell}{\partial \theta_1} =\ & \frac{n}{\theta_1} + \sum_{i=1}^{n}\ln(1+\delta_1 x_i + \sigma_1 x_i^2) - \sum_{i=1}^{n}(1+\delta_1 x_i + \sigma_1 x_i^2)^{\theta_1} \ln(1+\delta_1 x_i + \sigma_1 x_i^2) \\ & - \sum_{i=1}^{n}\frac{2\lambda(1+\delta_1 x_i + \sigma_1 x_i^2)^{\theta_1} \ln(1+\delta_1 x_i + \sigma_1 x_i^2)\,(1 - 2v(y_i))(1 - u(x_i))}{[1 + \lambda(1 - 2u(x_i))(1 - 2v(y_i))]},\end{aligned}$$

$$\begin{aligned}\frac{\partial \ell}{\partial \theta_2} =\ & \frac{n}{\theta_2} + \sum_{i=1}^{n}\ln(1+\delta_2 y_i + \sigma_2 y_i^2) - \sum_{i=1}^{n}(1+\delta_2 y_i + \sigma_2 y_i^2)^{\theta_2} \ln(1+\delta_2 y_i + \sigma_2 y_i^2) \\ & - \sum_{i=1}^{n}\frac{2\lambda(1+\delta_2 y_i + \sigma_2 y_i^2)^{\theta_2} \ln(1+\delta_2 y_i + \sigma_2 y_i^2)\,(1 - v(y_i))(1 - 2u(x_i))}{[1 + \lambda(1 - 2u(x_i))(1 - 2v(y_i))]},\end{aligned}$$

and

$$\frac{\partial \ell}{\partial \lambda} = \sum_{i=1}^{n} \frac{(1 - 2u(x_i))(1 - 2v(y_i))}{[1 + \lambda(1 - 2u(x_i))(1 - 2v(y_i))]}.$$

By equating the above partial derivatives with zero, we get a system of non-linear normal equations. This system needs a numerical solution such as a nonlinear optimization algorithm.

Now, for the censored sample case, the likelihood function under a Type-II censored sample is written as follows:

$$L(\omega) = \frac{n!}{(n-r)!}[1 - F(x_{r:n})]^{n-r} \prod_{i=1}^{r} f_{X,Y}(x_{i:n}, y_{i:n}). \tag{16}$$

Substituting Equations (6) and (9) into Equation (16), the log-likelihood function $\hat{L}(\omega)$ becomes

$$\hat{L}(\omega) = \frac{n!(\theta_1 \theta_2)^r}{(n-r)!}\left[e^{[1-(1+\delta_1 x_{r:n} + \sigma_1 x_{r:n}^2)^{\theta_1}]}\right]^{n-r} \prod_{i=1}^{r}(\delta_1 + 2\sigma_1 x_{i:n})(\delta_2 + 2\sigma_2 y_{i:n})(1 + \delta_1 x_{i:n} + \sigma_1 x_{i:n}^2)^{\theta_1 - 1}(1 + \delta_2 y_{i:n} + \sigma_2 y_{i:n}^2)^{\theta_2 - 1} \\ \prod_{i=1}^{r}[1 - u(x_{i:n})][1 - v(y_{i:n})] \prod_{i=1}^{r}[1 + \lambda(1 - v(y_{i:n}))(1 - 2u(x_{i:n}))], \tag{17}$$

where $u(x_{i:n}) = 1 - e^{[1-(1+\delta_1 x_{i:n} + \sigma_1 x_{i:n}^2)^{\theta_1}]}$ and $v(y_{i:n}) = 1 - e^{[1-(1+\delta_2 y_{i:n} + \sigma_2 y_{i:n}^2)^{\theta_2}]}$.
The log-likelihood function $\ell(\omega)$ can be expressed as follows:

$$\ell(\omega) \propto (n-r)[1 - (1 + \delta_1 x_{r:n} + \sigma_1 x_{r:n}^2)^{\theta_1}] + r \ln \theta_1 + r \ln \theta_2 + \sum_{i=1}^{r} \ln(\delta_1 + 2\sigma_1 x_{i:n}) + \sum_{i=1}^{r} \ln(\delta_2 + 2\sigma_2 y_{i:n}) + \\ (\theta_1 - 1)\sum_{i=1}^{r} \ln(1 + \delta_1 x_{i:n} + \sigma_1 x_{i:n}^2) + (\theta_2 - 1)\sum_{i=1}^{r} \ln(1 + \delta_2 y_{i:n} + \sigma_2 y_{i:n}^2) + \sum_{i=1}^{r}\left[1 - (1 + \delta_1 x_{i:n} + \sigma_1 x_{i:n}^2)^{\theta_1}\right] + \\ \sum_{i=1}^{r}\left[1 - (1 + \delta_2 y_{i:n} + \sigma_2 y_{i:n}^2)^{\theta_2}\right] + \sum_{i=1}^{r} \ln[1 + \lambda(1 - v(y_{i:n}))(1 - 2u(x_{i:n}))]. \tag{18}$$

Differentiating Equation (18) partially with respect to the model's parameters, we have:

$$\frac{\partial \ell}{\partial \delta_1} = -(n-r)\theta_1 x_{r:n}(1 + \delta_1 x_{r:n} + \sigma_1 x_{r:n}^2)^{\theta_1 - 1} + \sum_{i=1}^{r} \frac{1}{(\delta_1 + 2\sigma_1 x_{i:n})} \\ + (\theta_1 - 1)\sum_{i=1}^{r} \frac{x_{i:n}}{(1 + \delta_1 x_{i:n} + \sigma_1 x_{i:n}^2)} - \theta_1 \sum_{i=1}^{r} x_{i:n}(1 + \delta_1 x_{i:n} + \sigma_1 x_{i:n}^2)^{\theta_1 - 1} \\ - \sum_{i=1}^{r} \frac{2\lambda \theta_1 x_{i:n}(1 + \delta_1 x_{i:n} + \sigma_1 x_{i:n}^2)^{\theta_1 - 1}(1 - 2v(y_{i:n}))(1 - u(x_{i:n}))}{[1 + \lambda(1 - v(y_{i:n}))(1 - 2u(x_{i:n}))]},$$

$$\frac{\partial \ell}{\partial \delta_2} = \sum_{i=1}^{r} \frac{1}{(\delta_2 + 2\sigma_2 y_{i:n})} + (\theta_2 - 1)\sum_{i=1}^{r} \frac{y_{i:n}}{(1 + \delta_2 y_{i:n} + \sigma_2 y_{i:n}^2)} - \theta_2 \sum_{i=1}^{r} y_{i:n}(1 + \delta_2 y_{i:n} + \sigma_2 y_{i:n}^2)^{\theta_2 - 1} \\ - \sum_{i=1}^{r} \frac{2\lambda \theta_2 y_{i:n}(1 + \delta_2 y_{i:n} + \sigma_2 y_{i:n}^2)^{\theta_2 - 1}(1 - v(y_{i:n}))(1 - 2u(x_{i:n}))}{[1 + \lambda(1 - v(y_{i:n}))(1 - 2u(x_{i:n}))]},$$

$$\frac{\partial \ell}{\partial \sigma_1} = -(n-r)\theta_1 x_{r:n}^2(1 + \delta_1 x_{r:n} + \sigma_1 x_{r:n}^2)^{\theta_1 - 1} + \sum_{i=1}^{r} \frac{2 x_{i:n}}{(\delta_1 + 2\sigma_1 x_{i:n})} \\ + (\theta_1 - 1)\sum_{i=1}^{r} \frac{x_{i:n}^2}{(1 + \delta_1 x_{i:n} + \sigma_1 x_{i:n}^2)} - \theta_1 \sum_{i=1}^{r} x_{i:n}^2(1 + \delta_1 x_{i:n} + \sigma_1 x_{i:n}^2)^{\theta_1 - 1} \\ - \sum_{i=1}^{r} \frac{2\lambda \theta_1 x_{i:n}^2(1 + \delta_1 x_{i:n} + \sigma_1 x_{i:n}^2)^{\theta_1 - 1}(1 - 2v(y_{i:n}))(1 - u(x_{i:n}))}{[1 + \lambda(1 - v(y_{i:n}))(1 - 2u(x_{i:n}))]},$$

$$\frac{\partial \ell}{\partial \sigma_2} = \sum_{i=1}^{r} \frac{2 y_{i:n}}{(\delta_2 + 2\sigma_2 y_{i:n})} + (\theta_2 - 1)\sum_{i=1}^{r} \frac{y_{i:n}^2}{(1 + \delta_2 y_{i:n} + \sigma_2 y_{i:n}^2)} - \theta_2 \sum_{i=1}^{r} y_{i:n}^2(1 + \delta_2 y_{i:n} + \sigma_2 y_{i:n}^2)^{\theta_2 - 1} \\ - \sum_{i=1}^{r} \frac{2\lambda \theta_2 y_{i:n}^2(1 + \delta_2 y_{i:n} + \sigma_2 y_{i:n}^2)^{\theta_2 - 1}(1 - v(y_{i:n}))(1 - 2u(x_{i:n}))}{[1 + \lambda(1 - v(y_{i:n}))(1 - 2u(x_{i:n}))]},$$

$$\frac{\partial \ell}{\partial \theta_1} = -(n-r)(1+\delta_1 x_{r:n} + \sigma_1 x_{r:n}^2)^{\theta_1} \ln(1+\delta_1 x_{r:n} + \sigma_1 x_{r:n}^2) + \frac{r}{\theta_1}$$

$$+ \sum_{i=1}^{r} \ln(1+\delta_1 x_{i:n} + \sigma_1 x_{i:n}^2) - \sum_{i=1}^{r}(1+\delta_1 x_{i:n} + \sigma_1 x_{i:n}^2)^{\theta_1} \ln(1+\delta_1 x_{i:n} + \sigma_1 x_{i:n}^2)$$

$$- \sum_{i=1}^{r} \frac{2\lambda(1+\delta_1 x_{i:n} + \sigma_1 x_{i:n}^2)^{\theta_1} \ln(1+\delta_1 x_{i:n} + \sigma_1 x_{i:n}^2)(1-2v(y_{i:n}))(1-u(x_{i:n}))}{[1+\lambda(1-v(y_{i:n}))(1-2u(x_{i:n}))]},$$

$$\frac{\partial \ell}{\partial \theta_2} = \frac{r}{\theta_2} + \sum_{i=1}^{r} \ln(1+\delta_2 y_{i:n} + \sigma_2 y_{i:n}^2) - \sum_{i=1}^{r}(1+\delta_2 y_{i:n} + \sigma_2 y_{i:n}^2)^{\theta_2} \ln(1+\delta_2 y_{i:n} + \sigma_2 y_{i:n}^2)$$

$$- \sum_{i=1}^{r} \frac{2\lambda(1+\delta_2 y_{i:n} + \sigma_2 y_{i:n}^2)^{\theta_2} \ln(1+\delta_2 y_{i:n} + \sigma_2 y_{i:n}^2)(1-v(y_{i:n}))(1-2u(x_{i:n}))}{[1+\lambda(1-v(y_{i:n}))(1-2u(x_{i:n}))]}$$

and

$$\frac{\partial \ell}{\partial \lambda} = \sum_{i=1}^{r} \frac{(1-v(y_{i:n}))(1-2u(x_{i:n}))}{[1+\lambda(1-v(y_{i:n}))(1-2u(x_{i:n}))]}.$$

By equating these partial derivative equations to zeros, we get a system of non-linear (normal) equations.

Since there is no closed-form expression for the MLE $\hat{\theta}_1, \hat{\sigma}_1, \hat{\delta}_1, \hat{\theta}_2, \hat{\sigma}_2, \hat{\delta}_2$ and $\hat{\lambda}$, their values were computed numerically using a nonlinear optimization algorithm. All numerical calculations and their results were obtained and are summarized in Section 6.

4.2. Bayesian Estimation

The Bayesian method considers the parameters as random variables; the ambiguity of the parameters was defined as a joint prior distribution, which was observed before the collected failure times. The ability to collect prior knowledge in the analysis makes the Bayesian approach very helpful in reliability analysis. The limitation of data availability is one of the main challenges related to the reliability analysis; see [30]. Bayesian estimates of the unknown parameters $\omega = (\theta_1, \sigma_1, \delta_1, \theta_2, \sigma_2, \delta_2, \lambda)$ were achieved with respect to the squared error loss function (SEL).

The Bayesian estimation uses an appropriate choice of prior(s) for estimating each parameter. According to the Bayesian estimation theory, no prior distribution for a parameter is considered the best until it is tested and validated. Additionally, most prior distributions are selected according to one's subjective knowledge and beliefs. Hence, if one has enough knowledge of the parameter(s), it is wise to select an informative prior(s); otherwise, it is better to consider non-informative prior(s). For this research, we selected non-informative priors (uniform) and an informative prior (gamma). These assumed that prior distributions have been used widely by several authors such as [31–33]. The definitions of the above listed loss function are presented as follows:

- For informative prior, it is assumed that the parameters $\omega = (\theta_1, \sigma_1, \delta_1, \theta_2, \sigma_2, \delta_2, \lambda)$ are independent and follow the prior distributions,

$$\pi_j(\omega) = \omega^{-a_j-1} e^{-b_j \omega}, a_j, b_j > 0, j = 1, 2, ..., 7; \tag{19}$$

- For non-informative prior, it is assumed that the parameters ω are independent and follow the uniform distributions,

$$\pi_j(\omega) = \frac{1}{b_j - a_j}, \quad a_j < \omega < b_j, j = 1, 2, ..., 7; \tag{20}$$

- For non-prior, it is assumed that the parameters ω are independent and follow

$$\pi_j(\omega) = 1. \tag{21}$$

The posterior distribution of $\theta_1, \sigma_1, \delta_1, \theta_2, \sigma_2, \delta_2$ and λ, denoted by $\pi^*(\omega|\underline{x},\underline{y})$, can be obtained by combining the likelihood function with the priors and it can be written as

$$\pi^*(\omega|\underline{x},\underline{y}) = \frac{L(\omega|\underline{x},\underline{y})\,\pi_1(\theta_1)\,\pi_2(\sigma_1)\,\pi_3(\delta_1)\,\pi_4(\theta_2)\,\pi_5(\sigma_2)\,\pi_6(\delta_2)\,\pi_7(\lambda)}{\int_0^1\int_0^\infty\int_0^\infty\int_0^\infty\int_0^\infty\int_0^\infty\int_0^\infty L(\omega|\underline{x},\underline{y})\,\pi_1(\theta_1)\,\pi_2(\sigma_1)\,\pi_3(\delta_1)\,\pi_4(\theta_2)\,\pi_5(\sigma_2)\,\pi_6(\delta_2)\,\pi_7(\lambda)\,d\theta_1\,d\sigma_1 d\delta_1 d\theta_2\,d\sigma_2 d\delta_2\,d\lambda}.$$

A commonly-used loss function is the SEL, which is a symmetrical loss function that assigns equal losses to overestimation and underestimation. If ω is the parameter to be estimated by an estimator $\hat{\omega}$, then the square error loss function is defined as

$$v(\omega, \hat{\omega}) = (\hat{\omega} - \omega)^2.$$

Therefore, the Bayes estimate of any function of $\omega = (\theta_1, \sigma_1, \delta_1, \theta_2, \sigma_2, \delta_2, \lambda)$, say $g(\omega)$ under the SEL function, can be obtained as

$$\hat{g}_{BS}(\omega|\underline{x},\underline{y}) = E_{\omega|\underline{x},\underline{y}}(g(\omega)),$$

where

$$E_{\omega|\underline{x},\underline{y}}(g(\omega)) = \frac{\int_0^1\int_0^\infty\int_0^\infty\int_0^\infty\int_0^\infty\int_0^\infty\int_0^\infty g(\omega) L(\omega|\underline{x},\underline{y})\,\pi_1(\theta_1)\,\pi_2(\sigma_1)\,\pi_3(\delta_1)\,\pi_4(\theta_2)\,\pi_5(\sigma_2)\,\pi_6(\delta_2)\,\pi_7(\lambda)\,d\theta_1\,d\sigma_1 d\delta_1 d\theta_2\,d\sigma_2 d\delta_2\,d\lambda}{\int_0^1\int_0^\infty\int_0^\infty\int_0^\infty\int_0^\infty\int_0^\infty\int_0^\infty L(\omega|\underline{x},\underline{y})\,\pi_1(\theta_1)\,\pi_2(\sigma_1)\,\pi_3(\delta_1)\,\pi_4(\theta_2)\,\pi_5(\sigma_2)\,\pi_6(\delta_2)\,\pi_7(\lambda)\,d\theta_1\,d\sigma_1 d\delta_1 d\theta_2\,d\sigma_2 d\delta_2\,d\lambda}. \tag{22}$$

It was noticed that the ratio of multiple integrals in Equation (22) cannot be obtained in an explicit form. Thus, the MCMC technique generates samples from the joint posterior density function. To implement the MCMC technique, we considered the Gibbs within Metropolis–Hasting samplers procedure.

In the MCMC method, we estimated the posterior distribution and the multi-integrals via simulated samples generated from the posterior distribution. The MCMC technique was performed, where Gibbs sampling and the Metropolis–Hastings (M-H) algorithm were used for finding the MCMC method. That algorithm was first studied by [34,35]. Similar to the acceptance-rejection sampling, the M–H algorithm combines each iteration of the algorithm with a chosen value that can be generated from a certain distribution; so, the chosen value was accepted according to a suitable acceptance probability. This action assures the convergence of the Markov chain for the objective density. For more information about the applications of the M–H algorithm, one may refer to [36–38].

For the Type-II censoring data, Equation (16) can be used to replace Equation (14) to obtain the Bayes estimates of the unknown parameters $\hat{\theta}_1, \hat{\sigma}_1, \hat{\delta}_1, \hat{\theta}_2, \hat{\sigma}_2, \hat{\delta}_2$ and $\hat{\lambda}$. At this point, we concluded that the idea behind using the MCMC method over the MLE method is that it can obtain a reasonable interval estimate of the parameters by constructing the confidence intervals based on the empirical posterior distribution. This is regularly inaccessible in the MLE case. Chen and Shao [39] used a technique that extensively generates the highest posterior density intervals of unknown parameters of the distribution. In this study, the M–H algorithm was used to select samples that were used to generate failure times estimates.

5. Confidence Intervals

In this section, we introduce two methods for generating confidence intervals for the unknown parameters of the BMExE distribution, the asymptotic confidence interval, and the bootstrap confidence interval of $\theta_j, \sigma_j, \delta_j$, where $j = 1, 2$ and λ. There are two parts to the bootstrap method; bootstrap percentile and bootstrap-t.

5.1. Asymptotic Confidence Intervals

The asymptotic property of normal distribution for the MLEs is the most commonly used method for producing confidence bounds for the parameters. The Fisher information matrix $I(\omega)$ is created in combination with the asymptotic variance-covariance matrix of the MLE for the parameters by getting the negative second derivatives of the log-likelihood function at $\hat{\omega} = (\hat{\theta}_1, \hat{\sigma}_1, \hat{\delta}_1, \hat{\theta}_2, \hat{\sigma}_2, \hat{\delta}_2, \hat{\lambda})$.

Assume the parameter vector ω's asymptotic variance-covariance matrix is

$$I(\hat{\omega}) = \begin{bmatrix} I_{\hat{\theta}_1\hat{\theta}_1} & & & & & & \\ I_{\hat{\sigma}_1\hat{\theta}_1} & I_{\hat{\sigma}_1\hat{\sigma}_1} & & & & & \\ I_{\hat{\delta}_1\hat{\theta}_1} & I_{\hat{\delta}_1\hat{\sigma}_1} & I_{\hat{\delta}_1\hat{\delta}_1} & & & & \\ I_{\hat{\theta}_2\hat{\theta}_1} & I_{\hat{\theta}_2\hat{\sigma}_1} & I_{\hat{\theta}_2\hat{\delta}_1} & I_{\hat{\theta}_2\hat{\theta}_2} & & & \\ I_{\hat{\sigma}_2\hat{\theta}_1} & I_{\hat{\sigma}_2\hat{\sigma}_1} & I_{\hat{\sigma}_2\hat{\delta}_1} & I_{\hat{\sigma}_2\hat{\theta}_2} & I_{\hat{\sigma}_2\hat{\sigma}_2} & & \\ I_{\hat{\delta}_2\hat{\theta}_1} & I_{\hat{\delta}_2\hat{\sigma}_1} & I_{\hat{\delta}_2\hat{\delta}_1} & I_{\hat{\delta}_2\hat{\theta}_2} & I_{\hat{\delta}_2\hat{\sigma}_2} & I_{\hat{\delta}_2\hat{\delta}_2} & \\ I_{\hat{\lambda}\hat{\theta}_1} & I_{\hat{\lambda}\hat{\sigma}_1} & I_{\hat{\lambda}\hat{\delta}_1} & I_{\hat{\lambda}\hat{\theta}_2} & I_{\hat{\lambda}\hat{\sigma}_2} & I_{\hat{\lambda}\hat{\delta}_2} & I_{\hat{\lambda}\hat{\lambda}} \end{bmatrix},$$

where $V(\hat{\omega}) = I^{-1}(\hat{\omega})$ is the variance-covariance matrix.

A $100(1-\gamma)\%$ confidence interval for the parameter vector ω can be built based on the MLE's asymptotic normality. Using the asymptotic normality of the MLE, a $100(1-\gamma)\%$ confidence interval for parameter vector ω can be constructed as

$$\hat{\theta}_j \pm Z_\gamma \sqrt{I_{\hat{\theta}_j\hat{\theta}_j}},\ \hat{\sigma}_j \pm Z_\gamma \sqrt{I_{\hat{\sigma}_j\hat{\sigma}_j}},\ \hat{\delta}_j \pm Z_\gamma \sqrt{I_{\hat{\delta}_j\hat{\delta}_j}}\ \text{and}\ \hat{\lambda} \pm Z_\gamma \sqrt{I_{\hat{\lambda}\hat{\lambda}}},$$

where Z_γ is the percentile of the standard normal distribution with a right tail probability of $\frac{\gamma}{2}$.

5.2. Bootstrap Confidence Interval

Bootstrap is a re-sampling method used in statistical inference. It is frequently utilized to compute confidence intervals; for more information, see Efron [40]. In this subsection, we used the parametric bootstrap method to compute confidence intervals for the unknown parameters $\theta_k, \sigma_k, \delta_k$, where $k = 1, 2$ and λ. We provide two parametric bootstrap methods for confidence intervals, percentile bootstrap, and bootstrap-t confidence intervals.

5.2.1. Percentile Bootstrap Confidence Interval

The following steps summarize the algorithm for obtaining percentile bootstrap confidence intervals:

1. MLE and Bayesian estimators are measured for the BMExE distribution's parameters;
2. Create bootstrap samples with $\theta_k, \sigma_k, \delta_k$, and λ obtain the bootstrap estimate, respectively, as $\theta_k^b, \sigma_k^b, \delta_k^b$ and λ^b using the bootstrap sample;
3. Repeat step (2) $\mathcal{M}$ times to have $(\theta_k^{b(1)}, \theta_k^{b(2)}, \ldots, \theta_k^{b(\mathcal{M})}), (\sigma_k^{b(1)}, \sigma_k^{b(2)}, \ldots \sigma_k^{b(\mathcal{M})})$, $(\delta_k^{b(1)}, \delta_k^{b(2)}, \ldots, \delta_k^{b(\mathcal{M})})$, and $(\lambda^{b(1)}, \lambda^{b(2)}, \ldots, \lambda^{b(\mathcal{M})})$;
4. Arrange $(\theta_k^{b(1)}, \theta_k^{b(2)}, \ldots, \theta_k^{b(\mathcal{M})}), (\sigma_k^{b(1)}, \sigma_k^{b(2)}, \ldots, \sigma_k^{b(\mathcal{M})}), (\delta_k^{b(1)}, \delta_k^{b(2)}, \ldots, \delta_k^{b(\mathcal{M})})$ and $(\lambda^{b(1)}, \lambda^{b(2)}, \ldots, \lambda^{b(\mathcal{M})})$ in ascending order as $(\theta_k^{b[1]}, \theta_k^{b[2]}, \ldots, \theta_k^{b([\mathcal{M}])})$, $(\sigma_k^{b[1]}, \sigma_k^{b[2]}, \ldots, \sigma_k^{b[\mathcal{M}]}), (\delta_k^{b[1]}, \delta_k^{b[2]}, \ldots; \delta_k^{b[\mathcal{M}]})$ and $(\lambda^{b[1]}, \lambda^{b[2]}, \ldots, \lambda^{b[\mathcal{M}]})$.
5. The two side $100(1-\gamma)\%$ percentile bootstrap confidence intervals are given by $[\theta_k^{b([\mathcal{M}\frac{\gamma}{2}])}, \theta_k^{b([\mathcal{M}(1-\frac{\gamma}{2})])}], [\sigma_k^{b([\mathcal{M}\frac{\gamma}{2}])}, \sigma_k^{b([\mathcal{M}(1-\frac{\gamma}{2})])}], [\delta_k^{b([\mathcal{M}\frac{\gamma}{2}])}, \delta_k^{b([\mathcal{M}(1-\frac{\gamma}{2})])}]$ and $[\lambda^{b([\mathcal{M}\frac{\gamma}{2}])}, \lambda^{b([\mathcal{M}(1-\frac{\gamma}{2})])}]$.

5.2.2. Bootstrap-t Confidence Intervals

The following steps summarize the algorithm for obtaining bootstrap-t confidence intervals:

1. The steps (1, 2) are the same as in Boot-p;

2. Where $V(\hat{\omega}^b)$ is asymptotic variances of $\hat{\omega}^b$ and it can be obtained using the Fisher information matrix, the t-statistic of ω is computed as $T = \frac{\hat{\omega}^b - \hat{\omega}}{\sqrt{V(\hat{\omega}^b)}}$;

3. Repeat steps 2-3 $\mathcal{M}$ times and obtain $(T^{(1)}, T^{(2)}, \ldots, T^{(\mathcal{M})})$;
4. Arrange $(T^{(1)}, T^{(2)}, \ldots, T^{(\mathcal{M})})$ in ascending order as $(T^{[1]}, T^{[2]}, \ldots, T^{[\mathcal{M}]})$;
5. The two side $100(1 - \gamma)\%$ percentile bootstrap confidence intervals are given by

$$[\theta_k + T_k^{b([\mathcal{M}\frac{\gamma}{2}])}\sqrt{V(\theta_k^b)}, \theta_k + \sqrt{V(\theta_k^b)}T_k^{b([\mathcal{M}(1-\frac{\gamma}{2})])}],$$
$$[\sigma_k + T_k^{b([\mathcal{M}\frac{\gamma}{2}])}\sqrt{V(\sigma_k^b)}, \sigma_k + \sqrt{V(\sigma_k^b)}T_k^{b([\mathcal{M}(1-\frac{\gamma}{2})])}],$$
$$[\delta_k + T_k^{b([\mathcal{M}\frac{\gamma}{2}])}\sqrt{V(\delta_k^b)}, \delta_k + \sqrt{V(\delta_k^b)}T_k^{b([\mathcal{M}(1-\frac{\gamma}{2})])}] \text{ and}$$
$$[\lambda + T_k^{b([\mathcal{M}\frac{\gamma}{2}])}\sqrt{V(\lambda^b)}, \lambda + \sqrt{V(\lambda^b)}T_k^{b([\mathcal{M}(1-\frac{\gamma}{2})])}].$$

6. Simulation Study

In this section, a Monte Carlo simulation was performed using the copula function. The BMExE parameters were estimated using the R program. Nelsen [1] described how to generate a sample from a given joint distribution. We can generate a bivariate sample using the conditional approach by following the procedure below.

- By using uniform $(0,1)$ distribution, generate U and V independently;
- Set $x = Q_{MExE}(W) = \frac{1}{2\sigma_1}\sqrt{4\sigma_1(1 - \ln(1 - W))^{\frac{1}{\theta_1}} + (\delta_1^2 - 4\sigma_1)} - \delta_1$;
- Set $F(y|x) = V$ in Equation (12) to obtain y by numerical analysis;
- To obtain (x_i, y_i), $i = 1, \ldots, n$, repeat the above steps (n) times.

A simulation algorithm: simulation experiments were performed with the following data generated from the BMExE distribution, where X, Y are following the MExE distribution with θ_j, σ_j, and δ_j where $j = 1, 2$, thus the parameters $\theta_1, \sigma_1, \delta_1, \theta_2, \sigma_2, \delta_2$, and λ were obtained with the following cases for the random variables, generating:
Case-I: $\theta_1 = 2, \sigma_1 = 2.1, \delta_1 = 1.8, \theta_2 = 1.3, \sigma_2 = 1.5, \delta_2 = 1.3$ and $\lambda = -0.3$.
Case-II: $\theta_1 = 0.6, \sigma_1 = 0.7, \delta_1 = 0.8, \theta_2 = 0.9, \sigma_2 = 0.5, \delta_2 = 0.6$ and $\lambda = 0.8$.
Case-III: $\theta_1 = 1.6, \sigma_1 = 1.7, \delta_1 = 0.8, \theta_2 = 0.9, \sigma_2 = 1.5, \delta_2 = 0.6$ and $\lambda = 0.5$.

We chose several sample sizes, such as n = 40, 100, and 200, and different Type-II censored sample sizes, such as r = 30, and 40, for n = 40, r = 70, and 85 for n = 100, and r = 160, and 180 for n = 200. The bias, mean square error, length of asymptotic confidence intervals, length of bootstrap-p, and length of bootstrap-t were calculated for MLE, while bias, the mean square error, and length of credible confidence intervals were calculated for the Bayesian method. A comparison was performed between the different approaches of the resulting estimators with respect to the bias and MSE, where $L = 5000$ is the number of simulated samples.

The 95% of confidence intervals were produced using asymptotic MLE and credible intervals and they were evaluated against different criteria. The length of the typical confidence intervals was compared. Estimates of the parameters in the Bayes technique were generated from three cases as informative, non-informative priors, and non-prior in order to evaluate the type of prior. In the case of informative priors, the hyper-parameters were selected by elective hyper-parameters utilizing MLE information to display the outcomes of the estimated parameters.

In Tables A2–A4 and Figure 1, the simulation analysis yields the following findings, which we highlight as follows:

- When sample size grows and Bayes estimates are used, the bias, mean square error, and length of confidence intervals for computed MLE parameters have a downward trend;
- For bias, mean square error, and length of the asymptotic confidence interval under the Bayes estimates consistently outperform the MLEs;
- The performance improves as the censored sample size r rises, keeping the Type-II censored sample size n and time fixed;

- Compared to the confidence interval methods, the bootstrap confidence interval estimates are superior as they have the least confidence length;
- Compared to the estimation method, the Bayesian estimates are superior.

Tables A2–A4 are shown in Appendix A, where the notation MSE refers to mean square error, LACI is the length of asymptotic confidence intervals, LBP is the length of bootstrap-p, LBT is the length of bootstrap-t, and LCCI refers to the length of the credible confidence intervals.

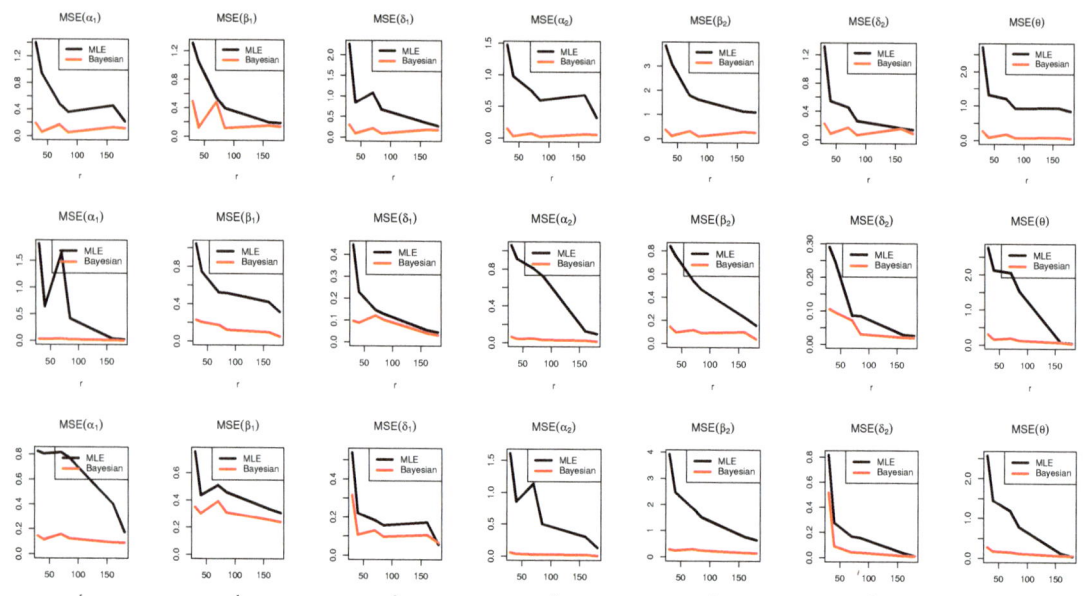

Figure 1. Comparison between MLE and Bayesian by MSE.

7. Application of Real Data

The data set contains n = 50 simulated primitive computer series systems, each with a processor and a memory. The computer works if both components of the system are in good working order. Let us assume that the system is going through a latent deterioration process. Over a small period of time, the deterioration progresses swiftly (in hours). The system becomes more susceptible to shocks as a result, making it possible for a lethal shock to randomly destroy either the first, the second, or both components. Because a fatal shock can kill both components at once, the independence premise could not be reliable, hence we used the Farlie–Gumbel–Morgenstern to examine this issue. Life of the processor can be denoted as X and memory lifetime can be denoted as Y. The data set is shown in Table A1 in Appendix A.

In order to check the outliers in the processor and memory data, Figure 2 was obtained, where a scatter plot with a boxplot was used to differentiate between the various groups. In contrast, Figure 3 presents the dependence plot of processor and memory, which have been reviewed to indicate independence in the data and the density distribution of numerical data. Figures 2 and 3 show that the data have right-skewed shapes, non-symmetric ships, and low negative correlation.

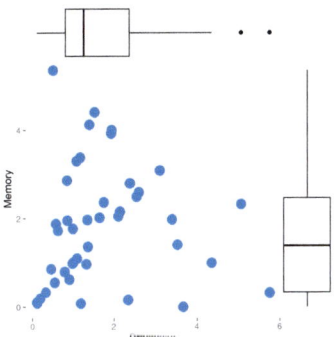

Figure 2. Scatter plot with boxplot.

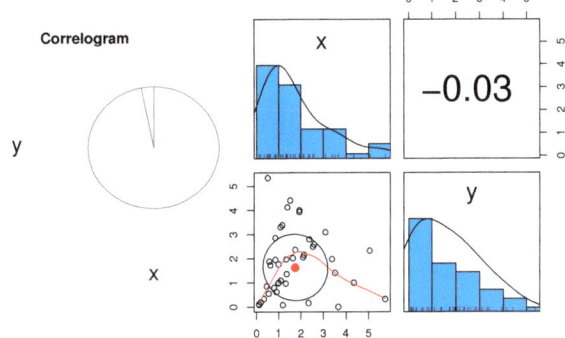

Figure 3. Correlation measure and its plot.

The MLE estimator of marginal parameters with standard error (SE) and Kolmogorov–Smirnov (KS) statistics with the *p*-value for all competing models, as MExE is included in Table 1. Figure 4 displays the estimated CDF with empirical CDF, pdf with histogram, and PP-plots for each sample. Table 1 includes the estimates' results along with various goodness-of-fit indices. We note that the data are consistent with this distribution.

Table 1. MLE of BMExE parameters with SE and KS test.

	Processor		Memory	
	Estimates	**SE**	**Estimates**	**SE**
θ	1.1278	0.0556	2.0966	0.0913
σ	0.0311	0.3501	0.0586	0.5922
δ	0.3843	0.0780	0.4618	0.1371
KS	1.1278		2.0966	
PVKS	0.9213		0.3040	

Table 2 discusses different measures of goodness-of-fit, including AIC, CAIC, the BIC, and (HQIC), as well as the MLE estimator for the marginal parameters with SE. The comparison was performed for three models: BMExE, Bivariate Lomax–Claim (BLC), which was introduced by [41], bivariate Fréchet (BF) by [6], and bivariate Lomax (BL) by [42].

Table 2 displays the MLE with SE for the BMExE model parameters. Comparisons of bivariate models based on the BMExE model, utilizing the AIC, CIAC, BIC, and HQIC metrics, are covered in Table 2. These findings lead us to the conclusion that the BMExE distribution outperforms other bivariate distributions in terms of the goodness-of-fit measures mentioned before. We found that Bayesian estimates outperform MLE in terms of the value of SE when comparing them.

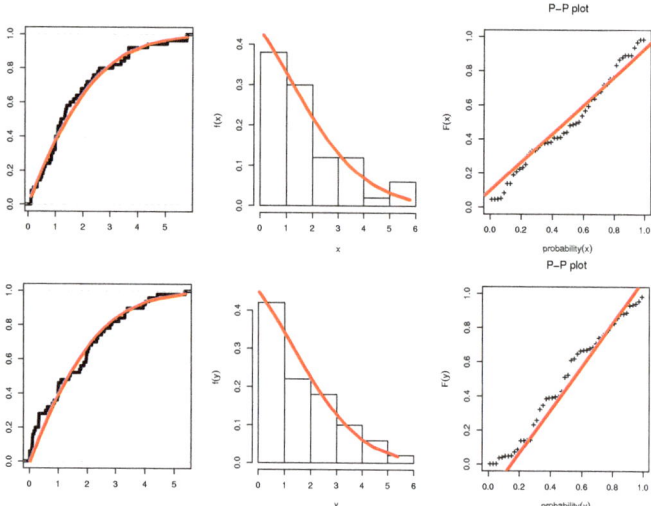

Figure 4. Marginal estimated plotting.

The distributions of the parameters from the computer data set were trace plotted in Figure 5 for $\theta_1, \sigma_1, \delta_1, \theta_2, \sigma_2$ and δ_2, to follow the posterior density of the MCMC outputs when r = 50 (complete sample). The marginal posterior density estimates of the parameters of the BMExE distribution, along with their histograms based on 10,000 chain values, are also displayed in Figure 6 for $\theta_1, \sigma_1, \delta_1, \theta_2, \sigma_2$ and δ_2. In the case of the copula parameter, we generated a plot of the MCMC, shown in Figure 7. All of the generated posteriors are symmetric with regard to the theoretical posterior density functions, according to the estimations, which are obvious in these figures.

Table 2. MLE with SE and different measures to compare between bivariate models.

	BMExE		BLC		BF		BL	
	Estimates	SE	Estimates	SE	Estimates	SE	Estimates	SE
θ_1	0.4829	0.2111	134.2571	56.9368	0.6353	0.1110	172.3502	464.2317
σ_1	0.6469	0.3098	0.0045	0.0018	0.8886	0.0889	103.0818	275.7959
δ_1	0.6425	0.2529	825.2748	5.6516				
θ_2	1.5264	0.7685	68.9989	56.7898	0.2456	0.0893	86.7328	164.7504
σ_2	0.0125	0.0309	0.0095	0.0078	0.4201	0.0389	55.3910	103.7675
δ_2	0.2967	0.1472	369.3705	256.5926				
λ	0.4050	0.3183	0.5054	0.3286	0.7265	0.4591	0.5006	0.3314
AIC	312.8515		319.6854		386.8855		313.2396	
CAIC	315.5182		322.3521		388.2492		317.6033	
BIC	322.2357		333.0696		396.4457		322.7998	
HQIC	317.9483		324.7822		390.5261		318.8802	

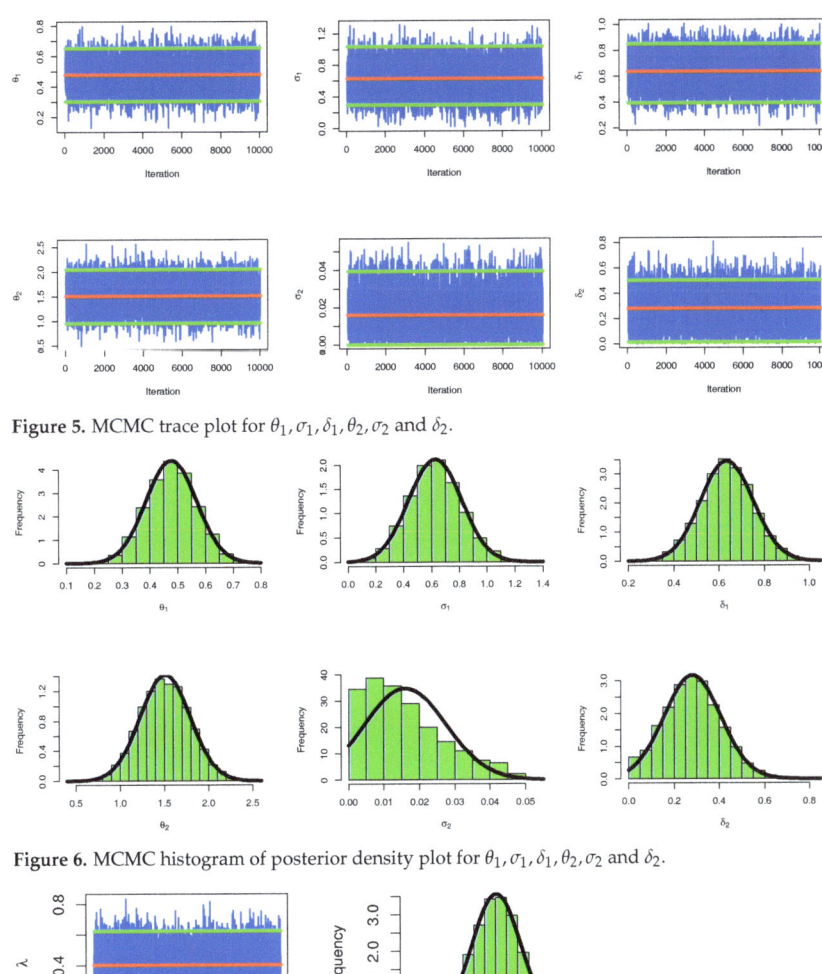

Figure 5. MCMC trace plot for $\theta_1, \sigma_1, \delta_1, \theta_2, \sigma_2$ and δ_2.

Figure 6. MCMC histogram of posterior density plot for $\theta_1, \sigma_1, \delta_1, \theta_2, \sigma_2$ and δ_2.

Figure 7. MCMC trace and MCMC histogram of posterior density plot for λ.

Table 3 discusses point and interval estimation for the unknown parameters using the MLE and Bayesian methods based on Type-II censored samples with different samples and a 95% confidence level. The results show how well the MCMC algorithm converges and how near the 95% asymptotic and higher posterior credible confidence interval boundaries are spaced. Also, the SE has a downward trend when r increases.

Table 3. MLE and Bayesian methods based on Type-II censored samples with different sample sizes.

r		MLE				Bayesian			
		Estimates	SE	Lower	Upper	Estimates	SE	Lower	Upper
50	θ_1	0.4829	0.2111	0.0692	0.8966	0.4769	0.0906	0.3038	0.6543
	σ_1	0.6469	0.3098	0.0397	1.2540	0.6268	0.1881	0.2954	1.0398
	δ_1	0.6425	0.2529	0.1468	1.1383	0.6347	0.1156	0.3928	0.8515
	θ_2	1.5264	0.7685	0.0201	3.0328	1.5087	0.2826	0.9539	2.0453
	σ_2	0.0125	0.0309	0.0048	0.0731	0.2805	0.0115	0.00001	0.0395
	δ_2	0.2967	0.1472	0.0082	0.5852	0.2805	0.1262	0.0136	0.5034
	λ	0.4050	0.3183	−0.2189	1.0289	0.4032	0.1122	0.1867	0.6240
35	α_1	0.4828	0.2121	0.0701	0.8955	0.4763	0.0902	0.3043	0.6537
	β_1	0.6473	0.3976	0.1320	1.4265	0.6266	0.1879	0.2582	1.0020
	δ_1	0.6424	0.3053	0.0440	1.2407	0.6344	0.1156	0.3926	0.8513
	α_2	1.5245	0.8049	0.0531	3.1021	1.5088	0.2806	0.9571	2.0395
	β_2	0.0126	0.0460	0.0078	0.1028	0.2815	0.0112	0.0000	0.0389
	δ_2	0.2971	0.2460	0.0185	0.7793	0.2815	0.1256	0.0134	0.5024
	λ	0.4051	0.3218	−0.2188	1.0289	0.4028	0.1121	0.1831	0.6203
30	α_1	1.5527	0.7124	0.1564	2.9490	1.5461	0.3553	0.8464	2.2203
	β_1	0.0048	0.3995	0.0000	0.1123	0.0406	0.0321	0.0000	0.1043
	δ_1	0.2941	0.3623	0.0482	0.7364	0.2974	0.1510	0.0001	0.5581
	α_2	1.5480	0.9020	0.2199	3.3160	1.5489	0.2984	0.9447	2.0986
	β_2	0.0119	0.0604	0.0011	0.1303	0.2936	0.0311	0.0000	0.1002
	δ_2	0.2921	0.2777	0.0115	0.6993	0.2936	0.1352	0.0085	0.5298
	λ	0.3895	0.3283	−0.2539	1.0330	0.3873	0.1138	0.1641	0.6081

8. Conclusions

In this paper, we proposed a new bivariate family called the BMExE, which is derived from the Farlie–Gumbel–Morgenster copula function and the modified extended exponential distribution. This family's statistical properties have been described; as a result, it can be used very efficiently in life testing data such as for medical data and computer survival times. The maximum likelihood method and the Bayesian method with the M–H algorithm were used to estimate parameters. Three types of prior distributions were assumed; the informative Gamma, the non-informative, and the non-prior distributions. As a result, we can conclude that Bayesian estimation is the best estimator for the BMExE distribution. Furthermore, the new bivariate model can be utilized in place of other classical bivariate distributions for a variety of applications. The BMExE model is efficient because the marginal functions are the same as the basic distribution; in addition, the moment-generating function and product moments have closed forms. For estimating unknown parameters, three types of confidence intervals were considered: asymptomatic, bootstrap, and Bayesian credible intervals. A real data analysis based on computers' failure times for their processors and memories was performed using some goodness-of-fit tests; the analysis shows the suitability of the new bivariate model with the failure times of computers. Some limitations may appear in modeling bivariate data, such as the skewness, the sample size, and the dependence issue between failure causes. Hence, more work is still needed in this field to improve more flexible bivariate models.

Author Contributions: Conceptualization, D.A.R. and H.H.A.; methodology, D.A.R. and E.M.A.; software, E.M.A.; validation, D.A.R. and H.H.A.; formal analysis, E.M.A.; investigation, D.A.R.; resources, H.H.A.; data curation, E.M.A.; writing—original draft preparation, D.A.R. and E.M.A.; writing—review and editing, D.A.R., H.H.A. and E.M.A.. All authors have read and agreed to the published version of the manuscript.

Funding: This work was supported by the Deanship of Scientific Research, Vice Presidency for Graduate Studies and Scientific Research, King Faisal University, Saudi Arabia [GRANT No. 3295].

Institutional Review Board Statement: Not applicable.

Informed Consent Statement: Not applicable.

Data Availability Statement: All data are available in the manuscript.

Acknowledgments: The authors would like to thank Editor in Chief, Associate Editor, and referees for their helpful and valuable suggestions and comments that greatly improved the paper.

Conflicts of Interest: The authors declare no conflict of interest.

Appendix A

Table A1. Computer works data.

i	X	Y	i	X	Y
1	1.9292	3.9291	26	0.1115	0.1115
2	3.6621	0.0026	27	0.8503	2.8578
3	3.6621	0.0026	28	0.1955	0.1955
4	3.6621	0.0026	29	0.4614	0.8584
5	1.0833	3.3059	30	3.3887	1.9796
6	1.0833	3.3059	31	0.1181	0.0884
7	0.3309	0.3309	32	5.0533	2.3238
8	0.3309	0.3309	33	1.6465	2.0197
9	0.5784	1.8795	34	0.9096	0.6214
10	0.552	0.552	35	1.7494	2.3643
11	1.9386	4.0043	36	0.1058	0.1058
12	2.1	2.0513	37	0.1058	0.1058
13	0.9867	0.9867	38	0.9938	1.7689
14	0.9867	0.9867	39	5.7561	0.3212
15	1.3989	4.1268	40	5.7561	0.3212
16	2.3757	2.7953	41	0.627	1.7289
17	3.5202	1.4095	42	0.7947	0.7947
18	2.3364	0.1624	43	0.5079	5.3535
19	0.8584	1.9556	44	2.5913	2.5913
20	4.3435	1.0001	45	2.5372	2.4923
21	1.1739	3.3857	46	1.1917	0.0801
22	1.3482	1.9705	47	1.5254	4.4088
23	3.0935	3.0935	48	1.0986	1.0986
24	2.1396	2.1548	49	1.0051	1.0051
25	1.3288	0.9689	50	1.364	1.364

Table A2. MLE and Bayesian for parameters of BMExE distribution: Case I.

							Bayesian						
Case I			MLE				Gamma Prior			Non-Informative			Non-Prior		
n	r		Bias	MSE	LACI	LBP	LBT	Bias	MSE	LCCI	Bias	MSE	LCCI	Bias	MSE	LCCI
40	30	θ_1	0.3412	1.3996	4.4426	0.1479	0.1473	0.0762	0.1886	1.5742	0.1217	0.2748	1.6517	0.0810	0.0842	0.9829
		σ_1	0.0480	1.3054	4.4770	0.1449	0.1450	−0.0374	0.4946	2.7226	0.4353	4.5401	7.8002	−0.0109	0.4360	2.5702
		δ_1	0.0471	2.2593	5.8922	0.1880	0.1880	0.0084	0.2977	1.9582	0.5365	1.8851	4.5058	0.0195	0.3334	2.1709
		θ_2	0.5006	1.4703	4.3314	0.1380	0.1372	0.1038	0.1481	1.1580	0.1151	0.3260	1.9067	0.0655	0.0881	1.1062
		σ_2	0.0612	3.8303	7.6720	0.2499	0.2486	−0.0612	0.3732	2.1782	0.3701	1.9305	4.7939	0.0355	0.3613	2.1449
		δ_2	−0.1086	1.3166	4.4799	0.1466	0.1479	0.0238	0.2260	1.8376	0.3670	0.9104	3.0914	0.0463	0.2627	1.8758
		λ	−0.0128	2.7066	2.2723	0.9092	0.9489	0.0531	0.2739	1.9791	−0.0151	0.1397	1.4803	0.0939	0.1211	1.1772
	40	θ_1	0.3789	0.9405	3.5011	0.1116	0.1115	0.0248	0.0601	0.9280	0.0741	0.2241	1.6587	0.0209	0.0248	0.5948
		σ_1	0.0017	1.0461	4.0114	0.1310	0.1309	−0.0054	0.1207	1.3744	0.6473	4.1495	6.8329	0.0028	0.1214	1.3236
		δ_1	−0.1353	0.8414	3.5582	0.1070	0.1072	−0.0141	0.0991	1.2076	0.5124	1.6294	4.0504	−0.0029	0.1025	1.1892
		θ_2	0.4732	0.9750	3.3989	0.1111	0.1099	0.0326	0.0340	0.6976	0.0881	0.2925	1.7683	0.0289	0.0343	0.7008
		σ_2	0.0424	3.0921	7.0710	0.2338	0.2528	−0.0216	0.1197	1.3350	0.4615	1.8002	4.7378	−0.0031	0.1128	1.3177
		δ_2	−0.2191	0.5427	2.7584	0.0859	0.0853	−0.0063	0.0858	1.0969	0.3833	0.8560	2.8654	0.0102	0.0899	1.1200
		λ	0.3572	1.3152	1.9026	0.6857	0.8163	−0.0014	0.0898	1.1513	0.0020	0.1154	1.2721	0.0315	0.0762	1.0194
100	70	θ_1	0.2005	0.4766	2.5908	0.0857	0.0861	0.0831	0.1709	1.5082	0.1478	0.3577	2.0246	0.0859	0.0853	0.9791
		σ_1	0.0383	0.5458	2.8936	0.0920	0.0912	−0.0185	0.4816	2.6493	0.2961	3.5878	7.1639	0.0040	0.4238	2.5265
		δ_1	0.0124	1.0822	4.0797	0.1297	0.1314	0.0143	0.2194	1.7979	0.4568	1.4503	3.8907	0.0154	0.2982	2.0084
		θ_2	0.3203	0.7531	3.1631	0.1007	0.1003	0.0645	0.0733	0.9884	0.0943	0.2421	1.5769	0.0764	0.0858	1.0045
		σ_2	−0.0371	1.7746	5.2226	0.1596	0.1691	−0.0579	0.3249	2.0676	0.4994	1.7361	4.3245	0.0156	0.3162	2.0386
		δ_2	−0.1170	0.4535	2.6011	0.0846	0.0877	0.0397	0.1725	1.5271	0.2139	0.4737	2.4359	0.0260	0.1933	1.6283
		λ	−0.1343	1.2022	2.5160	0.4022	0.4051	0.0731	0.1797	1.6045	0.0071	0.1023	1.2283	0.0089	0.1014	1.1652
	85	θ_1	0.1184	0.3587	2.3968	0.0800	0.0791	0.0089	0.0540	0.9089	0.1180	0.2578	1.6657	0.0314	0.0219	0.5475
		σ_1	−0.0124	0.3896	2.4476	0.0755	0.0755	−0.0332	0.1171	1.3321	0.4489	3.3804	6.6175	0.0018	0.1106	1.2615
		δ_1	0.0181	0.6702	3.2100	0.1048	0.1065	0.0219	0.0938	1.1289	0.3512	0.9778	3.3468	0.0009	0.1038	1.2222
		θ_2	0.3304	0.5922	2.7259	0.0834	0.0843	0.0222	0.0240	0.5846	0.0813	0.2374	1.7729	0.0222	0.0277	0.6041
		σ_2	0.0336	1.6090	4.4019	0.1413	0.1407	−0.0135	0.1055	1.2469	0.3656	1.5288	4.0971	−0.0044	0.1088	1.2830
		δ_2	−0.1607	0.2637	1.9129	0.0625	0.0621	−0.0059	0.0712	1.0236	0.1841	0.3709	2.0891	0.0282	0.0900	1.1145
		λ	0.2600	0.9221	1.1643	0.4015	0.3612	0.0221	0.0818	1.1452	0.0057	0.0881	1.1228	0.0062	0.0746	0.9865
200	160	θ_1	0.0379	0.4541	2.4622	0.0813	0.0814	0.0622	0.1314	1.3060	0.1110	0.2098	1.4561	0.0572	0.0497	0.7810
		σ_1	0.0070	0.1990	1.7497	0.0638	0.0632	−0.0654	0.1542	1.5111	0.2984	2.2272	5.2226	0.0084	0.3763	2.3492
		δ_1	−0.0161	0.3544	2.3345	0.0887	0.0883	0.0493	0.1919	1.6173	0.2432	0.6166	2.7866	0.0172	0.1929	1.6450
		θ_2	0.1676	0.6768	3.1594	0.1186	0.1206	0.0736	0.0620	0.8364	0.0894	0.1834	1.3994	0.0472	0.0610	0.8897
		σ_2	0.0555	1.1195	4.1447	0.1483	0.1484	−0.0267	0.2885	1.9571	0.3119	1.1073	3.5263	0.0364	0.3003	2.0330
		δ_2	−0.1162	0.1662	1.5330	0.0597	0.0590	0.0307	0.1616	1.1683	0.1220	0.2232	1.6960	0.0460	0.1202	1.2825
		λ	0.1469	0.9311	2.4837	0.3904	0.3903	0.0129	0.0866	1.1384	0.0010	0.0635	0.9623	0.0332	0.0701	0.9589
	180	θ_1	0.0615	0.2162	1.8077	0.0547	0.0565	0.0798	0.1190	1.2598	0.1030	0.1514	1.3370	0.0232	0.0145	0.4598
		σ_1	−0.0150	0.1893	1.7054	0.0549	0.0539	−0.0746	0.1413	1.4981	0.3180	2.1978	5.2201	0.0028	0.1174	1.3480
		δ_1	0.0134	0.2770	2.0636	0.0623	0.0625	0.0302	0.1821	1.5441	0.1674	0.4576	2.4139	−0.0110	0.0862	1.1300
		θ_2	0.2109	0.3179	2.0509	0.0639	0.0640	0.0702	0.0552	0.8278	0.0711	0.1694	1.3047	0.0198	0.0219	0.5728
		σ_2	0.0732	1.0995	4.1024	0.1347	0.1347	−0.0503	0.2711	1.9576	0.2984	1.0912	3.4578	0.0279	0.1174	1.3413
		δ_2	−0.1152	0.1436	1.4160	0.0423	0.0433	0.0238	0.0938	1.1715	0.1017	0.2015	1.6516	0.0111	0.0644	0.9630
		λ	0.0788	0.8367	1.2892	0.2609	0.2481	0.0159	0.0588	0.9287	0.0008	0.0477	0.8459	0.0173	0.0487	0.8320

Table A3. MLE and Bayesian for parameters of BMExE distribution: Case II.

Case II			MLE				Bayesian						
							Gamma Prior			Non-Informative			Non-Prior		
n	r		Bias	MSE	LACI	LBP	LBT	Bias	MSE	LCCI	Bias	MSE	LCCI	Bias	MSE	LCCI
40	30	θ_1	−0.0279	1.8203	5.2981	0.3369	0.3395	0.0555	0.0339	0.6017	0.0171	0.0348	0.6872	0.0437	0.0337	0.6453
		σ_1	0.1303	1.0415	3.9755	0.2502	0.2520	0.0512	0.2294	1.3993	0.2662	0.4775	2.0356	0.1279	0.2179	1.5395
		δ_1	−0.2400	0.4439	2.4410	0.1499	0.1494	0.0227	0.0962	1.2855	0.2716	0.5271	2.4130	0.1093	0.1740	1.4573
		θ_2	0.2542	1.0549	3.9085	0.2434	0.2450	0.0448	0.0647	0.8840	0.0836	0.1420	1.0932	0.0092	0.0686	0.8091
		σ_2	0.1391	0.8391	3.5561	0.2237	0.2334	0.0655	0.1460	1.1260	0.1365	0.1724	1.3382	0.1963	0.1677	1.2356
		δ_2	−0.2096	0.2894	1.9458	0.1249	0.1248	0.0495	0.1056	1.0977	0.1796	0.2184	1.3567	0.1562	0.1430	1.1551
		λ	1.2454	2.7575	3.0965	1.9012	1.8937	−0.0058	0.3108	2.0259	−0.1318	0.4159	2.6092	−0.3361	0.1794	0.8895
	40	θ_1	0.1300	0.6387	3.1053	0.2698	0.2819	0.0604	0.0294	0.6224	0.0227	0.0342	0.6132	0.0556	0.0325	0.5955
		σ_1	0.0063	0.7471	3.4038	0.2328	0.2234	0.0423	0.2041	1.2985	0.2997	0.4658	2.1937	0.0998	0.1830	1.4200
		δ_1	−0.1533	0.2277	1.7793	0.1372	0.1711	0.0265	0.0892	1.1587	0.2545	0.3877	1.9279	0.0873	0.1530	1.2701
		θ_2	0.1116	0.9097	3.8101	0.2401	0.2341	0.0369	0.0447	0.7428	0.0596	0.1174	1.0819	0.0318	0.0662	0.8851
		σ_2	0.1240	0.7488	3.4710	0.2144	0.2104	0.0277	0.0972	1.1349	0.1985	0.1523	1.5292	0.1601	0.1458	1.2523
		δ_2	−0.1690	0.2474	1.8421	0.1128	0.1179	0.0970	0.0965	1.1842	0.1511	0.1912	1.5130	0.1005	0.1231	1.1778
		λ	1.0185	2.1196	2.9589	1.7804	1.6423	−0.0121	0.1722	1.5851	−0.0319	0.2746	1.8937	−0.2774	0.1370	0.8157
100	70	θ_1	−0.1095	1.6453	5.0189	0.2909	0.3067	0.0703	0.0364	0.6788	0.0361	0.0484	0.6894	0.0706	0.0434	0.6853
		σ_1	0.1006	0.5213	2.8076	0.1706	0.1772	0.0064	0.1733	1.3901	0.2683	0.4557	2.1750	0.0895	0.1961	1.5373
		δ_1	−0.1176	0.1454	1.4243	0.0891	0.0874	0.0429	0.1210	1.1089	0.1717	0.2516	1.6126	0.0401	0.1116	1.2140
		θ_2	0.1693	0.7996	3.4480	0.2175	0.2115	0.0335	0.0484	0.8184	0.0628	0.1170	1.0616	0.0294	0.0602	0.8275
		σ_2	0.0734	0.5409	2.8736	0.1652	0.1655	0.0696	0.1176	1.1169	0.1597	0.1912	1.3807	0.1539	0.1561	1.3075
		δ_2	−0.0960	0.0858	1.0866	0.0645	0.0654	0.1075	0.0723	0.9160	0.1497	0.1355	1.2080	0.1353	0.1152	1.1706
		λ	1.0587	2.0526	2.3198	1.7366	1.8791	−0.0813	0.1934	1.6415	−0.1023	0.2866	2.0637	−0.2824	0.1390	0.8436
	85	θ_1	−0.0451	0.4194	4.2869	0.2499	0.2525	0.0581	0.0235	0.5013	0.0293	0.0345	0.6078	0.0471	0.0298	0.5517
		σ_1	0.0631	0.5162	2.5090	0.1684	0.1688	−0.0234	0.1227	1.2411	0.2067	0.2890	1.7492	0.0914	0.1828	1.4852
		δ_1	−0.1571	0.1288	1.2670	0.0759	0.0753	0.0452	0.1020	1.0411	0.1467	0.1734	1.3763	0.0574	0.1004	1.1454
		θ_2	0.0517	0.7289	3.3466	0.2166	0.2128	0.0215	0.0338	0.6750	0.0399	0.0955	0.9692	0.0255	0.0601	0.8916
		σ_2	0.1248	0.4635	2.6283	0.1591	0.1635	0.0796	0.0929	1.0841	0.1743	0.1770	1.3729	0.1753	0.1475	1.2905
		δ_2	−0.0647	0.0847	0.9813	0.0637	0.0627	0.0472	0.0315	0.6182	0.1136	0.1048	1.1304	0.0967	0.0787	0.9380
		λ	0.9029	1.5177	1.4766	1.6946	1.7172	−0.0502	0.1330	1.3737	−0.0832	0.1542	1.4465	−0.2439	0.1101	0.7945
200	160	θ_1	0.0630	0.0385	0.7289	0.0228	0.0228	0.0352	0.0140	0.4294	0.0520	0.0408	0.6540	0.0503	0.0276	0.5496
		σ_1	0.0951	0.4219	2.5202	0.0783	0.0781	0.0199	0.0980	1.2165	0.1641	0.2698	1.7806	0.0819	0.1601	1.3876
		δ_1	−0.0684	0.0552	0.8813	0.0262	0.0264	0.0417	0.0413	0.7265	0.0697	0.0938	1.1145	0.0299	0.0641	0.9341
		θ_2	0.1185	0.1265	1.3152	0.0407	0.0413	0.0284	0.0270	0.5823	0.0554	0.0818	0.9720	0.0168	0.0459	0.7905
		σ_2	0.0848	0.2269	1.8383	0.0583	0.0584	0.0613	0.1015	1.0018	0.1241	0.1246	1.1581	0.1467	0.1230	1.1677
		δ_2	−0.0572	0.0294	0.6338	0.0195	0.0197	0.0340	0.0210	0.5236	0.0686	0.0527	0.8170	0.0790	0.0489	0.7562
		λ	0.0165	0.0783	1.0955	0.0347	0.0344	−0.0495	0.0730	1.0388	−0.0844	0.0882	1.1066	−0.1891	0.0708	0.6723
	180	θ_1	0.0467	0.0280	0.6301	0.0201	0.0203	0.0210	0.0072	0.3010	0.0331	0.0277	0.5189	0.0316	0.0128	0.3792
		σ_1	0.0834	0.3167	2.1827	0.0671	0.0673	−0.0091	0.0535	0.8947	0.1589	0.2240	1.6564	0.0172	0.0729	0.9874
		δ_1	−0.0549	0.0440	0.7944	0.0246	0.0246	0.0226	0.0315	0.5860	0.0688	0.0688	0.9622	0.0090	0.0422	0.7927
		θ_2	0.1012	0.0962	1.1496	0.0369	0.0365	0.0233	0.0167	0.4803	0.0406	0.0617	0.8306	0.0145	0.0255	0.6001
		σ_2	0.0616	0.1578	1.5390	0.0464	0.0454	0.0222	0.0407	0.7185	0.1255	0.1174	1.1285	0.0645	0.0581	0.8185
		δ_2	−0.0509	0.0259	0.5991	0.0191	0.0190	0.0101	0.0208	0.4362	0.0396	0.0350	0.6830	0.0292	0.0265	0.5947
		λ	0.0007	0.0452	0.8334	0.0257	0.0260	−0.0221	0.0386	0.7574	−0.0620	0.0603	0.9743	−0.1025	0.0326	0.5204

Table A4. MLE and Bayesian for parameters of BMExE distribution: Case III.

Case III							Bayesian							
			MLE				Gamma Prior			Non-Informative			Non-Prior		
n	r		Bias	MSE	LACI	LBP	LBT	Bias	MSE	LCCI	Bias	MSE	LCCI	Bias	MSE	LCCI
40	30	θ_1	0.1923	0.8245	3.4851	0.2058	0.2119	0.0354	0.1460	1.4163	0.1806	0.7681	2.6878	0.0378	0.1398	1.3746
		σ_1	−0.0762	0.7538	3.3966	0.2041	0.2107	−0.0276	0.3490	2.1310	0.5137	2.5342	5.5758	0.0451	0.3549	2.1955
		δ_1	0.0512	0.5382	2.8740	0.1781	0.1822	0.1288	0.3162	1.4772	0.2081	0.3847	2.1325	0.1143	0.1913	1.4679
		θ_2	0.5891	1.6109	4.4149	0.2718	0.2707	0.0457	0.0574	0.7639	0.0522	0.1198	1.0622	0.0610	0.0552	0.7885
		σ_2	0.1573	3.9219	7.7528	0.4526	0.4510	0.0062	0.2904	2.0252	0.6265	1.9657	4.3585	0.0191	0.3469	2.1114
		δ_2	−0.2288	0.8162	3.4324	0.2139	0.2261	0.1089	0.5168	1.2836	0.1395	0.2368	1.8228	0.1469	0.1516	1.2540
		λ	0.3553	2.5694	2.0682	0.6956	0.7551	−0.0314	0.2871	1.9752	−0.0393	0.2521	1.9196	−0.1894	0.1355	1.1279
	40	θ_1	0.1634	0.8066	3.3908	0.1100	0.1117	0.0635	0.1150	1.2182	0.1653	0.5826	2.5263	0.0239	0.1217	1.2579
		σ_1	0.0420	0.4366	2.5802	0.1499	0.1507	−0.0350	0.3021	2.1559	0.4860	2.1191	4.6266	0.0450	0.3480	2.3639
		δ_1	−0.1177	0.2210	1.7848	0.0557	0.0558	0.0428	0.1087	1.2527	0.2019	0.3467	2.0030	0.1315	0.1729	1.4465
		θ_2	0.4600	0.8574	3.1518	0.1036	0.1045	0.0453	0.0384	0.6968	0.0478	0.1029	1.1596	0.0593	0.0547	0.8508
		σ_2	0.0400	2.4696	6.1614	0.1969	0.1977	−0.0199	0.2513	1.9473	0.5020	1.6866	4.1549	0.0132	0.3208	2.0043
		δ_2	−0.2304	0.2754	1.8493	0.0572	0.0568	0.0608	0.0931	1.1854	0.1411	0.2218	1.7167	0.1279	0.1454	1.2049
		λ	0.3120	1.4408	1.8553	0.1351	13.5553	−0.0350	0.1761	1.6262	−0.0408	0.2019	1.7513	−0.1349	0.1083	1.0553
100	70	θ_1	−0.0025	0.8166	3.2795	0.2046	0.2062	0.1295	0.1591	1.4675	0.1227	0.5089	2.3896	0.0400	0.1354	1.3707
		σ_1	−0.0166	0.5084	2.8069	0.2587	0.2749	−0.0637	0.3914	2.2389	0.4839	2.0677	4.7579	0.0310	0.3058	2.1595
		δ_1	−0.0186	0.1834	1.6847	0.1527	0.1538	0.0526	0.1304	1.1904	0.1948	0.2893	1.8306	0.1000	0.1356	1.2789
		θ_2	0.0955	1.1364	4.1806	0.3891	0.4389	0.0436	0.0275	0.5419	0.0701	0.1054	1.0796	0.0612	0.0468	0.7340
		σ_2	0.2628	1.8646	5.2762	0.4980	0.5034	0.0237	0.3007	1.9400	0.4479	1.4193	3.9201	−0.0098	0.2722	1.9890
		δ_2	−0.0983	0.1676	1.5648	0.1472	0.1670	0.0316	0.0435	0.8041	0.0930	0.1360	1.3986	0.1148	0.1139	1.1451
		λ	1.2078	1.1718	1.5663	0.4768	0.4969	−0.0867	0.1532	1.3772	−0.0524	0.1962	1.6983	−0.1524	0.1148	1.0683
	85	θ_1	0.1841	0.7692	3.0369	0.2031	0.2046	0.0700	0.1245	1.2476	0.1159	0.4823	2.2657	0.0162	0.0432	0.7756
		σ_1	0.0065	0.4573	2.7480	0.1985	0.2006	−0.0118	0.3087	2.0627	0.4858	1.9393	4.5295	0.0304	0.0985	1.1868
		δ_1	−0.0888	0.1563	1.5132	0.1022	0.1013	0.0709	0.0980	1.1302	0.1622	0.2192	1.5615	0.0276	0.0536	0.8725
		θ_2	0.1851	0.5012	2.6844	0.1752	0.1740	0.0503	0.0269	0.5860	0.0533	0.0813	0.9255	0.0311	0.0178	0.4668
		σ_2	0.1570	1.5032	4.7770	0.3179	0.3172	−0.0383	0.2477	1.9164	0.4413	1.2153	3.6372	−0.0157	0.1073	1.2644
		δ_2	−0.0915	0.1565	1.5120	0.0993	0.1022	0.0601	0.0414	0.7286	0.0857	0.1046	1.1234	0.0289	0.0507	0.8505
		λ	0.4384	0.7753	1.2669	0.4188	0.4053	−0.0451	0.1249	1.2968	−0.0434	0.1267	1.3767	−0.0490	0.0633	0.8571
200	160	θ_1	0.0552	0.4036	2.5008	0.3098	0.3415	−0.0264	0.0947	1.0220	0.0993	0.3258	2.0504	0.0307	0.1006	1.1943
		σ_1	−0.0073	0.3355	2.2887	0.2979	0.3002	0.1020	0.2557	1.7316	0.3783	1.3526	3.9226	0.0251	0.2767	1.9947
		δ_1	0.0115	0.1721	1.5813	0.2253	0.2810	0.1026	0.1060	1.2579	0.1176	0.1351	1.2482	0.0828	0.0889	1.0708
		θ_2	0.1567	0.3046	2.0911	0.2613	0.2628	0.0384	0.0232	0.4418	0.0651	0.0810	0.9165	0.0441	0.0355	0.6420
		σ_2	−0.1110	0.7601	3.4170	0.4364	0.4354	−0.0211	0.1627	1.4987	0.3031	0.8765	3.1774	0.0195	0.2520	1.9188
		δ_2	−0.0502	0.0368	0.7315	0.0892	0.0911	0.0310	0.0172	0.4795	0.0582	0.0614	0.8532	0.0847	0.0606	0.8430
		λ	0.0776	0.1163	1.3121	0.1593	0.1600	−0.0188	0.0712	0.9691	−0.0481	0.0919	1.1572	−0.0820	0.0626	0.8729
	180	θ_1	0.0746	0.1750	1.6404	0.2769	0.2826	0.0159	0.0915	1.2362	0.0812	0.3046	1.9759	0.0218	0.0376	0.7306
		σ_1	0.1237	0.3061	2.0782	0.2577	0.2577	0.0078	0.2389	1.6017	0.3646	1.2295	3.8526	0.0064	0.0961	1.1813
		δ_1	−0.0026	0.0553	0.9370	0.1612	0.1579	0.0975	0.0644	0.8327	0.0805	0.0972	1.1153	0.0240	0.0428	0.7688
		θ_2	0.2025	0.1345	1.2187	0.2093	0.2126	0.0184	0.0086	0.2980	0.0422	0.0626	0.7998	0.0203	0.0112	0.3941
		σ_2	−0.2822	0.6379	2.9773	0.3517	0.3516	−0.0693	0.1471	1.1522	0.3478	0.8709	3.1675	−0.0140	0.0921	1.1367
		δ_2	−0.0278	0.0136	0.4516	0.0833	0.0840	0.1051	0.0146	0.3813	0.0377	0.0513	0.8087	0.0346	0.0363	0.7069
		λ	−0.0208	0.0434	0.8256	0.1461	0.1502	−0.0746	0.0615	0.8677	−0.0325	0.0609	0.9586	−0.0288	0.0441	0.7842

References

1. Nelsen, R.B. *An Introduction to Copulas*; Springer Science Business Media: New York, NY, USA, 2007.
2. Flores, A.Q. Testing Copula Functions as a Method to Derive Bivariate Weibull Distributions. In Proceedings of the American Political Science Association (APSA), Annual Meeting 2009, Toronto, ON, Canada, 3–6 September 2009; Volume 4.
3. Verrill, S.P.; Evans, J.W.; Kretschmann, D.E.; Hatfield, C.A. Asymptotically efficient estimation of a bivariate Gaussian–Weibull distribution and an introduction to the associated pseudo-truncated Weibull. *Commun. Stat. Theory Methods* **2015**, *44*, 2957–2975. [CrossRef]
4. El-Sherpieny, E.S.; Almetwally, E.M. Bivariate Generalized Rayleigh Distribution Based on Clayton Copula. In Proceedings of the Annual Conference on Statistics (54rd), Computer Science and Operation Research, Faculty of Graduate Studies for Statistical Research, Giza, Egypt, 9–11 December 2019; pp. 1–19.
5. Qura, M.E.; Fayomi, A.; Kilai, M.; Almetwally, E.M. Bivariate power Lomax distribution with medical applications. *PLoS ONE* **2023**, *18*, e0282581. [CrossRef] [PubMed]
6. Almetwally, E.M.; Muhammed, H.Z. On a Bivariate Frechet Distribution. *J. Stat. Appl. Prob.* **2020**, *9*, 1–21.
7. Almetwally, E.M.; Muhammed, H.Z.; El-Sherpieny, E.S.A. Bivariate Weibull Distribution: Properties and Different Methods of Estimation. *Ann. Data Sci.* **2020**, *7*, 163–193. [CrossRef]
8. Samanthi, R.G.M.; Sepanski, J. On bivariate Kumaraswamy-distorted copulas. *Commun. Stat. Theory Methods* **2020**, 1–19. [CrossRef]
9. Muhammed, H.Z. On a bivariate generalized inverted Kumaraswamy distribution. *Phys. A Stat. Mech. Appl.* **2020**, *553*, 124281. [CrossRef]
10. Alotaibi, R.M.; Rezk, H.R.; Ghosh, I.; Dey, S. Bivariate exponentiated half logistic distribution: Properties and application. *Commun. Stat. Theory Methods* **2020**, *50*, 1–23. [CrossRef]
11. Eliwa, M.; Alhussain, Z.; Ahmed, E.; Salah, M.; Ahmed, H.; El-Morshedy, M. Bivariate Gompertz generator of distributions: Statistical properties and estimation with application to model football data. *J. Natl. Sci. Found. Sri Lanka* **2020**, *48*, 149–162. [CrossRef]
12. Muhammed, H.Z. Bivariate inverse Weibull distribution. *J. Stat. Comput. Simul.* **2016**, *86*, 2335–2345. [CrossRef]
13. El-Morshedy, M.; Alhussain, Z.A.; Atta, D.; Almetwally, E.M.; Eliwa, M.S. Bivariate Burr X generator of distributions: Properties and estimation methods with applications to complete and type-II censored samples. *Mathematics* **2020**, *8*, 131. [CrossRef]
14. Eliwa, M.S.; El-Morshedy, M. Bivariate odd Weibull-G family of distributions: Properties, Bayesian and non-Bayesian estimation with bootstrap confidence intervals and application. *J. Taibah Univ. Sci.* **2020**, *14*, 331–345. [CrossRef]
15. Rafiei, M.; Iranmanesh, A. A Bivariate Gamma Distribution Whose Marginals are Finite Mixtures of Gamma Distributions. *Statistics. Optim. Inf. Comput.* **2020**, *8*, 950–971. [CrossRef]
16. Bekker, A.; Ferreira, J. Bivariate gamma type distributions for modeling wireless performance metrics. *Stat. Optim. Inf. Comput.* **2018**, *6*, 335–353. [CrossRef]
17. Genest, C.; Quessy, J.F.; Rémillard, B. Goodness-of-fit procedures for copula models based on the probability integral transformation. *Scand. J. Stat.* **2006**, *33*, 337–366. [CrossRef]
18. Suzuki, A.K.; Louzada-Neto, F.; Cancho, V.G.; Barriga, G.D. The FGM bivariate lifetime copula model: A bayesian approach. *Adv. Appl. Stat.* **2011**, *21*, 55–76.
19. El-Sherpieny, E.-S.A.; Almetwally, E.M.; Muhammed, H.Z. Bivariate Weibull-G Family Based on Copula Function: Properties, Bayesian and non-Bayesian Estimation and Applications. *Stat. Optim. Inf. Comput.* **2022**, *10*, 678–709. [CrossRef]
20. Pabaghi, Z.; Bazrafshan, O.; Zamani, H.; Shekari, M.; Singh, V.P. Bivariate Analysis of Extreme Precipitation Using Copula Functions in Arid and Semi-Arid Regions. *Atmosphere* **2023**, *14*, 275. [CrossRef]
21. Sklar, A. Random variables, joint distribution functions, and copulas. *Kybernetika* **1973**, *9*, 449–460.
22. Gumbel, E.J. Bivariate exponential distributions. *J. Am. Stat. Assoc.* **1960**, *55*, 698–707. [CrossRef]
23. El-Damcese, M.A.; Ramadan, D.A. Studies on properties and estimation problems for modified extension of exponential distribution. *Int. J. Comput. Appl.* **2015**, *125*, 21–28. [CrossRef]
24. Mansour, M.; Ramadan, D.A. Statistical inference of the parameters of the modified extended exponential distribution under the type-II hybrid censoring scheme. *J. Appl. Prob. Stat.* **2020**, *15*, 19–44.
25. Ramadan, D.A.; Aboshady, M.S.; Mansour, M.M.M. Inference for modified extended exponential distribution based on progressively Type-I hybrid censored data with application to some mechanical models. *J. Appl. Prob. Stat.* **2022**, *9*, 1510.
26. Mahmoud, M.A.W.; Ramadan, D.A.; Mansour, M.M.M. Estimation of lifetime parameters of the modified extended exponential distribution with application to a mechanical model. *Commun. Stat. Simul. Comput.* **2022**, *51*, 7005–7018. [CrossRef]
27. Osmetti, S.A.; Chiodini, P.M. A method of moments to estimate bivariate survival functions: The copula approach. *Statistica* **2011**, *71*, 469–488.
28. Basu, A.P. Bivariate failure rate. *J. Am. Stat. Assoc.* **1971**, *66*, 103–104. [CrossRef]
29. Kim, G.; Silvapulle, M.J.; Silvapulle, P. Comparison of semiparametric and parametric methods for estimating copulas. *Comput. Stat. Data Anal.* **2007**, *51*, 2836–2850. [CrossRef]
30. Bourguignon, M.; Silva, R.B.; Cordeiro, G.M. The Weibull-G family of probability distributions. *J. Data Sci.* **2014**, *12*, 53–68. [CrossRef]

31. Dey, S. Bayesian estimation of the shape parameter of the generalized exponential distribution under different loss functions. *Pak. J. Stat. Oper. Res.* **2010**, *6*, 163–174. [CrossRef]
32. Aliyu, Y.; Yahaya, A. Bayesian estimation of the shape parameter of generalized Rayleigh distribution under non-informative prior. *Int. J. Adv. Stat. Prob.* **2016**, *4*, 1–10. [CrossRef]
33. Ahmad, K.; Ahmad, S.P.; Ahmed, A. On parameter estimation of erlang distribution using bayesian method under different loss functions. In Proceedings of the International Conference on Advances in Computers, Communication, and Electronic Engineering, Dehradun, India, 1–2 May 2015; pp. 200–206.
34. Metropolis, N.; Rosenbluth, A.W.; Rosenbluth, M.N.; Teller, A.H.; Teller, E. Equation of state calculations by fast computing machines. *J. Chem. Phys.* **1953**, *21*, 1087–1091. [CrossRef]
35. Hastings, W.K. Monte Carlo sampling methods using Markov chains and their applications. *Biometrika* **1970**, *57*, 97–109. [CrossRef]
36. Robert, C.P.; Casella, G. *Monte Carlo Statistical Methods*; Springer: New York, NY, USA, 2004.
37. Casella, G. Monte Carlo Statistical Methods. Ph.D. Thesis, University of Florida, Gainesville, FL, USA, 2004.
38. El-Sherpieny, A.; Muhammed, H.Z.; Almetwally, E.M. Progressive Type-II Censored Samples for Bivariate Weibull Distribution with Economic and Medical Applications. *Ann. Data Sci.* **2022**, *9*, 1–35. [CrossRef]
39. Chen, M.H.; Shao, Q.M. Monte Carlo estimation of Bayesian credible and HPD intervals. *J. Comput. Graph. Stat.* **1999**, *8*, 69–92.
40. Efron, B. Bootstrap methods: Another look at the jackknife. In *Breakthroughs in Statistics*; Springer: New York, NY, USA, 1992; pp. 569–593.
41. Zhao, J.; Faqiri, H.; Ahmad, Z.; Emam, W.; Yusuf, M.; Sharawy, A.M. The lomax-claim model: Bivariate extension and applications to financial data. *Complexity* **2021**, *2021*, 9993611. [CrossRef]
42. Philip, A.; Thomas, P.Y. On concomitants of order statistics and its application in defining ranked set sampling from Farlie-Gumbel-Morgenstern bivariate Lomax distribution. *JIRSS* **2017**, *16*, 67–95.

Disclaimer/Publisher's Note: The statements, opinions and data contained in all publications are solely those of the individual author(s) and contributor(s) and not of MDPI and/or the editor(s). MDPI and/or the editor(s) disclaim responsibility for any injury to people or property resulting from any ideas, methods, instructions or products referred to in the content.

Article

A Three-Stage Nonparametric Kernel-Based Time Series Model Based on Fuzzy Data

Gholamreza Hesamian [1], Arne Johannssen [2,*] and Nataliya Chukhrova [3]

1. Department of Statistics, Payame Noor University, Tehran 19395-3697, Iran; gh.hesamian@pnu.ac.ir
2. Faculty of Business Administration, University of Hamburg, 20146 Hamburg, Germany
3. HafenCity University of Hamburg, 20457 Hamburg, Germany; nataliya.chukhrova@hcu-hamburg.de
* Correspondence: arne.johannssen@uni-hamburg.de

Abstract: In this paper, a nonlinear time series model is developed for the case when the underlying time series data are reported by *LR* fuzzy numbers. To this end, we present a three-stage nonparametric kernel-based estimation procedure for the center as well as the left and right spreads of the unknown nonlinear fuzzy smooth function. In each stage, the nonparametric Nadaraya–Watson estimator is used to evaluate the center and the spreads of the fuzzy smooth function. A hybrid algorithm is proposed to estimate the unknown optimal bandwidths and autoregressive order simultaneously. Various goodness-of-fit measures are utilized for performance assessment of the fuzzy nonlinear kernel-based time series model and for comparative analysis. The practical applicability and superiority of the novel approach in comparison with further fuzzy time series models are demonstrated via a simulation study and some real-life applications.

Keywords: fuzzy regression; fuzzy time series model; nonparametric time series analysis; time series analysis

MSC: 03E72; 37M10; 62A86

1. Introduction

The field of time series analysis comprises methods used to analyze the characteristics of a response variable with respect to time. It takes into consideration the fact that observations made over time may have an internal structure (such as autocorrelations, trends, seasonal and/or cyclic variations) that should be accounted for. The main aims of time series analysis are as follows:

- *Trend analysis*: to identify the underlying pattern or trend in the data over time, such as an upward or downward trend.
- *Seasonality analysis*: to identify if the data exhibit a repeating pattern over a set period, such as daily, weekly, or yearly.
- *Forecasting*: to forecast future values using historical data.
- *Anomaly detection*: to identify any unusual or unexpected observations in the data that deviate from the normal pattern.
- *Model selection*: to choose an appropriate model to represent the underlying relationships between variables in the data.
- *Noise reduction*: to remove any unwanted variability or random fluctuations from the data to improve the accuracy of predictions and make the underlying patterns more clear.

These aims can inform decision makers, provide insight into the underlying patterns and relationships in the data, and support the development of data-driven strategies in various fields such as economics, engineering, finance, and more (see, e.g., [1–8]).

Common time series models rely on exact observations and ensure crisp predictions. However, due to various uncertainty factors, it is sometimes preferable to make predictions using imprecise values. For instance, we usually observe imprecise observations in carbon emissions, social benefits and oil reserves, among others [9]. Traditional statistical time series models fail to address prediction problems based on ambiguous or vague information represented by fuzzy data. This shortcoming can be overcome by time series models that use techniques of fuzzy statistics. In general, fuzzy statistics is a branch of statistics that deals with uncertainty and imprecision, e.g., in the data. It includes, for instance, the fields of fuzzy estimation, fuzzy regression, fuzzy clustering, and fuzzy hypothesis testing [10–12].

Fuzzy time series models were originally introduced in 1993 [13], and since then they have replaced conventional (crisp) time series approaches when observations are uncertain. When considering fuzzy time series models, the prediction of future values requires three principal steps. In step 1, the exact data are reported. In step 2, through the identification of fuzzy logical relations [14,15], the predictions are transformed into fuzzy quantities. Finally, step 3 provides a defuzzification approach [16–22] to transform the fuzzy values into crisp ones. The techniques used to identify fuzzy logical relations in step 2 primarily involve fuzzy logical relation groups and matrices [13,23–34], soft computing methods [35–44], and statistical approaches in interaction with fuzzy logic [21,24,45–47]. Step 2 is an essential part of the predictive power of the presented model. Fuzzy time series models that rely on imprecise observations have attracted substantial attention in recent years, mainly due to their high applicability to real-life problems.

In fact, a lot of researchers have focused on time series models using imprecise observations. The soft computing techniques employed in this framework are mostly combinations of artificial neural networks, evolutionary algorithms, fuzzy and rough sets. These approaches are widely used for crisp or fuzzy forecasts based on crisp past observations such as electricity load, stock index prices and temperature (for a review of these techniques, we refer to [48–56]). In addition, various methods combine techniques of time series and fuzzy regression analysis [57]. For some recent advances in fuzzy regression analysis, see [58–63].

The reliability of forecasting methods generally requires exact observations in the sample. But there is often only vague information that is given in terms of imprecise quantities. Moreover, there are various real-world problems related to biological, economic, environmental, medical and sociological data where we face inaccurate instead of accurate data. In many real-life applications, e.g., monthly Co_2 emission, annual sea surface temperature or the water level of a lake, conventional observations are often reported as mean values. In such cases, the data obtained are not sufficient informative since some information contained in the range of the data is neglected. To overcome this shortcoming, one alternative would be to report such kind of data as interval valued (comparable to conventional confidence intervals). However, a potential shortcoming of interval-valued data is the fact that all values within the interval have the same importance. To avoid this issue of interval-valued data, the reported data can alternatively be represented with help of fuzzy numbers [64]. These fuzzy quantities can be modeled via experts opinion, or as simple alternative, they can be constructed via a method proposed by Buckley [65]. In this approach, conventional confidence intervals are employed to construct fuzzy numbers around the conventional mean values.

In addition to the abovementioned methods, there are also fuzzy time series models that rely on fuzzy data, but comparatively few overall. In this regard, Hesamian and Akbari [66] first suggested a fuzzy semi-parametric time series model (**FSPTSM**) based on fuzzy data, non-fuzzy coefficients, and fuzzy smooth functions. Secondly, Zarei et al. [67] used a specific variant of the **FSPTSM** [66] for triangular fuzzy data and different distance measures for fuzzy data. And thirdly, Hesamian et al. [68] introduced a forward additive time series model (**FATSM**) for fuzzy observations.

In this paper we develop a fuzzy nonparametric time series model (**FNPTSM**) for fuzzy observations that is inspired by nonparametric regression models and kernel smoothing methods [57]. As an initial idea, note that in nonparametric regression analysis, the *Nadaraya-Watson estimator* [69,70] is fairly common. Now, let us consider the issue of parameter estimation in the nonlinear regression model $x_t = f(x_{t-1}, x_{t-2}, \ldots, x_{t-p}) + \epsilon_t$ with $f : \mathbb{R}^p \to \mathbb{R}$. Based on this general model, a simple nonparametric way of estimating the function f is to employ the kernel-based Nadaraya–Watson estimator

$$\widehat{f}(x_t) = \sum_{j=p+1}^{T^*} w^h(t,j) x_j \qquad (1)$$

with

$$w^h(t,j) = \frac{\sum_{i=1}^{p} K\left(\frac{x_{t-i} - x_{j-i}}{h}\right)}{\sum_{j=p+1}^{T^*} \sum_{i=1}^{p} K\left(\frac{x_{t-i} - x_{j-i}}{h}\right)},$$

where K is a kernel function and $h > 0$ the bandwidth parameter. Note that the estimator (1) is a weighted average of $x_1, x_2, \ldots, x_T$ using the weights $w^h(t,j)$. As for determining the optimal bandwidth h, the Generalized Cross Validation (GCV) criterion

$$\widehat{h} = \arg\min_{h>0} \text{GCV}(h) = \arg\min_{h>0} \frac{1}{T^* - p} \sum_{t=p+1}^{T^*} \left(\frac{x_t - \sum_{j=p+1}^{T^*} w^h(t,j) x_j}{1 - \frac{\text{tr}(W_h)}{T^* - p}}\right)^2$$

can be utilized, where $\text{tr}(W_h)$ is the trace of the matrix $W_h = [w^h(t,j)]$. It is a matter of fact that the estimated values of f are ensured to be within the range of the response variable. This beneficial property is one of the reasons why we apply the Nadaraya–Watson kernel-based estimator for our fuzzy time series model. By utilizing this idea, the proposed **FNPTSM** provides an estimation procedure of the unknown (nonlinear) relationship between the fuzzy observations in three stages. The advantage of this methodology is that it considerably decreases the complexity in the estimation procedure. While the other fuzzy time series models [66–68] are based on estimating the unknown components of the model by unifying the centers and spreads of fuzzy data and their corresponding predicted values, our proposed method provides a smooth estimation procedure according to three separate stages. In the framework of a simulation study and two real-data examples, the efficiency and appropriateness of the **FNPTSM** is assessed in comparison with previous time series models for fuzzy data by utilizing four approved goodness-of-fit criteria.

The paper is organized as follows. First, we recall some necessary concepts related to fuzzy numbers in Section 2. In Section 3, the three-stage nonparametric kernel-based time series model using fuzzy data is presented. In Section 4, various application examples are given. Concluding remarks are provided in Section 5.

2. Fuzzy Numbers

In this section, we introduce basic definitions of fuzzy numbers that are needed to develop our proposed method.

A *fuzzy set* $\widetilde{A}$ is a mapping on $\mathbb{X}$ that assigns a specific degree of membership $0 \leq \mu_{\widetilde{A}}(x) \leq 1$ to each $x \in \mathbb{X}$. In addition, a *fuzzy number* (**FN**) $\widetilde{A}$ is a convex normalized fuzzy set on the real line $\mathbb{R}$ with an upper semi-continuous membership function of bounded support [71]. In many real applications, vague data a can be reported as $\widetilde{A}$: "about a". Such fuzzy data can often be represented via a special case of **FN**s, so called *LR*-**FN**s, which split $\mu_{\widetilde{A}}$ into two curves: a part on the left and a part on the right of the modal value. So, when considering real-life applications in fuzzy environments, *LR*-**FN**s play an important role. The membership function of an *LR*-**FN** $\mu_{\widetilde{A}}(x) = (a; l_a, r_a)_{LR}$ can be defined by:

$$\mu_{\tilde{A}}(x) = \begin{cases} L\left(\dfrac{a-x}{l_a}\right) & \text{if } x \leq a \\ R\left(\dfrac{x-a}{r_a}\right) & \text{if } x > a \end{cases} \qquad (2)$$

In (2), L and R are continuous and strictly decreasing functions from $[0,1]$ to $[0,1]$ satisfying $L(0) = R(0) = 1$ and $L(1) = R(1) = 0$. In addition, $a \in \mathbb{R}$ represents the *modal value*, while $l_a > 0$ and $r_a > 0$ are the *left spreads* and *right spreads* of $\tilde{A}$, respectively. The set of all LR-FNs is represented by $\mathcal{F}_{LR}(\mathbb{R})$. A special case of an LR-FN is the so-called *triangular fuzzy number* (**TFN**), whose membership function has the following form:

$$\mu_{\tilde{A}}(x) = \begin{cases} \dfrac{x-(a-l_a)}{l_a} & a - l_a \leq x \leq a \\ \dfrac{a+r_a-x}{r_a} & a < x \leq a + r_a \\ 0 & \text{otherwise} \end{cases}$$

There are various operations that can be defined between two LR-FNs, i.e., between $\tilde{A} = (a; l_a, r_a)_{LR}$ and $\tilde{B} = (b; l_b, r_b)_{LR}$. For instance, as we need both operations in this paper, we define *Addition* and *Scalar multiplication* of $\tilde{A}$ and $\tilde{B}$ in the following [72]:

- Addition: $\tilde{A} \oplus \tilde{B} = (a + b; l_a + l_b, r_a + r_b)_{LR}$
- Scalar multiplication:

$$\lambda \otimes \tilde{A} = \begin{cases} (\lambda a; \lambda l_a, \lambda r_a)_{LR} & \text{if } \lambda > 0 \\ (\lambda a; -\lambda r_a, -\lambda l_a)_{RL} & \text{if } \lambda < 0 \end{cases}$$

Moreover, there are numerous concepts used to define distances between two LR-FNs $\tilde{A} = (a; l_a, r_a)_{LR}$ and $\tilde{B} = (b; l_b, r_b)_{LR}$ [71]. Here, we utilize the *squared error distance measure* D for performance evaluation of the **FNPTSM** in comparison with other models. It is defined as

$$D(\tilde{A}, \tilde{B}) = \left(((a-b)^2 + c_1(l_a - l_b)^2 + c_2(r_a - r_b)^2)/3\right)^{0.5}$$

with $c_1 = \int_0^1 L^{-1}(\alpha) d\alpha$ and $c_2 = \int_0^1 R^{-1}(\alpha) d\alpha$ [73].

3. Nonparametric Kernel-Based Time Series Model for Fuzzy Data

In this section, the **FNPTSM** is developed along with the suggested parameter estimation method.

3.1. The Model

First, we recall the definition of fuzzy time series data.

Definition 1. *Let $\tilde{x}_T = \{\tilde{x}_1, \tilde{x}_2, \ldots, \tilde{x}_T\}$ be a set of FNs of size T. Then, $\tilde{x}_T$ is called fuzzy time series data if $\{\tilde{x}_1, \tilde{x}_2, \ldots, \tilde{x}_T\}$ is the vague concept of ordinary time series data $\{x_1, x_2, \ldots, x_T\}$ [68,74].*

As discussed in the Introduction, there are many situations where it is preferable to report exact data x by an FN $\tilde{x}$ as "about x". Then, $\tilde{x}$ is the respective vague concept of x.

Definition 2. *Let $\tilde{x}_T = \{\tilde{x}_1, \tilde{x}_2, \ldots, \tilde{x}_T\}$ be fuzzy time series data. The FNPTSM for fuzzy time series data $\tilde{x}_T$ is then defined by*

$$\tilde{x}_t = \tilde{f}(\tilde{x}_{t-1}, \tilde{x}_{t-2}, \ldots, \tilde{x}_{t-p}) \oplus \tilde{\epsilon}_t, \qquad (3)$$

where

1. $\tilde{x}_t = (x_t; l_{x_t}, r_{x_t})_{LR}$,
2. $\tilde{f}(\tilde{x}_{t-1}, \tilde{x}_{t-2}, \ldots, \tilde{x}_{t-p}) = (f(x_{t-1}, x_{t-2}, \ldots, x_{t-p});$ $l_{f(l_{x_{t-1}}, l_{x_{t-2}}, \ldots, l_{x_{t-p}})}, r_{f(r_{x_{t-1}}, r_{x_{t-2}}, \ldots, r_{x_{t-p}})})_{LR}$,

3. $\tilde{\epsilon}_t = (\epsilon_t; l_{\epsilon_t}, r_{\epsilon_t})_{LR}$'s are fuzzy errors, where $\epsilon_t \in \mathbb{R}$ and $l_{\epsilon_t}, r_{\epsilon_t} \in \mathbb{R}^+$.

Remark 1. *Note that (3) provides an FN in the form $\tilde{x}_t^* = (x_t^*; l_{x_t^*}, r_{x_t^*})_{LR}$ with $x_t^* = f(x_{t-1}, x_{t-2}, \ldots, x_{t-p}) + \epsilon_t$, $l_{x_t^*} = l_{f(l_{x_{t-1}}, \ldots, l_{x_{t-p}}) + l_{\epsilon_t}}$ and $r_{x_t^*} = r_{f(r_{x_{t-1}}, \ldots, r_{x_{t-p}}) + r_{\epsilon_t}}$ with $t = 1, 2, \ldots, T$. According to Definition 1, as $\{x_1, x_2, \ldots, x_T\}$ is ordinary time series data, $\tilde{x}_T^* = \{\tilde{x}_1^*, \tilde{x}_2^*, \ldots, \tilde{x}_T^*\}$ is also a vague concept of ordinary time series data $\{x_1^*, x_2^*, \ldots, x_T^*\}$. Thus, the proposed fuzzy time series model (3) generates new fuzzy time series data.*

3.2. Three-Stage Estimation Method for the Nonlinear Fuzzy Smooth Function

Below, we suggest a three-stage method to estimate the unknown fuzzy smooth function $\tilde{f}$ in (3). For this purpose, the fuzzy predictions are obtained based on a within-sample forecast $x_{T^*} = (x_1, x_2, \ldots, x_{T^*})^\top$ with $T^* < T$. From (3), one can get three ordinary nonlinear time series models as (1) $x_t = f(x_{t-1}, x_{t-2}, \ldots, x_{t-p}) + \epsilon_t$, (2) $l_{x_t} = l_{f(l_{x_{t-1}}, \ldots, l_{x_{t-p}}) + l_{\epsilon_t}}$, and (3) $r_{x_t} = r_{f(r_{x_{t-1}}, \ldots, r_{x_{t-p}}) + r_{\epsilon_t}}$ for $t = 1, 2, \ldots, T^*$. Therefore, to estimate the fuzzy smooth function at $\tilde{x} = (x; l_x, r_x)_T$ with $x = (x_1, x_2, \ldots, x_p)^\top$, $l_x = (l_{x_1}, l_{x_2}, \ldots, l_{x_p})^\top$ and $r_x = (r_{x_1}, r_{x_2}, \ldots, r_{x_p})^\top$, we follow the three-stage procedure below:

- **Stage (1):** Consider the nonlinear regression model $l_{x_t} = l_{f(x_{t-1}, x_{t-2}, \ldots, x_{t-p})} + l_{\epsilon_t}$. Based on the time series data $l_{x_t} = (l_{x_{t-1}}, \ldots, l_{x_{t-p}})^\top$, we employ the weighted Nadaraya–Watson estimator to estimate l_f for a within-sample forecast $T^* \leq T$ at $l_x = (l_{x_1}, \ldots, l_{x_p})^\top$ as

$$l_{\widehat{f}(l_{x_t})} = \sum_{j=p+1}^{T^*} w^{h_l}(t,j) l_{x_j},$$

where

$$w^{h_l}(t,j) = \frac{\sum_{i=1}^p K\left(\frac{l_{x_{t-i}} - l_{x_{j-i}}}{h_l}\right)}{\sum_{j=p+1}^{T^*} \sum_{i=1}^p K\left(\frac{l_{x_{t-i}} - l_{x_{j-i}}}{h_l}\right)} \quad (4)$$

with kernel function $K(.)$ and bandwidth parameter $h_l > 0$. The optimal value of h_l can be estimated by implementing the GCV criterion,

$$\hat{h}_l = \arg\min_{h_l > 0} \text{GCV}(h) = \arg\min_{h_l > 0} \frac{1}{T^* - p} \sum_{t=p+1}^{T^*} \left(\frac{l_{x_t} - \sum_{j=p+1}^{T^*} w^{h_l}(t,j) l_{x_j}}{1 - \frac{\text{tr}(W_{h_l})}{T^* - p}}\right)^2, \quad (5)$$

where $\text{tr}(W_{h_l})$ is the trace of the matrix $W_{h_l} = [w^{h_l}(t,j)]$ with $w^{h_l}(t,j)$ as defined in (4).

- **Stage (2):** Consider the nonlinear regression model $r_{x_t} = r_{f(x_{t-1}, x_{t-2}, \ldots, x_{t-p})} + r_{\epsilon_t}$. Based on the within-sample time series forecast data $r_{x_t} = (r_{x_{t-1}}, \ldots, r_{x_{t-p}})^\top$, $t = 1, 2, \ldots, T^*$, the weighted Nadaraya–Watson estimation of r_f at $r_x = (r_{x_1}, \ldots, r_{x_p})^\top$ can be established via

$$r_{\widehat{f}(r_{x_t})} = \sum_{j=p+1}^{T^*} w^{h_r}(t,j) r_{x_j},$$

where

$$w^{h_r}(t,j) = \frac{\sum_{i=1}^p K\left(\frac{r_{x_{t-i}} - r_{x_{j-i}}}{h_r}\right)}{\sum_{j=p+1}^{T^*} \sum_{i=1}^p K\left(\frac{r_{x_{t-i}} - r_{x_{j-i}}}{h_r}\right)} \quad (6)$$

and $h_r > 0$ is a bandwidth parameter. The optimal value of h_r can be estimated using the GCV criterion,

$$\widehat{h}_r = \arg\min_{h_r>0} \text{GCV}(h) = \arg\min_{h_r>0} \frac{1}{T^* - p} \sum_{t=p+1}^{T^*} \left(\frac{r_{x_t} - \sum_{j=p+1}^{T^*} w^{h_r}(t,j) r_{x_j}}{1 - \frac{\text{tr}(W_{h_r})}{T^* - p}} \right)^2, \quad (7)$$

where $\text{tr}(W_{h_r})$ is the trace of the matrix $W_{h_r} = [w^{h_r}(t,j)]$ with $w^{h_r}(t,j)$ as defined in (6).

- **Stage (3):** Consider the nonlinear regression model $x_t = f(x_{t-1}, x_{t-2}, \ldots, x_{t-p}) + \epsilon_t$. Based on the within-sample time series forecast data $(x_t = (x_{t-1}, x_{t-2}, \ldots, x_{t-p})^\top)$, $t = 1, 2, \ldots, T^*$, a nonparametric estimator f can be achieved as

$$\widehat{f}(x_t) = \sum_{j=p+1}^{T^*} w^h(t,j) x_j,$$

where

$$w^h(t,j) = \frac{\sum_{i=1}^p K\left(\frac{x_{t-i} - x_{j-i}}{h}\right)}{\sum_{j=p+1}^{T^*} \sum_{i=1}^p K\left(\frac{x_{t-i} - x_{j-i}}{h}\right)} \quad (8)$$

and bandwidth parameter $h > 0$. Similar to the previous stages, the optimal value of h is estimated with the help of the GCV criterion,

$$\widehat{h} = \arg\min_{h>0} \text{GCV}(h) = \arg\min_{h>0} \frac{1}{T^* - p} \sum_{t=p+1}^{T^*} \left(\frac{x_t - \sum_{j=p+1}^{T^*} w^h(t,j) x_j}{1 - \frac{\text{tr}(W_h)}{T^* - p}} \right)^2, \quad (9)$$

where $\text{tr}(W_h)$ is the trace of the matrix $W_h = [w^h(t,j)]$ with $w^h(t,j)$, as defined in (8).

Therefore, the forecast $\widetilde{x}_{T^*+k}$ with time lag $k \in \mathbb{N}$ can be achieved by an *LR*-**FN** via $\widetilde{x}_{T^*+k} = (\widehat{x}_{T^*+k}; l_{\widehat{x}_{T^*+k}}, r_{\widehat{x}_{T^*+k}})_{LR}$ with

$$\widehat{x}_{T^*+k} = \sum_{j=p+1}^{T^*+k-1} \frac{\sum_{i=1}^p K\left(\frac{x_{t-i} - x_{j-i}}{\widehat{h}}\right)}{\sum_{j=p+1}^{T^*+k-1} \sum_{i=1}^p K\left(\frac{x_{t-i} - x_{j-i}}{\widehat{h}}\right)} \cdot x_j,$$

$$l_{\widehat{x}_{T^*+k}} = \sum_{j=p+1}^{T^*+k-1} \frac{\sum_{i=1}^p K\left(\frac{l_{x_{t-i}} - l_{x_{j-i}}}{\widehat{h}_l}\right)}{\sum_{j=p+1}^{T^*+k-1} \sum_{i=1}^p K\left(\frac{l_{x_{t-i}} - l_{x_{j-i}}}{\widehat{h}_l}\right)} \cdot l_{x_j},$$

$$r_{\widehat{x}_{T^*+k}} = \sum_{j=p+1}^{T^*+k-1} \frac{\sum_{i=1}^p K\left(\frac{r_{x_{t-i}} - r_{x_{j-i}}}{\widehat{h}_r}\right)}{\sum_{j=p+1}^{T^*+k} \sum_{i=1}^p K\left(\frac{r_{x_{t-i}} - r_{x_{j-i}}}{\widehat{h}_r}\right)} \cdot r_{x_j}.$$

According to Stages (2) and (3), it can be seen that the spreads of the fuzzy prediction $\widetilde{x}_{T^*+k}$ are always non-negative.

Remark 2. *Since the proposed time series model relies on fuzzy data, let us recall the previous time series models based on fuzzy data [66–68]. First, Hesamian and Akbari [66] proposed a fuzzy semi-parametric autoregressive integrated moving average (ARIMA) model as follows:*

$$\widetilde{x}_i = \bigoplus_{l=1}^p (\theta_l \otimes \widetilde{x}_{i-l} \oplus \widetilde{f}(t_i) \oplus \widetilde{\epsilon}_i), \quad i = p+1, \ldots, T.$$

The parameters of the model are estimated by employing a hybrid method including a non-parametric kernel-based method and least absolute deviations. For a second time series model based on fuzzy data, Zarei et al. [67] applied the method [66] to estimate the model parameters and the fuzzy smooth function based on a specific distance, kernel and triangular fuzzy numbers. Finally, Hesamian et al. [68] proposed the fuzzy nonlinear time series model

$$\widetilde{x}_t = \widetilde{f}(\widetilde{x}_{t-1}, \widetilde{x}_{t-2}, \ldots, \widetilde{x}_{t-p}) \oplus \widetilde{\epsilon}_t, \qquad t = 1, 2, \ldots, T,$$

where

$$\widetilde{f}(\widetilde{x}_{t-1}, \widetilde{x}_{t-2}, \ldots, \widetilde{x}_{t-p}) = \bigoplus_{l=1}^{p} f_l(\widetilde{x}_{t-l}).$$

As for the estimation of the unknown fuzzy smooth functions $\widetilde{f}_l$, they applied a forward additive nonparametric technique.

Remark 3. We have extended some common performance measures used to compare the predictive accuracy of different time series models that we implement in Section 4. For this purpose, a time series model is first estimated based on a within-sample fuzzy time series dataset of size $T^* < T$ and then the performance of the model is evaluated via the remaining fuzzy time series dataset of size $T - T^*$.

1. Mean Forecast Error:

$$MFE = \frac{\sum_{t=T^*+1}^{T} D^2(\widehat{\widetilde{x}}_t, \widetilde{x}_t)}{T - T^*}$$

2. Mean Absolute Scaled Error:

$$MASE = \frac{\sum_{t=T^*+1}^{T} q_t}{T - T^*}$$

with

$$q_t = \frac{D(\widehat{\widetilde{x}}_t, \widetilde{x}_t)}{\frac{1}{T-T^*} \sum_{t=T^*+1}^{T} D^2(\widetilde{x}_t, \widetilde{x}_{t-1})}$$

3. Basis of the Index of Agreement:

$$BIA = 1 - \frac{\sum_{t=T^*+1}^{T} D^2(\widetilde{x}_t, \widehat{\widetilde{x}}_t)}{\sum_{t=T^*+1}^{T} (D(\widetilde{x}_t, \overline{\widetilde{x}}) + D(\overline{\widetilde{x}}, \widehat{\widetilde{x}}_t))^2}$$

with

$$\overline{\widetilde{x}} = \frac{\sum_{t=T^*+1}^{T} \widetilde{x}_t}{T - T^*}$$

4. Mean Similarity Measure:

$$MSM = \frac{1}{T - T^*} \sum_{t=T^*+1}^{T} \frac{\int \min\{\widehat{\widetilde{x}}_t(x), \widetilde{x}_t(x)\} dx}{\int \max\{\widehat{\widetilde{x}}_t(x), \widetilde{x}_t(x)\} dx}$$

Let A and B be two fuzzy time series models. As $MSM : \mathcal{F}_{LR}(\mathbb{R}) \times \mathcal{F}_{LR}(\mathbb{R}) \to [0,1]$ is a similarity measure, values of MSM above 0.5 show a good degree of similarity between the fuzzy responses and their fuzzy predictions. If we observe $MSM_B < MSM_A$, then model A outperforms model B. Further, if $MFE_A < MFE_B$, $MASE_A < MASE_B$ or $BIA_A < BIA_B$, then model A acts better in terms of prediction accuracy compared to model B.

Remark 4. While the proposed estimation procedure does not depend on the shape functions L and R corresponding to fuzzy data, the performance measures MFE, MASE and MSM depend on these shape functions. Therefore, the selected type of the shape functions L and R may affect the prediction criteria. For instance, assume that the data have reported by $\widetilde{x}_t = (x_t, l_{x_t}, r_{x_t})_{LR}$ with $L(x) = 1 - x$ and $R(x) = \sqrt{1-x}$. That is, $c_1 = \frac{1}{2}$ and $c_2 = \frac{2}{3}$. Therefore, the distance between

$\tilde{x}_t$ and its prediction is $D^2(\tilde{x}_t, \widehat{\tilde{x}}_t) = (x_t - \hat{x}_t)^2 + \frac{1}{2}(l_{x_t} - l_{\hat{x}_t})^2 + \frac{2}{3}(r_{x_t} - r_{\hat{x}_t})^2$. This implies that the MFE criterion is more sensitive to right spreads than to left spreads in this case. Considering $L(x) = R(x) = 1 - x$, it can be seen that $D^2(\tilde{x}_t, \widehat{\tilde{x}}_t)$ would be equally dependent from the left and right spreads. However, when we compare the performance of fuzzy time series models, it is reasonable that the shape functions L and R are assumed to be the same for all the considered models. Thus, following this approach, the performance criteria are not sensitive to the selection of L and R since c_1 and c_2 remain fixed for each model.

3.3. Selection of Autoregressive Order and Optimal Bandwidths

When implementing the **FNPTSM** (3), it is necessary to select the optimal bandwidths h, h_l and h_r, to choose the kernel function and to determine the autoregressive order p. The procedure used to select the autoregressive order and the optimal bandwidths is proposed as follows:

(1) Let $p = 1$.
(2)
   (2.1) Compute $\hat{h}_l^p$ based on (5).
   (2.2) Compute $\hat{h}_r^p$ based on (7).
   (2.3) Compute $\hat{h}^p$ based on (9).
(3) Let $p = p + 1$ and return to (2) until

$$\hat{p} = \arg\min_{p} \text{RMSE}_p,$$

where

$$\text{RMSE}_p = \sqrt{\frac{\sum_{i=p+1}^{T^*} D^2(\widehat{\tilde{x}}_i, \tilde{x}_i)}{T^* - p}}.$$

Then, $\hat{p}$, $\hat{h}_p$, $\hat{h}_l^p$ and $\hat{h}_r^p$ are the optimal values.

4. Numerical Examples

In this section, the effectiveness of the **FNPTSM** is investigated considering a simulation study and application examples that rely on fuzzy data. Recall that there are three other time series models that are based on fuzzy data (see Remark 2), i.e., the models introduced by Hesamian and Akbari [66], Zarei et al. [67] and Hesamian et al. [68]. However, as the method of Zarei et al. [67] is based on Hesamian and Akbari's method [66] (with a different distance measure), we omit this technique in the comparisons below. Thus, we compare our proposed method with the models suggested by Hesamian and Akbari (**FSPTSM**) [66] and Hesamian et al. (**FATSM**) [68] via three different kernel functions (Gaussian, Epanechnikov, and triweight).

Example 1. *In this example, 10 fuzzy datasets, each of size 300, are generated by the following FNPTSM:*

$$\tilde{x}_t = f(\tilde{x}_{t-1}, \tilde{x}_{t-2}, \tilde{x}_{t-3}) \oplus \tilde{\epsilon}_t, \quad t = 4, 5, \ldots, 300,$$

where

1.
$$f(\tilde{x}_1, \tilde{x}_2, \tilde{x}_3) = \left(x_1 - \cos(x_2) - \exp\left(\frac{x_3}{1 + |x_3|}\right); \cos^2\left(0.9 \prod_{j=1}^{3} l_{x_j}\right), \exp\left(0.002 \prod_{j=1}^{3} r_{x_j}\right)\right)_{LR}$$

2. $\tilde{x}_j = (x_j; l_{x_j}, r_{x_j})_{LR}$, $j = 1, 2, 3$ are the initial values with $x_j \sim N(0, 1)$, and l_{x_j} and r_{x_j} are random variables following $U(0, 0.2)$ and $U(0, 0.9)$, respectively,
3. $\tilde{\epsilon}_t = (\epsilon_t; l_{\epsilon_t}, r_{\epsilon_t})_{LR}$ with $\epsilon_t \sim N(0, 4)$, l_{ϵ_t} and r_{ϵ_t} are random variables following $U(0, 0.4)$ and $U(0, 0.5)$, respectively, and
4. $L(x) = 1 - x^2$ and $R(x) = 1 - x$.

The kernels Gaussian, Epanechnikov, and triweight are applied to predict $\widetilde{x}_t$. Based on the 10 sample fuzzy datasets (each of size 300), the mean values of the goodness-of-fit measures and their corresponding bandwidth mean values are summarized in Table 1. Consulting the results for the **FNPTSM**, it is evident that the best results among various kernels are obtained via the Gaussian kernel (lowest values of $\overline{MFE}$, $\overline{MASE}$ and largest values of $\overline{BIA}$, $\overline{MSM}$). In addition, the results of the **FSPTSM** and **FATSM** can also be found in Table 1. Comparing these results with the results of the **FNPTSM**, it is obvious that the **FNPTSM** provides more accurate predictions compared to both other methods for all three kernels, as all the considered goodness-of-fit measures show better results for the **FNPTSM**. That is, we observe the lowest values of $\overline{MFE}$, $\overline{MASE}$ and the largest values of $\overline{BIA}$, $\overline{MSM}$ for the **FNPTSM**.

Table 1. The mean performance measures of the **FNPTSM**, **FSPTSM** and **FATSM** corresponding to some specific kernels in Example 1.

Method	Kernel	Results	Goodness-of-Fit Criteria
FNPTSM	Gaussian	$\overline{\widehat{h}} = 0.45$ $\overline{\widehat{h^l}} = 0.04$ $\overline{\widehat{h^r}} = 0.22$	$\overline{MFE} = 1.0452$ $\overline{MASE} = 1.6089$ $\overline{BIA} = 0.9996$ $\overline{MSM} = 0.4167$
	Epanechnikov	$\overline{\widehat{h}} = 0.66$ $\overline{\widehat{h^l}} = 0.05$ $\overline{\widehat{h^r}} = 0.39$	$\overline{MFE} = 1.1728$ $\overline{MASE} = 1.6478$ $\overline{BIA} = 0.9992$ $\overline{MSM} = 0.3953$
	triweight	$\overline{\widehat{h}} = 1.89$ $\overline{\widehat{h^l}} = 0.08$ $\overline{\widehat{h^r}} = 0.54$	$\overline{MFE} = 1.2482$ $\overline{MASE} = 1.6339$ $\overline{BIA} = 0.9991$ $\overline{MSM} = 0.3721$
FSPTSM	Gaussian	$\overline{h}_{opt} = 0.07$ $\widehat{\theta}_1 = 0.5575$ $\widehat{\theta}_2 = -0.0956$ $\widehat{\theta}_3 = -0.1247$	$\overline{MFE} = 8.9728$ $\overline{MASE} = 4.2652$ $\overline{BIA} = 0.9536$ $\overline{MSM} = 0.2207$
	Epanechnikov	$\overline{h}_{opt} = 0.02$ $\widehat{\theta}_1 = -0.6424$ $\widehat{\theta}_2 = -0.5168$ $\widehat{\theta}_3 = -0.4258$	$\overline{MFE} = 5.2530$ $\overline{MASE} = 4.5383$ $\overline{BIA} = 0.9603$ $\overline{MSM} = 0.2920$
	triweight	$\overline{h}_{opt} = 0.13$ $\widehat{\theta}_1 = 0.3757$ $\widehat{\theta}_2 = -0.1576$ $\widehat{\theta}_3 = -0.2568$	$\overline{MFE} = 11.5231$ $\overline{MASE} = 3.8643$ $\overline{BIA} = 0.9738$ $\overline{MSM} = 0.2133$

Table 1. Cont.

Method	Kernel	Results	Goodness-of-Fit Criteria
FATSM	Gaussian	$\overline{\widetilde{h}}_1 = 1.8$ $\overline{\widetilde{h}}_2 = 0.7$ $\overline{\widetilde{h}}_3 = 0.03$	$\overline{MFE} = 1.392$ $\overline{MASE} = 18.028$ $\overline{BIA} = 0.972$ $\overline{MSM} = 0.321$
	Epanechnikov	$\overline{\widetilde{h}}_1 = 1.75$ $\overline{\widetilde{h}}_2 = 1.20$ $\overline{\widetilde{h}}_3 = 0.6$	$\overline{MFE} = 1.353$ $\overline{MASE} = 17.548$ $\overline{BIA} = 0.976$ $\overline{MSM} = 0.339$
	triweight	$\overline{\widetilde{h}}_1 = 2$ $\overline{\widetilde{h}}_2 = 1.5$ $\overline{\widetilde{h}}_3 = 0.5$	$\overline{MFE} = 1.348$ $\overline{MASE} = 16.459$ $\overline{BIA} = 0.979$ $\overline{MSM} = 0.349$

Example 2. *Three models, the **FNPTSM**, **FSPTSM** and **FATSM**, are implemented to analyze the dataset in Table 2 taken from [67].*

Table 2. Fuzzy time series data in Example 2.

t	$\widetilde{x}_t$	t	$\widetilde{x}_t$
1	$(1.7337; 0.8051)_T$	15	$(2.9145; 1.1507)_T$
2	$(2.3302; 0.9228)_T$	16	$(2.6085; 1.1335)_T$
3	$(1.3199; 0.7742)_T$	17	$(3.0432; 0.4489)_T$
4	$(5.0507; 0.8948)_T$	18	$(6.8010; 0.9588)_T$
5	$(1.4206; 1.0540)_T$	19	$(4.9351; 0.8115)_T$
6	$(4.0273; 0.9331)_T$	20	$(3.5672; 0.6054)_T$
7	$(2.8624; 1.0480)_T$	21	$(3.8828; 1.1579)_T$
8	$(4.7107; 1.0647)_T$	22	$(0.5183; 0.9652)_T$
9	$(4.1098; 1.1028)_T$	23	$(3.6846; 0.8175)_T$
10	$(4.4843; 1.0881)_T$	24	$(3.5117; 0.4248)_T$
11	$(1.3249; 1.0064)_T$	25	$(2.8294; 0.7956)_T$
12	$(3.2249; 0.5503)_T$	26	$(2.3836; 1.0101)_T$
13	$(3.2916; 0.5049)_T$	27	$(3.9454; 0.7558)_T$
14	$(3.5508; 0.9268)_T$	28	$(3.6012; 1.0266)_T$

Eighty percent of the data were used for parameter estimation and the rest were applied to fit the model. The goodness-of-fit values are given in Table 3 for the three kernels. The best results among various kernels are obtained by employing the triweight kernel (lowest values of MFE, MASE, and largest value of MSM). The results of the **FSPTSM** and **FATSM** are also given in Table 3. As for the **FSPTSM**, the best results are obtained based on the triweight kernel with $MFE = 1.6252$, $MASE = 1.3371$, $MSM = 0.1205$ and $BIA = 0.9443$. The best results of **FATSM** are also obtained based on the triweight kernel with $MFE = 0.254$, $MASE = 0.733$, $MSM = 0.358$ and $BIA = 0.958$. However, all goodness-of-fit measures related to the **FNPTSM** show a better performance compared to both the **FSPTSM** and **FATSM**, i.e., the lowest values of MFE, MASE and the largest values of BIA, MSM are observed for the **FNPTSM**. The results show that the newly presented **FNPTSM** is more efficient than the **FSPTSM** and **FATSM** for the data in Table 3. The plot of the fuzzy data and corresponding estimates based on the triweight kernel is given in Figure 1 for all methods (**FNPTSM**, **FSPTSM**, **FATSM**).

Table 3. The performance measures of the **FNPTSM**, **FSPTSM** and **FATSM** corresponding to some specific kernels in Example 2.

Method	Kernel	Results	Goodness-of-Fit Criteria
FNPTSM	Gaussian	$\hat{p}=2$ $\hat{h}=0.64$ $\hat{h^l}=0.22$ $\hat{h^r}=0.22$	MFE = 0.1127 MASE = 0.4030 BIA = 0.9629 MSM = 0.3791
	Epanechnikov	$\hat{p}=2$ $\hat{h}=1.35$ $\hat{h^l}=0.09$ $\hat{h^r}=0.09$	MFE = 0.1133 MASE = 0.4016 BIA = 0.9639 MSM = 0.3836
	triweight	$\hat{p}=3$ $\hat{h}=1.44$ $\hat{h^l}=0.65$ $\hat{h^r}=0.65$	MFE = 0.1087 MASE = 0.3722 BIA = 0.9626 MSM = 0.4439
FSPTSM	Gaussian	$\hat{h}=0.13, \hat{p}=3$ $\hat{\theta}_1=-0.2305$ $\hat{\theta}_2=-0.0613$ $\hat{\theta}_3=-0.4164$	MFE = 3.6124 MASE = 2.1452 BIA = 0.9666 MSM = 0.0061
	Epanechnikov	$\hat{h}=0.16$ $\hat{p}=1$ $\hat{\theta}_1=-0.3192$	MFE = 1.5749 MASE = 1.3135 BIA = 0.9438 MSM = 0.1080
	triweight	$\hat{h}=0.23$ $\hat{p}=1$ $\hat{\theta}_1=-0.3437$	MFE = 1.6252 MASE = 1.3371 BIA = 0.9443 MSM = 0.1205
FATSM	Gaussian	$\hat{p}=2$ $\hat{h}_1=0.2$ $\hat{h}_2=0.3$	MFE = 0.327 MASE = 0.943 BIA = 0.954 MSM = 0.341
	Epanechnikov	$\hat{p}=2$ $\hat{h}_1=0.3$ $\hat{h}_2=1.5$	MFE = 0.386 MASE = 1.302 BIA = 0.948 MSM = 0.3208
	triweight	$\hat{p}=3$ $\hat{h}_1=0.5$ $\hat{h}_2=0.1$ $\hat{h}_3=0.05$	MFE = 0.254 MASE = 0.733 BIA = 0.958 MSM = 0.358

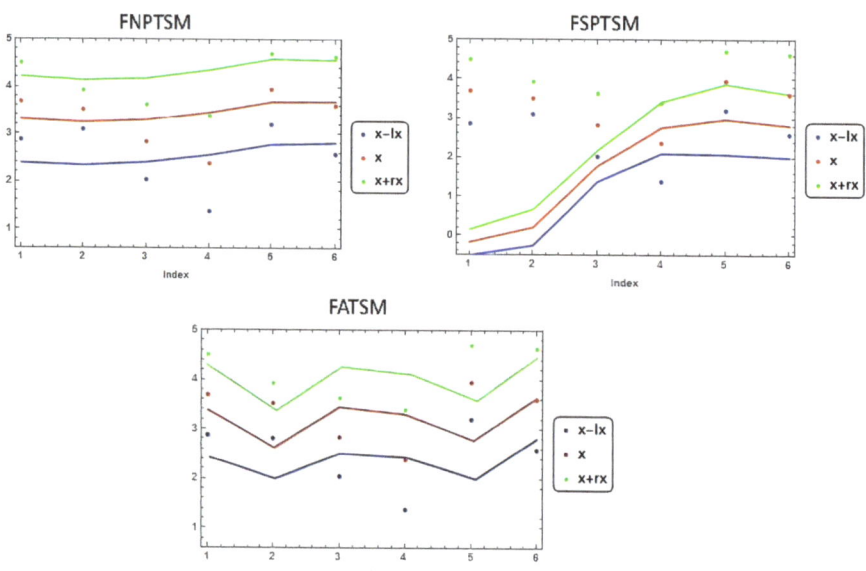

Figure 1. Plot of $x-l, x, x-r$ and $\hat{x}-\hat{r}, \hat{x}, \hat{x}+\hat{r}$ for **FNPTSM**, **FSPTSM** and **FATSM** (based on the triweight kernel) in Example 2.

Example 3. *In this example, we employ the FNPTSM and both the FSPTSM and FATSM to predict the global land–ocean temperature [75]. For this purpose, we use the global land–ocean temperature from January 2000 to December 2020, as shown in Figure 2. The data are reported as average values for each month. Therefore, the data can also be interpreted as "mean of each month" and appropriately modeled via triangular fuzzy numbers. Inspired by Buckley [65], this dataset can be used to evaluate the global land-ocean temperature with the help of a TFN $\widetilde{x}_t = (\overline{x}_t; Z_{0.005} s_t / \sqrt{n_t}, 0.15 Z_{0.025} s_t / \sqrt{n_t})_T$, where n_t, $\overline{x}_t$, and s_t denote the number of days, mean, and standard deviation of the global land–ocean temperature in the tth month, respectively, and Z_α is the α-quantile of the standard normal distribution. However, since we do not have daily values of the global land–ocean temperature, we model the monthly global land–ocean temperature for a month t via $\widetilde{x}_t = (x_t; 0.2 x_t, 0.15 x_t)_T$.*

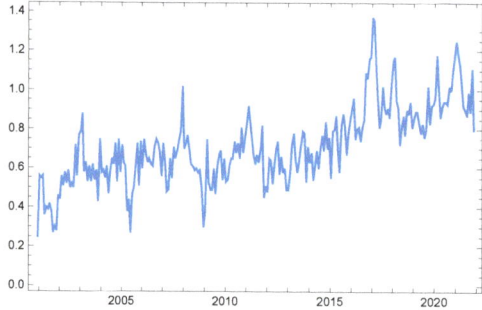

Figure 2. Time series on global temperature in Example 3.

In this example, 200 observations were used to estimate the parameters. A further 52 observations were used to fit the model. The goodness-of-fit values that correspond to the **FNPTSM**, **FSPTSM** and **FATSM** are given in Table 4. The results reveal that the **FNPTSM** outperforms the **FSPTSM** and **FATSM** for the global land–ocean temperature dataset. Note that the best results of the proposed **FNPTSM** are obtained based on the Gaussian kernel and the best results of both the **FSPTSM** and **FATSM** are given when implementing the triweight kernel. The fuzzy data, along with the corresponding estimations related to the **FNPTSM** (based on the Gaussian kernel) as well as of the **FSPTSM** and **FATSM** (based on the triweight kernel), are visualized in Figure 3. In comparison to the **FSPTSM** and **FATSM**, the values predicted by the **FNPTSM** are closer to the fuzzy observations, which reveals that the proposed **FNPTSM** performs better for the global land–ocean temperature dataset.

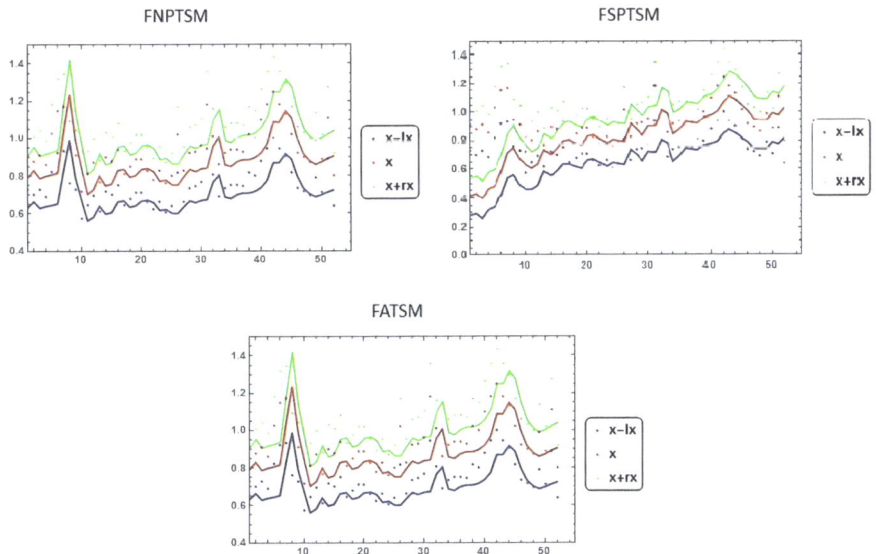

Figure 3. Plot of $x - l$, x, $x - r$ and $\hat{x} - \hat{r}$, $\hat{x}$, $\hat{x} + \hat{r}$ for **FNPTSM**, **FSPTSM** and **FATSM** in Example 3.

Table 4. The performance measures of the **FNPTSM**, **FSPTSM** and **FATSM** corresponding to some specific kernels in Example 3.

Method	Kernel	Results	Goodness-of-Fit Criteria
FNPTSM	Gaussian	$\hat{p} = 2$ $\hat{h} = 0.030$ $\hat{h^l} = 0.006$ $\hat{h^r} = 0.005$	MFE = 0.0061 MASE = 14.6082 BIA = 0.9783 MSM = 0.3781
	Epanechnikov	$\hat{p} = 2$ $\hat{h} = 0.051$ $\hat{h^l} = 0.013$ $\hat{h^r} = 0.007$	MFE = 0.0061 MASE = 14.6360 BIA = 0.9782 MSM = 0.3787
	triweight	$\hat{p} = 2$ $\hat{h} = 0.032$ $\hat{h^l} = 0.011$ $\hat{h^r} = 0.010$	MFE = 0.0067 MASE = 14.8885 BIA = 0.9761 MSM = 0.3921

Table 4. Cont.

Method	Kernel	Results	Goodness-of-Fit Criteria
FSPTSM	Gaussian	$\hat{h} = 0.05, \hat{p} = 3$ $\hat{\theta}_1 = 0.3912$ $\hat{\theta}_2 = 0.2045$ $\hat{\theta}_3 = -0.1916$	MFE = 0.0153 MASE = 20.4387 BIA = 0.9462 MSM = 0.3714
	Epanechnikov	$\hat{h} = 0.10, \hat{p} = 3$ $\hat{\theta}_1 = 0.4011$ $\hat{\theta}_2 = 0.2095$ $\hat{\theta}_3 = -0.1894$	MFE = 0.0148 MASE = 20.0997 BIA = 0.9483 MSM = 0.3711
	triweight	$\hat{h} = 0.11, \hat{p} = 3$ $\hat{\theta}_1 = 0.3643$ $\hat{\theta}_2 = 0.1853$ $\hat{\theta}_3 = -0.2175$	MFE = 0.0157 MASE = 19.7872 BIA = 0.9457 MSM = 0.3954
FATSM	Gaussian	$\hat{p} = 3$ $\hat{h}_1 = 1.5$ $\hat{h}_2 = 1.75$ $\hat{h}_3 = 0.05$	MFE = 0.011 MASE = 19.625 BIA = 0.964 MSM = 0.310
	Epanechnikov	$\hat{p} = 3$ $\hat{h}_1 = 1.25$ $\hat{h}_2 = 1.73$ $\hat{h}_3 = 0.1$	MFE = 0.010 MASE = 17.625 BIA = 0.966 MSM = 0.33
	triweight	$\hat{p} = 3$ $\hat{h}_1 = 1.5$ $\hat{h}_2 = 1.77$ $\hat{h}_3 = 0.13$	MFE = 0.0098 MASE = 16.745 BIA = 0.960 MSM = 0.345

5. Conclusions

Nonparametric statistical inference deals with situations where the functional relationships of the involved distribution functions are unspecified. In this regard, nonparametric time series models were broadly utilized to identify the "best fit" curve for a given time series of data. However, there are numerous situations where the available data are fuzzy rather than exact. In this paper, a nonparametric kernel-based time series model that relies on fuzzy data was introduced. The Nadaraya–Watson estimator was utilized to provide a fuzzy time series model within a three-stage procedure. Some popular goodness-of-fit measures have been implemented to investigate the performance of the fuzzy nonparametric time series model based on different kernel functions. The effectiveness and feasibility of the proposed time series model were also compared with the performance of existing time series models based on fuzzy data. Considering three common kernel functions (Gaussian, Epanechnikov, and triweight), the results indicated the superior performance of our proposed method in comparison to previous approaches. In addition to the performance aspect, the handling of the new nonparametric kernel-based time series model is much simpler than that of the previous methods, as we implemented an estimation procedure that is divided into three independent stages. In addition, our proposed time series model can be employed for arbitrary shapes of LR fuzzy numbers. However, the model can be applied only for LR fuzzy numbers, and thus it could be a promising future direction to develop a more general methodology that can handle arbitrary fuzzy numbers. Future studies could also focus on extending our approach to cases where the underlying time series data contain outliers. Finally, extending the proposed methodology for other nonlinear models such as wavelet-based or neural network-based time series models are further ideas for future research.

Author Contributions: Conceptualization, G.H.; methodology, G.H., A.J. and N.C.; software, G.H.; validation, G.H.; formal analysis, A.J.; investigation, A.J. and N.C.; writing—original draft, G.H.; writing—review & editing, A.J. and N.C.; supervision, A.J. and N.C.; project administration, N.C. All authors have read and agreed to the published version of the manuscript.

Funding: This research received no external funding.

Data Availability Statement: The data that support the findings of this study are available from the respective references as mentioned in the main text.

Acknowledgments: The authors thank the three anonymous reviewers for their valuable feedback and suggestions, which were important and helpful for improving the paper.

Conflicts of Interest: The authors declare no conflict of interest.

References

1. Brockwell, P.J.; Davis, R.A. *Time Series: Theory and Methods*; Springer: New York, NY, USA, 2009.
2. Shumway, R.H.; Stoffer, D.S. *Time Series Analysis and Its Applications*; Springer: London, UK, 2017.
3. Box, G.E.P.; Jenkins, G.M. *Time Series Analysis: Forecasting and Control*; Holden-Day: San Francisco, CA, USA, 1976.
4. Chukhrova, N.; Johannssen, A. State Space Models and the Kalman Filter in Stochastic Claims Reserving: Forecasting, Filtering and Smoothing. *Risks* **2017**, *5*, 30. [CrossRef]
5. Chukhrova, N.; Johannssen, A. Stochastic Claims Reserving Methods with State Space Representations—A Review. *Risks* **2021**, *9*, 198. [CrossRef]
6. Palma, W. *Time Series Analysis*; Wiley: Hoboken, NJ, USA, 2016.
7. Tong, H. *Nonlinear Time Series: A Dynamical System Approach*; Oxford University Press: Oxford, UK, 1990.
8. Woodward, W.A.; Gray, H.L.; Elliott, A.C. *Applied Time Series Analysis*; CRC Press: Boca Raton, FL, USA, 2012.
9. Yang, X.; Liu, B. Uncertain time series analysis with imprecise observations. *Fuzzy Optim. Decis. Mak.* **2019**, *18*, 263–278. [CrossRef]
10. Chukhrova, N.; Johannssen, A. Generalized One-Tailed Hypergeometric Test with Applications in Statistical Quality Control. *J. Qual. Technol.* **2020**, *52*, 14–39. [CrossRef]
11. Chukhrova, N.; Johannssen, A. Non-parametric fuzzy hypothesis testing for quantiles applied to clinical characteristics of COVID-19. *Int. J. Intell. Syst.* **2021**, *36*, 2922–2963. [CrossRef]
12. Chukhrova, N.; Johannssen, A. Employing fuzzy hypothesis testing to improve modified p charts for monitoring the process fraction nonconforming. *Inf. Sci.* **2023**, *633*, 141–157. [CrossRef]
13. Song, Q.; Chissom, B.S. Fuzzy time series and its models. *Fuzzy Sets Syst.* **1993**, *54*, 269–277. [CrossRef]
14. Sun, C.; Li, H. Parallel fuzzy relation matrix factorization towards algebraic formulation, universal approximation and interpretability of MIMO hierarchical fuzzy systems. *Fuzzy Sets Syst.* **2022**, *450*, 68–86. [CrossRef]
15. Sun, C.; Li, H. Construction of universal approximations for multi-input single-output Hierarchical Fuzzy Systems. *IEEE Trans. Fuzzy Syst.* **2023**, in press.
16. Yu, H.K. Weighted fuzzy time-series models for TAIEX forecasting. *Phys. A Stat. Mech. Appl.* **2005**, *349*, 609–624. [CrossRef]
17. Chen, S.M.; Tanuwijaya, K. Multivariate fuzzy forecasting based on fuzzy time series and automatic clustering techniques. *Expert Syst. Appl.* **2011**, *38*, 10594–10605. [CrossRef]
18. Huang, Y.L.; Horng, S.J.; He, M.; Fan, P.; Kao, T.W.; Khan, M.K.; Lai, J.L.; Kuo, I.H. A hybrid forecasting model for enrollments based on aggregated fuzzy time series and particle swarm optimization. *Expert Syst. Appl.* **2011**, *38*, 8014–8023. [CrossRef]
19. Li, S.T.; Kuo, S.C.; Cheng, Y.C.; Chen, C.C. Deterministic vector long-term forecasting for fuzzy time series. *Fuzzy Sets Syst.* **2010**, *161*, 1852–1870. [CrossRef]
20. Peng, H.W.; Wu, S.F.; Wei, C.C.; Lee, S.J. Time series forecasting with a neuro-fuzzy modeling scheme. *Appl. Soft Comput.* **2015**, *32*, 481–493. [CrossRef]
21. Duru, O.; Bulut, E. A nonlinear clustering method for fuzzy time series: Histogram damping partition under the optimized cluster paradox. *Appl. Soft Comput.* **2014**, *24*, 742–748. [CrossRef]
22. Bose, M.; Mali, K. A novel data partitioning and rule selection technique for modeling high-order fuzzy time series. *Appl. Soft Comput.* **2018**, *63*, 87–96. [CrossRef]
23. Uslu, V.R.; Bas, E.; Yolcu, U.; Egrioglu, E. A fuzzy time series approach based on weights determined by the number of recurrences of fuzzy relations. *Swarm Evol. Comput.* **2014**, *15*, 19–26.
24. Bulut, E. Modeling seasonality using the fuzzy integrated logical forecasting (FILF) approach. *Expert Syst. Appl.* **2014**, *41*, 1806–1812. [CrossRef]
25. Chen, M.Y.; Chen, B.T. Online fuzzy time series analysis based on entropy discretization and a fast Fourier transform. *Appl. Soft Comput.* **2014**, *14*, 156–166. [CrossRef]
26. Singh, P.; Borah, B. Forecasting stock index price based on M-factors fuzzy time series and particle swarm optimization. *Int. J. Approx. Reason.* **2014**, *55*, 812–833. [CrossRef]
27. Chen, S.M.; Chen, S.W. Fuzzy forecasting based on two-factors second-order fuzzy-trend logical relationship groups and the probabilities of trends of fuzzy logical relationships. *IEEE Trans. Cyber.* **2015**, *45*, 405–417.

28. Cheng, S.H.; Chen, S.M.; Jian, W.S. Fuzzy time series forecasting based on fuzzy logical relationships and similarity measures. *Inf. Sci.* **2016**, *327*, 272–287. [CrossRef]
29. Sadaei, H.J.; Enayatifar, R.; Abdullah, A.H.; Gani, A. Short-term load forecasting using a hybrid model with a refined exponentially weighted fuzzy time series and an improved harmony search. *Int. J. Electr. Power Energy Syst.* **2014**, *62*, 118–129. [CrossRef]
30. Ye, F.; Zhang, L.; Zhang, D.; Fujita, H.; Gong, Z. A novel forecasting method based on multi-order fuzzy time series and technical analysis. *Inf. Sci.* **2016**, *367–368*, 41–57. [CrossRef]
31. Efendi, R.; Ismail, Z.; Deris, M.M. A new linguistic out-sample approach of fuzzy time series for daily forecasting of Malaysian electricity load demand. *Appl. Soft Comput.* **2015**, *28*, 422–430. [CrossRef]
32. Talarposhtia, F.M.; Hossein, J.S.; Rasul, E.; Guimaraesc, F.G.; Mahmud, M.; Eslami, T. Stock market forecasting by using a hybrid model of exponential fuzzy time series. *Int. J. Approx. Reason.* **2016**, *70*, 79–98.
33. Wang, W.; Liu, X. Fuzzy forecasting based on automatic clustering and axiomatic fuzzy set classification. *Inf. Sci.* **2015**, *294*, 78–94. [CrossRef]
34. Sadaei, H.J.; Enayatifar, R.; Lee, M.H.; Mahmud, M. A hybrid model based on differential fuzzy logic relationships and imperialist competitive algorithm for stock market forecasting. *Appl. Soft Comput.* **2016**, *40*, 132–149. [CrossRef]
35. Aladag, C.H.; Yolcu, U.; Egrioglu, E. A high order fuzzy time series forecasting model based on adaptive expectation and artificial neural network. *Math. Comput. Simul.* **2010**, *81*, 875–882. [CrossRef]
36. Chen, M.Y. A high-order fuzzy time series forecasting model for internet stock trading. *Future Gener. Comput. Syst.* **2014**, *37*, 461–467. [CrossRef]
37. Egrioglu, E.; Aladag, C.H.; Yolcu, U. Fuzzy time series forecasting with a novel hybrid approach combining fuzzy c-means and neural networks. *Expert Syst. Appl.* **2013**, *40*, 854–857. [CrossRef]
38. Yolcu, O.C.; Yolcu, U.; Egrioglu, E.; Aladag, C.H. High order fuzzy timeseries forecasting method based on an intersection operation. *Appl. Math. Model.* **2016**, *40*, 8750–8765. [CrossRef]
39. Singh, P.; Borah, B. High-order fuzzy-neuro expert system for daily temperature forecasting. *Knowl. Based Syst.* **2013**, *46*, 12–21. [CrossRef]
40. Yolcu, O.C.; Lam, H.K. A combined robust fuzzy time series method for prediction of time series. *Neurocomputing* **2017**, *247*, 87–101. [CrossRef]
41. Yolcu, O.C.; Alpaslan, F. Prediction of TAIEX based on hybrid fuzzy time series model with single optimization process. *Appl. Soft Comput.* **2018**, *66*, 18–33. [CrossRef]
42. Aladag, C.H. Using multiplicative neuron model to establish fuzzy logic relationships. *Expert Syst. Appl.* **2013**, *40*, 850–853. [CrossRef]
43. Gaxiola, F.; Melin, P.; Valdez, F.; Castillo, O. Interval type-2 fuzzy weight adjustment for back propagation neural networks with application in time series prediction. *Inf. Sci.* **2014**, *260*, 1–14. [CrossRef]
44. Wei, L.Y. A hybrid ANFIS model based on empirical mode decomposition for stock time series forecasting. *Appl. Soft Comput.* **2016**, *42*, 368–376. [CrossRef]
45. Sadaei, H.J.; Enayatifar, R.; Guimaraes, F.G.; Mahmud, M.; Alzamil, Z.A. Combining ARFIMA models and fuzzy time series for the forecast of long memory time series. *Neurocomputing* **2016**, *175*, 782–796. [CrossRef]
46. Torbat, S.; Khashei, M.; Bijari, M. A hybrid probabilistic fuzzy ARIMA model for consumption forecasting in commodity markets. *Econ. Anal. Policy* **2018**, *58*, 22–31. [CrossRef]
47. Kocak, C. ARMA(p,q)-type high order fuzzy time series forecast method based on fuzzy logic relations. *Appl. Soft Comput.* **2017**, *58*, 92–103. [CrossRef]
48. Abhishekh, S.S.G.; Singh, S.R. A score function-based method of forecasting using intuitionistic fuzzy time series. *New Math. Nat. Comput.* **2018**, *14*, 91–111. [CrossRef]
49. Cheng, C.H.; Chen, C.H. Fuzzy time series model based on weighted association rule for financial market forecasting. *Expert Syst.* **2018**, *35*, 23–30. [CrossRef]
50. Guan, H.; Dai, Z.; Zhao, A.; He, J. A novel stock forecasting model based on High-order-fuzzy-fluctuation trends and back propagation neural network. *PLoS ONE* **2018**, *13*, e0192366. [CrossRef] [PubMed]
51. Gupta, C.; Jain, G.; Tayal, D.K.; Castillo, O. ClusFuDE: Forecasting low dimensional numerical data using an improved method based on automatic clustering, fuzzy relationships and differential evolution. *Eng. Appl. Artif. Intell.* **2018**, *71*, 175–189. [CrossRef]
52. Gautam, S.S.; Singh, S. A refined method of forecasting based on high-order intuitionistic fuzzy time series data. *Prog. Artif. Intell.* **2018**, *7*, 339–350.
53. Li, R. Water quality forecasting of Haihe River based on improved fuzzy time series model. *Desal. Water Treat.* **2018**, *106*, 285–291. [CrossRef]
54. Novak, V. Detection of structural breaks in time series using fuzzy techniques. *Int. J. Fuzzy Logic Intell. Syst.* **2018**, *18*, 1–12. [CrossRef]
55. Phan, T.T.H.; Big, A.; Caillault, E.P. A new fuzzy logic-based similarity measure applied to large gap imputation for uncorrelated multivariate time series. *Appl. Comput. Intel. Soft Comput.* **2018**, *2018*, 1–15. [CrossRef]
56. Rahim, N.F.; Othman, M.; Sokkalingam, R.; Kadir, E.A. Forecasting crude palm oil prices using fuzzy rule-based time series method. *IEEE Access* **2018**, *6*, 32216–32224. [CrossRef]

57. Chukhrova, N.; Johannssen, A. Fuzzy regression analysis: Systematic review and bibliography. *Appl. Soft Comput.* **2019**, *84*, 105708. [CrossRef]
58. Akbari, M.G.; Hesamian, G. Linear model with exact inputs and interval-valued fuzzy outputs. *IEEE Trans. Fuzzy Syst.* **2017**, *26*, 518–530. [CrossRef]
59. Hesamian, G.; Akbari, M.G. Semi-parametric partially logistic regression model with exact inputs and intuitionistic fuzzy outputs. *Appl. Soft Comput.* **2017**, *58*, 517–526. [CrossRef]
60. Hesamian, G.; Akbari, M.G.; Asadollahi, M. Fuzzy semi-parametric partially linear model with fuzzy inputs and fuzzy outputs. *Expert Syst. Appl.* **2017**, *71*, 230–239. [CrossRef]
61. Akbari, M.G.; Hesamian, G. Elastic net oriented to fuzzy semiparametric regression model with fuzzy explanatory variables and fuzzy responses. *IEEE Trans. Fuzzy Syst.* **2019**, *27*, 2433–2442. [CrossRef]
62. Hesamian, G.; Akbari, M.G. A fuzzy additive regression model with exact predictors and fuzzy responses. *Appl. Soft Comput.* **2020**, *95*, 106507. [CrossRef]
63. Hesamian, G.; Torkian, F.; Johannssen, A.; Chukhrova, N. A fuzzy nonparametric regression model based on an extended center and range method. *J. Comput. Appl. Math.* **2023**, *2023*, 115377. [CrossRef]
64. Viertl, R. *Statistical Methods for Fuzzy Data*; Wiley: New York, NY, USA, 2011.
65. Buckley, J.J. *Fuzzy Statistics, Studies in Fuzziness and Soft Computing*; Springer: Berlin, Germany, 2006.
66. Hesamian, G.; Akbari, M.G. A semi-parametric model for time series based on fuzzy data. *IEEE Trans. Fuzzy Syst.* **2018**, *26*, 2953–2966. [CrossRef]
67. Zarei, R.; Akbari, M.G.; Chachi, J. Modeling autoregressive fuzzy time series data based on semi-parametric methods. *Soft Comput.* **2020**, *24*, 7295–7304. [CrossRef]
68. Hesamian, G.; Torkian, F.; Yarmohammadi, M. A fuzzy nonparametric time series model based on fuzzy data. *Iran. J. Fuzzy Syst.* **2022**, *19*, 61–72.
69. Golub, G.H.; Heath, M.; Wahba, G. Generalized cross-validation as a method for choosing a good ridge parameter. *Technometrics* **1979**, *21*, 215–223. [CrossRef]
70. Craven, P.; Wahba, G. Smoothing noisy data with spline functions: Estimating the correct degree of smoothing by the method of generalized cross-validation. *Numer. Math.* **1979**, *31*, 377–403. [CrossRef]
71. Chukhrova, N.; Johannssen, A. Fuzzy hypothesis testing: Systematic review and bibliography. *Appl. Soft Comput.* **2021**, *106*, 107331. [CrossRef]
72. Lee, K.H. *First Course on Fuzzy Theory and Applications*; Springer: Berlin, Germany, 2005.
73. Coppi, R.; D'Urso, P.; Giordani, P.; Santoro, A. Least squares estimation of a linear regression model with LR-fuzzy response. *Comput. Stat. Data Anal.* **2006**, *51*, 267–286. [CrossRef]
74. Grzegorzewski, P. Testing statistical hypotheses with vague data. *Fuzzy Sets Syst.* **2000**, *11*, 501–510. [CrossRef]
75. Mills, T.C. *Applied Time Series Analysis: A Practical Guide to Modelling and Forecasting*; Academic Press: London, UK, 2019.

Disclaimer/Publisher's Note: The statements, opinions and data contained in all publications are solely those of the individual author(s) and contributor(s) and not of MDPI and/or the editor(s). MDPI and/or the editor(s) disclaim responsibility for any injury to people or property resulting from any ideas, methods, instructions or products referred to in the content.

Article
Process Capability and Performance Indices for Discrete Data

Vasileios Alevizakos

Department of Mathematics, National Technical University of Athens, Zografou, 15773 Athens, Greece; basalebiz@yahoo.gr

Abstract: Process capability and performance indices (PCIs and PPIs) are used in industry to provide numerical measures for the capability and performance of several processes. The majority of the literature refers to PCIs and PPIs for continuous data. The aim of this paper is to compute the classical indices for discrete data following Poisson, binomial or negative binomial distribution using various transformation techniques. A simulation study under different situations of a process and comparisons with other existing PCIs for discrete data are also presented. The methodology of computing the indices is easy to use, and as a result, one can have an assessment of the process capability and performance without difficulty. Three examples are further provided to illustrate the application of the transformation techniques.

Keywords: discrete data; binomial distribution; negative binomial distribution; Poisson distribution; process capability indices

MSC: 62P30

Citation: Alevizakos, V. Process Capability and Performance Indices for Discrete Data. *Mathematics* **2023**, *11*, 3457. https://doi.org/10.3390/math11163457

Academic Editors: Arne Johanssen and Nataliya Chukhrova

Received: 17 July 2023
Revised: 29 July 2023
Accepted: 1 August 2023
Published: 9 August 2023

Copyright: © 2023 by the author. Licensee MDPI, Basel, Switzerland. This article is an open access article distributed under the terms and conditions of the Creative Commons Attribution (CC BY) license (https://creativecommons.org/licenses/by/4.0/).

1. Introduction

Process capability indices (PCIs) are widely used in industry as they are considered practical tools for continuous quality improvement. They provide numerical measures of the process ability to produce output within specification limits on a quality characteristic of interest. The four most broadly used PCIs are C_p (Juran [1]), C_{pk} (Kane [2]), C_{pm} (Chan, Cheng, and Spiring [3], Hsiang and Taguchi [4]) and C_{pmk} (Choi and Owen [5], Pearn, Kotz, and Johnson [6]) and are defined as follows:

$$C_p = \frac{U - L}{6\sigma},$$

$$C_{pk} = \frac{min\{U - \mu, \mu - L\}}{3\sigma},$$

$$C_{pm} = \frac{U - L}{6\sqrt{\sigma^2 + (\mu - T)^2}},$$

$$C_{pmk} = \frac{min\{U - \mu, \mu - L\}}{3\sqrt{\sigma^2 + (\mu - T)^2}},$$

where U and L are the upper and the lower specification limits, respectively; T is the target value of the process; μ is the process mean; and σ is the process standard deviation. If μ and σ are unknown, one can replace them with the sample mean $\bar{X}$ and the sample standard deviation S, respectively. The properties, the estimation methods and the confidence limits of PCIs have been examined by many authors, such as Rodriguez [7], Kotz and Johnson [8,9], Kotz and Lovelace [10], and Pearn and Kotz [11].

In 1991, the Automotive Industry Action Group (AIAG) recommended using the PCIs C_p and C_{pk} when the process is in control with the process standard deviation estimated using the formula $\hat{\sigma} = \bar{R}/d_2$ (s-short-term). On the other hand, when the process is not in control, AIAG recommended using the process performance indices (PPIs) P_p and P_{pk}, given by

$$P_p = \frac{U - L}{6s},$$

$$P_{pk} = \frac{min\{U - \mu, \mu - L\}}{3s},$$

where s is computed using $s = \sqrt{\frac{\sum_{i=1}^{n}(X_i - \bar{X})^2}{n-1}}$ (s-long-term). The main difference between PCIs and PPIs is that PCIs are used to predict the capability of a process to produce parts conforming to specifications, while PPIs are used to evaluate the behaviour of a process. Note that for a stable process, the differences between C_p (C_{pk}) and P_p (P_{pk}) are negligible (Montgomery [12]).

The use of the four well-known PCIs presupposes that the process is in control as well as that the quality characteristic of interest is normally distributed. However, there are many processes in which the distribution of the quality characteristic is continuous but non-normal. Many authors have studied PCIs on continuous non-normal distributions. Clements [13] used a percentile method (known as "Clements' method") in order to calculate the C_p and C_{pk} indices for non-normal Pearsonian populations. Pearn and Kotz [14] applied Clements' method for computing the C_{pm} and C_{pmk} indices. Rivera, Hubele and Lawrence [15] studied the effect of several transformation methods on the estimate of C_{pk} index for non-normal data. Pearn and Chen [16] proposed a modification of Clements' method in order to calculate the four well-known PCIs for non-normal Pearsonian populations. Castagliola [17] proposed a method based on Burr's distributions to estimate the C_p and C_{pk} indices independently of the real process distribution. Hosseinifard et al. [18] proposed a root transformation technique to estimate PCIs of right skewed processes. Hosseinifard, Abbasi and Niaki [19] studied and compared various methods for estimating PCIs of non-normal processes.

In practice, there are many processes that the quality characteristic of interest is discrete and, unfortunately, the above studies cannot be used. Few studies have been focused on PCIs for discrete quality characteristics. Bothe [20] computed the C_{pk} index based on the percentage of nonconforming parts of a process with attribute data. Yeh and Bhattcharya [21] proposed a PCI that is related to the probability of non-conformance of a process and can be applied to continuous as well as discrete distributions. Borges and Ho [22] proposed a PCI based on the process fraction defective. Perakis and Xekalaki [23,24] proposed a PCI based on the proportion of conformance of the process, which can be used for continuous or discrete processes. Maiti, Saha and Nanda [25] defined a generalized index, named C_{py}, based on the ratio of the proportion of specification conformance to the proportion of desired conformance, which can be used for continuous as well as discrete processes. Moreover, they introduced two modifications of the C_{py} index, the C_{pyk} and C_{pTk} indices, for studying off-centered and off-target processes, respectively. Maravelakis [26] proposed the Q transformation technique to estimate the four well-known PCIs for Poisson and binomial data. Dey and Saha [27] considered three bootstrap confidence intervals of the C_{pyk} index using different methods of estimation for the logistic exponential distribution. Pal and Gauri [28] studied several approaches for estimating the PCIs C_{pu} and C_{pl} of a binomial process. In addition, Gauri and Pal [29] assessed the relative goodness of some generalized PCIs in quantifying the capability of a Poisson or binomial process.

Motivated by the work of Maravelakis [26] and of Pal and Gauri [28], in this paper, we expand the transformation techniques for Poisson and binomial data and we also apply them to negative binomial data in order to compute the four classical PCIs and the two PPIs. Moreover, we compare these indices with other existing PCIs for discrete data to make conclusions about the proposed methodology. Finally, we illustrate the application of the transformation techniques to real data from Poisson, binomial and negative binomial processes.

The paper is organized as follows. In Section 2, we present the transformation techniques for Poisson, binomial and negative binomial data, while a brief overview of existing PCIs for discrete data is given in Section 3. Section 4 presents the results of a simulation study that was performed to compare the indices using different transformation methods. In Section 5, we present three illustrative examples, one for each discrete distribution. Finally, we conclude the paper in Section 6.

2. Transformation Techniques to Compute PCIs

In this section, we present several transformation techniques for Poisson, binomial and negative binomial data, in order to transform them into being approximately independent and normally distributed.

2.1. Transformation Techniques for Poisson Data

Let X be the number of defects or nonconformities in an inspection unit with a standard size. If the process is stable, i.e., the defects rate c is constant, and each inspection is performed independently, then the variable X follows the Poisson distribution with probability mass function (pmf) given by

$$f(x;c) = \frac{e^{-c}c^x}{x!}, x = 0,1,2,\ldots, \tag{1}$$

where $c > 0$ is the parameter of the Poisson distribution. The mean and the variance of X are equal to c.

2.1.1. Anscombe's Transformation

Let x_i, $i = 1,2,\ldots$ be observations of the Poisson variable X. Bartlett [30] used two nonlinear transformations based on the square root in order to transform Poisson data into normal data. Anscombe [31] proposed a variation of Balett's transformation afterwards, and he demonstrated that the statistics $y_i = \sqrt{x_i + 3/8}$, $i = 1,2,\ldots$ are approximately normally distributed with mean and variance given by $\sqrt{c + 3/8} - 1/(8c)$ and 0.25, respectively. In this paper, we use a modification of this transformation in order to obtain a variance close to 1, that is

$$y_i = 2\sqrt{x_i + \frac{3}{8}}, \text{ for } i = 1,2,\ldots \tag{2}$$

2.1.2. Freeman–Tukey's Transformation

Freeman and Tukey [32] proposed another square root transformation for approximately normalizing Poisson data. They displayed that the statistics

$$y_i = \sqrt{x_i} + \sqrt{x_i + 1}, \ i = 1,2,\ldots \tag{3}$$

are approximately normal with mean and variance given by $2c$ and 1, respectively. Ryan and Schwertman [33] applied this transformation to the classical Shewhart c-chart, so as to study its performance as well as to investigate the optimal control limits.

2.1.3. The Q Transformation

Quesenberry [34,35] defined the Q statistic and introduced the Poisson Q-chart for data with known and unknown parameter c. For $c = c_0$ known a priori, define

$$u_i = e^{-c_0}\sum_{k=0}^{x_i}\frac{c_0^k}{(k!)},$$
$$Q_i = \Phi^{-1}(u_i), \text{ for } i = 1,2,\ldots \tag{4}$$

where $\Phi^{-1}(\cdot)$ is the inverse function of the standard normal distribution function. Then, the values $Q_1, Q_2, \ldots$ are approximately standard normal.

If the defects rate c is unknown, then we consider that the data are in a sequence of values (n_i, x_i), for $i = 1, 2, \ldots$ Let $t_i = \sum_{j=1}^{i} x_j$, $N_i = \sum_{j=1}^{i} n_j$ and define

$$u_i = \sum_{k=0}^{x_i} \binom{t_i}{k} \left(\frac{n_i}{N_i}\right)^k \left(1 - \frac{n_i}{N_i}\right)^{(n_i - k)} \tag{5}$$

$$Q_i = \Phi^{-1}(u_i), \text{ for } i = 2, 3, \ldots$$

Then, the values $Q_2, Q_3, \ldots$ are approximately standard normal. The Poisson Q-chart is obtained by plotting the statistics Q with control limits UCL = 3, CL = 0 and LCL = -3. We notice that for the case where c is unknown, the first plotting statistic is the Q_2. Moreover, according to Quesenberry [36], this approximation performs satisfactorily after the first observations, as in an effort to approximate c, past observations are used to update the Q statistic.

2.2. Transformation Techniques for Binomial Data

Let X be the number of noncomforming products found when n products are inspected. If the process is stable, i.e., the probability p that any product does not conform to specifications is constant, and each inspection is performed independently, then the variable X follows the binomial distribution with parameters n and p and the pmf given by

$$f(x; n, p) = \binom{n}{x} p^x (1-p)^{n-x}, \; x = 0, 1, \ldots, n \tag{6}$$

The mean and the variance of X are equal to np and $np(1-p)$, respectively.

2.2.1. Freeman–Tukey's Transformation

Let x_i be the number of nonconforming items produced in a sample of size n with known and constant probability $p = p_0$ of observing a nonconforming product. Freeman and Tukey [32] used the average angular transformation to transform binomial data to normal. The transformation is defined as

$$y_i = \sin^{-1}\left(\sqrt{\frac{x_i}{n+1}}\right) + \sin^{-1}\left(\sqrt{\frac{x_i+1}{n+1}}\right), \text{ for } i = 1, 2, \ldots \tag{7}$$

The statistics y_i are normally distributed with mean and variance equal to $\sin^{-1}\sqrt{p}$ and $1/(n+0.5)$, respectively. Ryan [37] used this transformation to construct a control chart for monitoring p.

2.2.2. Chen's Transformation

Johnson and Kotz [38] showed that the statistics

$$y_i = \sin^{-1}\left(\sqrt{\frac{x_i + \frac{3}{8}}{n + \frac{3}{4}}}\right) \text{ for } i = 1, 2, \ldots \tag{8}$$

are approximately normally distributed with mean and variance equal to $\sin^{-1}\sqrt{p}$ and $1/(4n)$, respectively. Ryan [37] used this transformation to construct a control chart for monitoring p. Chen [39] used a variation of Johnson and Kotz's transformation for monitoring p, given by

$$y_i = 2\sqrt{n_i}\left[\sin^{-1}\left(\sqrt{\frac{x_i + \frac{3}{8}}{n + \frac{3}{4}}}\right) - \sin^{-1}(\sqrt{p})\right], \text{ for } i = 1, 2, \ldots, \tag{9}$$

for the case where the parameter p is known. If the parameter p is unknown, then we use the transformation

$$y_i = 2\sqrt{n_i}\left[\sin^{-1}\left(\sqrt{\frac{x_i + \frac{3}{8}}{n + \frac{3}{4}}}\right) - \sin^{-1}(\sqrt{\hat{p}_{i-1}})\right], \text{ for } i = 2, 3, \ldots, \tag{10}$$

where $\hat{p}_i = (x_1 + x_2 + \ldots + x_i)/(n_1 + n_2 + \ldots + n_i)$. The arcsine-$p$ chart, proposed by Chen [39], is obtained by plotting the statistics y_i's with control limits at UCL = 3, CL = 0 and LCL = -3. We notice that for the case where p is unknown, the first plotted statistic is y_2.

2.2.3. The Q Transformation

Quesenberry [40] introduced the binomial Q-chart for data with known and unknown p. For $p = p_0$ known a priori, we define

$$u_i = \sum_{k=0}^{x_i} \binom{n_i}{k} p_0^k (1 - p_0)^{(n_i - k)},$$

$$Q_i = \Phi^{-1}(u_i), \text{ for } i = 1, 2, \ldots \tag{11}$$

According to Quesenberry [40], the values $Q_1, Q_2, \ldots$, are approximately standard normal. For p unknown, let $t_i = \sum_{j=1}^{i} x_j$ and $N_i = \sum_{j=1}^{i} n_j$ and define

$$u_i = \sum_{k=h_i}^{x_i} \binom{n_i}{k} \frac{\binom{t_i - k}{N_{i-1}}}{\binom{n_i + N_{i-1}}{t_i}},$$

$$Q_i = \Phi^{-1}(u_i), \text{ for } i = 1, 2, \ldots \tag{12}$$

where $h_i = \max(0, t_i - N_{i-1})$. Then, the values $Q_2, Q_3, \ldots$ are approximately standard normal. The binomial Q-chart is obtained by plotting the statistics Q on a chart with control limits UCL = 3, CL = 0 and LCL = -3. We notice that for the case where p is unknown, the first plotting statistic is Q_2. Furthermore, the approximation has adequate performance after the first observations, as in the case of Poisson data.

2.3. Transformation Techniques for Negative Binomial Data

Let X be the total number of products inspected until a nonconforming r ($r \geq 1$) is counted and p be the probability of observing a nonconforming one. Then, the variable X follows the negative binomial distribution with parameters r and p with a pmf given by

$$f(x; r, p) = \binom{x - 1}{r - 1} p^r (1 - p)^{x - r}, \; x = r, r + 1, \ldots \tag{13}$$

The mean and variance of X are equal to r/p and $r(1-p)/p^2$, respectively. Note that the geometric distribution is a special case of the negative binomial distribution for $r = 1$. The CCC-r chart, proposed by Xie et al. [41], is used to monitor observations of the variable X and it is suggested when the parameter p is low. For more details, the reader can refer to the work of Xie, Goh and Kuralmani [42] and that of Joekes, Smrekar and Righetti [43].

2.3.1. Anscombe's Transformation

Let x_i be the total number of products inspected until a nonconforming r ($r \geq 1$) is counted. Anscombe [31] showed that the transformation

$$y_i = \ln(x_i + \frac{1}{2}r), \text{ for } i = 1, 2, \ldots \tag{14}$$

for a large mean $m = r/p$ and $r \geq 1$ is approximately normal with variance approximated by $1/(r - 0.5)$ for a large r.

2.3.2. The Box–Cox Transformation

Box and Cox [44] proposed a family of power transformations on a positive variable X given by

$$Y^{(L)} = \begin{cases} \frac{X^L - 1}{L}, & \text{for } L \neq 0, \\ \ln(X), & \text{for } L = 0, \end{cases} \quad (15)$$

where L is a single parameter that can be estimated using the maximum likelihood estimation (MLE), so that $Y^{(L)}$ is approximately normally distributed. Applying this transformation to observations x_i, we have the transformed data y_i, which follow approximately the normal distribution.

3. A Brief Overview of PCIs for Discrete Distributions

Yeh and Bhattcharga [21] proposed an index based on the proportion of nonconforming products. This index is defined as follows:

$$C_f = min\left(\frac{p_0^L}{p_L}, \frac{p_0^U}{p_U}\right), \quad (16)$$

where $p_L = P(X < L)$, $p_U = P(X > U)$ and p_0^L, p_0^U are the expected proportions of nonconforming products that the manufacturer can tolerate on L and U, respectively.

Perakis and Xekalaki [23] proposed an index that can be used for continuous as well as for discrete distributions and is defined as follows:

$$C_{pc} = \frac{1 - p_0}{1 - p}, \quad (17)$$

where $p = P(L < X < U)$ and p_0 is the minimum allowable proportion of conformance, usually equal to 0.9973. Perakis and Xekalaki [24] investigated the above index for discrete data where only one specification limit is set. More specifically, if only the U is set, then the index is denoted by C_{pcu} while if only the L is set, the index is denoted by C_{pcl}. The two indices are defined as follows:

$$C_{pcu} = \frac{0.0027}{1 - P(X < U)}, C_{pcl} = \frac{0.0027}{1 - P(X > L)}. \quad (18)$$

Alternative forms of the two indices for a Poisson process can be obtained by using the chi-square distribution. Thus, they can be written as follows:

$$C_{pcu} = \frac{0.0027}{P(x_{2U}^2 < 2c)}, C_{pcl} = \frac{0.0027}{P(x_{2(L+1)}^2 > 2c)}. \quad (19)$$

Maiti, Saha and Nanda [25] proposed an index based on the process yield, which is defined as follows:

$$C_{py} = \frac{p}{p_0}, \quad (20)$$

where $p = P(L \leq X \leq U)$ and p_0 is the desired yield equal to $1 - p_0^L - p_0^U$.

For an off-centered process, i.e., $F(U) + F(L) \neq 1$, where $F(\cdot)$ is the cumulative density function (cdf), the index is defined as follows:

$$C_{pyk} = min\left(\frac{F(U) - F(M)}{\frac{1}{2} - p_0^U}, \frac{F(M) - F(L)}{\frac{1}{2} - p_0^L}\right) = min\left(\frac{F(U) - \frac{1}{2}}{\frac{1}{2} - p_0^U}, \frac{\frac{1}{2} - F(L)}{\frac{1}{2} - p_0^L}\right), \quad (21)$$

where M is the median of the distribution and $F(M) = \frac{F(U) + F(L)}{2} = \frac{1}{2}$ for a centered process. We notice that for a centered process, $C_{py} = C_{pyk}$, while if $C_{pyk} < C_{py}$, the process median is off-centered.

For generalized asymmetric tolerance, i.e., $T \neq M$, the index is defined as follows:

$$C_{pTk} = min\left(\frac{F(U) - F(T)}{\frac{1}{2} - p_0^U}, \frac{F(T) - F(L)}{\frac{1}{2} - p_0^L}\right). \tag{22}$$

We point out that $C_{pTk} = C_{pyk}$ when $M = T$.

The above PCIs have been studied for various non-normal as well as for Poisson distributions. However, these indices can be easily extended to other discrete data following a binomial or negative binomial distribution using the appropriate cdf.

When the value of the parameter of distribution is unknown, then it has to be estimated in order to compute the indices. In most cases, the maximum likelihood estimator (MLE) method is used. If $X_1, X_2, \ldots, X_m$ are independent observations from a Poison process, the MLE of c is given by $\hat{c} = \frac{1}{m}\sum_{i=1}^{m} X_i$, while if the observations are from a binomial process, the MLE of p is given by $\hat{p} = \frac{1}{mn}\sum_{i=1}^{m} X_i$. Finally, for a negative binomial process, the MLE of p is given by $\hat{p} = \frac{mr}{\sum_{i=1}^{m} X_i}$.

4. A Simulation Study

In this section, a simulation study is performed to evaluate the PCIs and PPIs for different Poisson, binomial and negative binomial processes. Six different processes are considered for each distribution: two centered, two off-centered and two off-target processes. For the centered processes, the process target is approximately equal to the median, and the U and L are chosen so that the condition $F(U) + F(L) = 1$ is satisfied. For the off-centered processes, the values of specification limits are chosen so that $F(U) + F(L) \neq 1$. Finally, under off-target processes, the process target is such that $F(T) \neq (F(U) + F(L))/2$. These situations of a process are also considered in Maiti, Saha and Nanda [25].

In our simulation study, random variables of each discrete distribution are generated for specified values of parameters and for sample of sizes $m = 50$ and 100 to compute the mean and the standard deviation of each index. In order to estimate them, first, the appropriate transformation method is applied to the data. Then, the specification limits and the target value are transformed via the transformation applied on the data. Next, the obtained specification limits, the target value, as well as the mean and the standard deviation of the transformed data are used to estimate each index. As C_p has an indirect association with C_{pc} and C_{py}, C_{pk} with C_{pyk} and C_{pmk} with C_{pTk}, comparisons among them and the classical PCIs are also presented. The estimators of the C_{pc}, C_{py}, C_{pyk} and C_{pTk} are obtained using the MLE method. Moreover, in order to make the latter indices comparable with the classical PCIs, we use $p_0 = 0.9973$ and $p_0^L = p_0^U = 0.00135$. All the computations are carried out using the R version 4.3.0. software with 10,000 iterations.

The results of the simulation study are presented in Tables 1–6. The first two processes in each table are centered, the other two are off-centered and the last two are off-target. The first columns in these tables show the values of the parameters of each distribution, the specification limits and the target value of the processes. In each process, the first row shows the mean of the index estimators and the second row shows their standard deviations.

As can be observed in the aforementioned tables, we observe that for each process, the differences in the mean values of PCI and PPI estimators applying different transformation methods are minor. More specifically, the differences are very small when applying Anscombe's and Freeman–Tukey's transformations for Poisson processes and Freeman–Tukey's and Chen's transformations for binomial processes. Moreover, for all processes, the values of $\hat{C}_p$ are very close to the corresponding values of $\hat{C}_{py}$, while at almost all of them, the values of $\hat{C}_{pc}$ differ from those of $\hat{C}_p$ and $\hat{C}_{py}$. This can be explained as follows: The C_{pc} index is based on the ratio of the allowable proportion of nonconforming products to the observed proportion of nonconforming products, while the C_{py} index is based on the ratio of the observed proportion of conformance to the allowable proportion of conformance. In other words, the C_{py} index is the reciprocal of the complementary proportions of the C_{pc} index. For off-centered processes, the values of $\hat{C}_{pk}$ indices are slightly smaller than those of $\hat{C}_{pyk}$. Furthermore, for binomial and negative binomial off-target processes, the values of $\hat{C}_{pmk}$ indices are very close to the corresponding $\hat{C}_{pTk}$ indices, while for Poisson processes, the differences are larger but both of them indicate incapability processes. Finally, it is observed that the values of $\hat{P}_p$ and $\hat{P}_{pk}$ are approximately similar to those of $\hat{C}_p$ and $\hat{C}_{pk}$, respectively. Thus, the processes can be considered stable.

In addition, we study the impact of sample size and parameters of distribution on the results. For a certain process, the differences in the mean values of each index estimator are negligible for different values of sample sizes. This entails the fact that one should use a small value of sample size (say $m = 50$) to obtain safe conclusions about the process capability/performance. On the other hand, the value of the standard deviation of each index estimator for $m = 50$ is always higher than the corresponding value for $m = 100$. It should be noticed that the standard deviations of $\hat{C}_{py}$, $\hat{C}_{pyk}$ and $\hat{C}_{pTk}$ are smaller than that of the classical PCIs estimators. Moreover, for certain values of U, L and T, the capability of the process depends on the parameters of distribution. For example, consider the second and the fifth processes in Table 1, where $(U, L, T) = (15, 0, 6)$, while $c = 6.5$ for the second process and $c = 4.5$ for the fifth one. It is recalled that the second process is centered and the fifth process is off-target. As can be seen in Table 1, we observe that the differences in the mean and standard deviation values of the $\hat{C}_p$ and $\hat{P}_p$ indices for the two processes are negligible, while the corresponding differences are significant for the other three classical PCIs and $\hat{P}_{pk}$. This was expected as the C_p and P_p indices do not depend on the mean value of the process.

Table 1. The results of the simulation study for Poisson processes (sample size $m = 50$).

	$\hat{C}_p$			$\hat{C}_{pc}$	$\hat{C}_{py}$	$\hat{C}_{pk}$			$\hat{C}_{pyk}$	$\hat{C}_{pm}$		
	Ansc	F-T	Q			Ansc	F-T	Q		Ansc	F-T	Q
(U, L) = (10, 0)	0.8994	0.9477	0.8425	0.1439	0.9954	0.7364	0.7345	0.7808	0.9773	0.8823	0.9299	0.8220
(c, T) = (4.5, 4)	0.1238	0.1328	0.1132	0.0101	0.0035	0.1068	0.1072	0.1133	0.0061	0.1168	0.1252	0.1081
(U, L) = (15, 0)	1.1350	1.1868	1.0493	0.9365	1.0014	0.9229	0.9241	0.9871	0.9987	1.1164	1.1673	1.0284
(c, T) = (6.5, 6)	0.1540	0.1615	0.1400	0.0984	0.0008	0.1293	0.1298	0.1380	0.0012	0.1466	0.1537	0.1343
(U, L) = (20, 0)	1.3452	1.3932	1.3868	0.2560	1.0027	1.0624	1.1609	0.8949	0.9795	1.3196	1.3670	1.3531
(c, T) = (4.5, 4)	0.1852	0.1953	0.1864	0.0798	$4 \cdot 10^{-8}$	0.1603	0.1772	0.1289	0.0071	0.1747	0.1840	0.1779
(U, L) = (12, 0)	0.9967	1.0481	0.8918	0.1601	0.9856	0.6462	0.6467	0.6783	0.9684	0.9803	1.0308	0.8741
(c, T) = (6.5, 6)	0.1352	0.1426	0.1190	0.0507	0.0069	0.0956	0.0959	0.1033	0.0138	0.1287	0.1358	0.1141
(U, L) = (15, 0)	1.1408	1.1888	1.1309	0.2545	1.0027	1.0602	1.1437	0.8949	0.9795	0.9001	0.9388	0.8952
(c, T) = (4.5, 6)	0.1571	0.1666	0.1520	0.0779	$2 \cdot 10^{-5}$	0.1582	0.1673	0.1289	0.0071	0.1072	0.1142	0.0981
(U, L) = (13, 0)	1.0445	1.0960	0.9368	0.3149	0.9950	0.7418	0.7426	0.7786	0.9871	0.7359	0.7723	0.6619
(c, T) = (6.5, 9)	0.1417	0.1491	0.0863	0.0847	0.0035	0.1070	0.1073	0.0797	0.0070	0.0779	0.0819	0.0454

	$\hat{C}_{pmk}$			$\hat{C}_{pTk}$	$\hat{P}_p$			$\hat{P}_{pk}$		
	Ansc	F-T	Q		Ansc	F-T	Q	Ansc	F-T	Q
(U, L) = (10, 0)	0.7233	0.7216	0.7650	0.8857	0.8855	0.9298	0.8335	0.7250	0.7206	0.7748
(c, T) = (4.5, 4)	0.1075	0.1076	0.1153	0.0745	0.0964	0.1069	0.0855	0.0855	0.0873	0.0899
(U, L) = (15, 0)	0.9085	0.9097	0.9683	0.9007	1.1221	1.1727	1.0408	0.9124	0.9131	0.9792
(c, T) = (6.5, 6)	0.1288	0.1292	0.1391	0.0746	0.1214	0.1286	0.1076	0.1037	0.1048	0.1096
(U, L) = (20, 0)	1.0413	1.1381	0.8720	0.8934	1.3243	1.3669	1.3721	1.0459	1.1390	0.8854
(c, T) = (4.5, 4)	0.1450	0.1609	0.1144	0.0754	0.1442	0.1571	0.1407	0.1307	0.1482	0.1014
(U, L) = (12, 0)	0.6364	0.6368	0.6657	0.8827	0.9854	1.0356	0.8847	0.6388	0.6390	0.6729
(c, T) = (6.5, 6)	0.0965	0.0967	0.1054	0.0726	0.1066	0.1136	0.0915	0.0787	0.0793	0.0853
(U, L) = (15, 0)	0.8387	0.9042	0.7111	0.3438	1.1231	1.1664	1.1188	1.0438	1.1222	0.8854
(c, T) = (4.5, 6)	0.1262	0.1227	0.1067	0.0772	0.1223	0.1341	0.1148	0.1285	0.1377	0.1014
(U, L) = (13, 0)	0.5204	0.5210	0.5489	0.2355	1.0327	1.0830	0.9381	0.7334	0.7337	0.7796
(c, T) = (6.5, 9)	0.0359	0.0360	0.0257	0.0557	0.1118	0.1188	0.0970	0.0870	0.0878	0.0945

Table 2. The results of the simulation study for Poisson processes (sample size $m = 100$).

	$\hat{C}_p$			$\hat{C}_{pc}$	$\hat{C}_{py}$	$\hat{C}_{pk}$			$\hat{C}_{pyk}$	$\hat{C}_{pm}$		
	Ansc	F-T	Q			Ansc	F-T	Q		Ansc	F-T	Q
(U, L) = (10, 0)	0.8903	0.9378	0.8344	0.1477	0.9957	0.7296	0.7275	0.7807	0.9791	0.8788	0.9259	0.8186
(c, T) = (4.5, 4)	0.0854	0.0917	0.0780	0.0058	0.0023	0.0738	0.0740	0.0798	0.0042	0.0817	0.0876	0.0754
(U, L) = (15, 0)	1.1239	1.1751	1.0394	0.9713	1.0015	0.9143	0.9155	0.9824	0.9992	1.1120	1.1626	1.0244
(c, T) = (6.5, 6)	0.1055	0.1108	0.0958	0.0610	0.0005	0.0888	0.0891	0.0961	0.0007	0.1020	0.1070	0.0932
(U, L) = (20, 0)	1.3316	1.3787	1.3735	0.2496	1.0027	1.0511	1.1481	0.8862	0.9800	1.3143	1.3612	1.3475
(c, T) = (4.5, 4)	0.1277	0.1348	0.1284	0.0540	$2 \cdot 10^{-8}$	0.1106	0.1224	0.0888	0.0049	0.1222	0.1287	0.1241
(U, L) = (12, 0)	0.9869	1.0377	0.8835	0.1571	0.9860	0.6403	0.6408	0.6720	0.9694	0.9765	1.0267	0.8707
(c, T) = (6.5, 6)	0.0927	0.0978	0.0814	0.0365	0.0047	0.0657	0.0659	0.0710	0.0095	0.0895	0.0945	0.0792
(U, L) = (15, 0)	1.1293	1.1764	1.1200	0.2489	1.0027	1.0508	1.1403	0.8862	0.9800	0.8957	0.9338	0.8920
(c, T) = (4.5, 6)	0.1083	0.1150	0.1047	0.0532	10^{-5}	0.1104	0.1178	0.0888	0.0049	0.0744	0.0795	0.0680
(U, L) = (13, 0)	1.0343	1.0852	0.9368	0.3147	0.9953	0.7350	0.7357	0.7786	0.9878	0.7331	0.7693	0.6619
(c, T) = (6.5, 9)	0.0971	0.1023	0.0863	0.0642	0.0024	0.0735	0.0737	0.0797	0.0048	0.0540	0.0569	0.0454

	$\hat{C}_{pmk}$			$\hat{C}_{pTk}$		$\hat{P}_p$			$\hat{P}_{pk}$		
	Ansc	F-T	Q			Ansc	F-T	Q	Ansc	F-T	Q
(U, L) = (10, 0)	0.7205	0.7187	0.7664	0.9064	0.8777	0.9207	0.8270	0.7192	0.7142	0.7737	
(c, T) = (4.5, 4)	0.0751	0.0751	0.0823	0.0570	0.0665	0.0741	0.0588	0.0592	0.0605	0.0639	
(U, L) = (15, 0)	0.9050	0.9061	0.9687	0.9232	1.1124	1.1621	1.0326	0.9049	0.9054	0.9760	
(c, T) = (6.5, 6)	0.0896	0.0899	0.0982	0.0566	0.0837	0.0890	0.0737	0.0716	0.0725	0.0768	
(U, L) = (20, 0)	1.0370	1.1331	0.8689	0.9156	1.3127	1.3536	1.3613	1.0361	1.1272	0.8783	
(c, T) = (4.5, 4)	0.1018	0.1131	0.0801	0.0575	0.0995	0.1090	0.0969	0.0904	0.1030	0.0699	
(U, L) = (12, 0)	0.6339	0.6343	0.6627	0.9018	0.9768	1.0263	0.8777	0.6337	0.6337	0.6677	
(c, T) = (6.5, 6)	0.0672	0.0674	0.0735	0.0557	0.0735	0.0786	0.0627	0.0544	0.0549	0.0589	
(U, L) = (15, 0)	0.8346	0.9058	0.7072	0.3417	1.1132	1.1550	1.1100	1.0359	1.1196	0.8783	
(c, T) = (4.5, 6)	0.0897	0.0894	0.0739	0.0545	0.0844	0.0930	0.0790	0.0901	0.0978	0.0699	
(U, L) = (13, 0)	0.5199	0.5204	0.5489	0.2338	1.0237	1.0732	0.9307	0.7275	0.7276	0.7736	
(c, T) = (6.5, 9)	0.0250	0.0251	0.0257	0.0395	0.0770	0.0822	0.0665	0.0601	0.0608	0.0653	

Table 3. The results of the simulation study for binomial processes (sample size $m = 50$).

	$\hat{C}_p$			$\hat{C}_{pc}$	$\hat{C}_{py}$	$\hat{C}_{pk}$			$\hat{C}_{pyk}$	$\hat{C}_{pm}$		
	F-T	Chen	Q			F-T	Chen	Q		F-T	Chen	Q
(U, L, T) = (13, 0, 5)	1.1239	1.0724	0.9983	0.5475	1.0009	0.8964	0.8956	0.9479	0.9957	1.0919	1.0420	0.9641
(n, p) = (113, 0.05)	0.1570	0.1490	0.1359	0.0497	0.0010	0.1288	0.1283	0.1343	0.0020	0.1447	0.1374	0.1268
(U, L, T) = (22, 1, 10)	1.2768	1.2779	1.2115	5.5306	1.0024	1.0878	1.0863	1.1417	1.0020	1.2570	1.2581	1.1902
(n, p) = (105, 0.1)	0.1731	0.1733	0.1627	0.8569	0.0002	0.1515	0.1512	0.1589	0.0003	0.1660	0.1662	0.1566
(U, L, T) = (10, 0, 5)	0.9652	0.9141	0.8204	0.0949	0.9751	0.5790	0.5789	0.6030	0.9476	0.9377	0.8881	0.7922
(n, p) = (113, 0.05)	0.1348	0.1270	0.1117	0.0256	0.0100	0.0893	0.0891	0.0955	0.0199	0.1243	0.1171	0.1042
(U, L, T) = (17, 1, 10)	1.0634	1.0649	0.9795	0.1782	0.9855	0.6609	0.6603	0.6813	0.9684	1.0468	1.0484	0.9623
(n, p) = (105, 0.1)	0.1442	0.1444	0.1316	0.0693	0.0065	0.0983	0.0982	0.1031	0.0131	0.1382	0.1385	0.1266
(U, L, T) = (11, 0, 6)	1.0201	0.9689	0.8813	0.1895	0.9908	0.6888	0.6885	0.7248	0.9787	0.9821	0.9326	0.8530
(n, p) = (113, 0.05)	0.1425	0.1346	0.1200	0.0417	0.0050	0.1025	0.1022	0.1100	0.0097	0.1357	0.1280	0.1123
(U, L, T) = (18, 1, 11)	1.1078	1.1092	1.0273	0.3650	0.9943	0.7497	0.7490	0.7769	0.9858	1.0706	1.0719	0.9962
(n, p) = (105, 0.1)	0.1502	0.1504	0.1380	0.1483	0.0036	0.1090	0.1089	0.1146	0.0072	0.1429	0.1431	0.1300

Table 3. Cont.

	$\hat{C}_{pmk}$			$\hat{C}_{pTk}$	$\hat{P}_p$			$\hat{P}_{pk}$		
	F-T	Chen	Q		F-T	Chen	Q	F-T	Chen	Q
(U, L, T) = (13, 0, 5) (n, p) = (113, 0.05)	0.8722 0.1283	0.8715 0.1278	0.9165 0.1331	0.9099 0.0663	1.1087 0.1230	1.0591 0.1149	0.9895 0.1022	0.8843 0.1024	0.8845 0.1009	0.9396 0.1039
(U, L, T) = (22, 1, 10) (n, p) = (105, 0.1)	1.0716 0.1505	1.0702 0.1503	1.1224 0.1591	0.9082 0.0700	1.2687 0.1343	1.2696 0.1344	1.2056 0.1248	1.0808 0.1197	1.0793 0.1195	1.1361 0.1248
(U, L, T) = (10, 0, 5) (n, p) = (113, 0.05)	0.5638 0.0913	0.5637 0.0912	0.5837 0.0986	0.8883 0.0624	0.9522 0.1056	0.9027 0.0980	0.8131 0.0840	0.5711 0.0733	0.5717 0.0727	0.5977 0.0785
(U, L, T) = (17, 1, 10) (n, p) = (105, 0.1)	0.6513 0.0992	0.6508 0.0991	0.6701 0.1048	0.8907 0.0679	1.0565 0.1118	1.0580 0.1120	0.9748 0.1009	0.6566 0.0805	0.6560 0.0804	0.6780 0.0848
(U, L, T) = (11, 0, 6) (n, p) = (113, 0.05)	0.6616 0.0865	0.6613 0.0861	0.7002 0.0933	0.6508 0.0948	1.0064 0.1116	0.9569 0.1038	0.8735 0.0902	0.6795 0.0829	0.6800 0.0820	0.7184 0.0886
(U, L, T) = (18, 1, 11) (n, p) = (105, 0.1)	0.7232 0.0932	0.7224 0.0930	0.7521 0.0984	0.7019 0.0985	1.1006 0.1165	1.1020 0.1167	1.0223 0.1058	0.7449 0.0882	0.7441 0.0881	0.7731 0.0931

Table 4. The results of the simulation study for binomial processes (sample size $m = 100$).

	$\hat{C}_p$			$\hat{C}_{pc}$	$\hat{C}_{py}$	$\hat{C}_{pk}$			$\hat{C}_{pyk}$	$\hat{C}_{pm}$		
	F-T	Chen	Q			F-T	Chen	Q		F-T	Chen	Q
(U, L, T) = (13, 0, 5) (n, p) = (113, 0.05)	1.1115 0.1080	1.0608 0.1024	0.9879 0.0935	0.5646 0.0310	1.0010 0.0006	0.8871 0.0887	0.8865 0.0884	0.9447 0.0939	0.9962 0.0014	1.0865 0.1007	1.0369 0.0957	0.9593 0.0884
(U, L, T) = (22, 1, 10) (n, p) = (105, 0.1)	1.2640 0.1195	1.2651 0.1197	1.1996 0.1124	5.7655 0.6005	1.0024 0.0001	1.0774 0.1049	1.0759 0.1047	1.1336 0.1110	1.0021 0.0002	1.2512 0.1160	1.2524 0.1161	1.1849 0.1095
(U, L, T) = (10, 0, 5) (n, p) = (113, 0.05)	0.9545 0.0927	0.9041 0.0873	0.8119 0.0769	0.0938 0.0188	0.9758 0.0070	0.5732 0.0617	0.5731 0.0616	0.5969 0.0661	0.9488 0.0140	0.9331 0.0865	0.8838 0.0815	0.7883 0.0727
(U, L, T) = (17, 1, 10) (n, p) = (105, 0.1)	1.0527 0.0995	1.0542 0.0997	0.9699 0.0909	0.1717 0.0466	0.9860 0.0045	0.6547 0.0683	0.6542 0.0682	0.6749 0.0716	0.9693 0.0091	1.0420 0.0966	1.0436 0.0968	0.9580 0.0886
(U, L, T) = (11, 0, 6) (n, p) = (113, 0.05)	1.0088 0.0980	0.9583 0.0925	0.8721 0.0826	0.1905 0.0321	0.9912 0.0034	0.6818 0.0707	0.6816 0.0705	0.7174 0.0762	0.9797 0.0069	0.9758 0.0945	0.9268 0.0892	0.8487 0.0783
(U, L, T) = (18, 1, 11) (n, p) = (105, 0.1)	1.0966 0.1037	1.0981 0.1039	1.0172 0.0953	0.3514 0.1019	0.9946 0.0025	0.7427 0.0756	0.7419 0.0755	0.7695 0.0796	0.9864 0.0049	1.0649 0.0997	1.0662 0.0999	0.9915 0.0908

	$\hat{C}_{pmk}$			$\hat{C}_{pTk}$	$\hat{P}_p$			$\hat{P}_{pk}$		
	F-T	Chen	Q		F-T	Chen	Q	F-T	Chen	Q
(U, L, T) = (13, 0, 5) (n, p) = (113, 0.05)	0.8678 0.0895	0.8672 0.0892	0.9179 0.0949	0.9350 0.0476	1.0965 0.0848	1.0480 0.0789	0.9801 0.0701	0.8751 0.0705	0.8758 0.0693	0.9372 0.0730
(U, L, T) = (22, 1, 10) (n, p) = (105, 0.1)	1.0669 0.1054	1.0654 0.1052	1.1200 0.1123	0.9321 0.0524	1.2559 0.0928	1.2569 0.0929	1.1941 0.0863	1.0705 0.0829	1.0690 0.0827	1.1284 0.0877
(U, L, T) = (10, 0, 5) (n, p) = (113, 0.05)	0.5609 0.0639	0.5609 0.0638	0.5803 0.0692	0.9111 0.0452	0.9417 0.0728	0.8932 0.0672	0.8054 0.0576	0.5654 0.0506	0.5662 0.0501	0.5921 0.0543
(U, L, T) = (17, 1, 10) (n, p) = (105, 0.1)	0.6484 0.0697	0.6479 0.0695	0.6669 0.0737	0.9118 0.0520	1.0459 0.0773	1.0474 0.0775	0.9655 0.0697	0.6505 0.0559	0.6500 0.0558	0.6718 0.0590
(U, L, T) = (11, 0, 6) (n, p) = (113, 0.05)	0.6587 0.0603	0.6584 0.0600	0.6975 0.0653	0.6509 0.0683	0.9952 0.0770	0.9467 0.0713	0.8652 0.0618	0.6725 0.0571	0.6733 0.0564	0.7117 0.0612
(U, L, T) = (18, 1, 11) (n, p) = (105, 0.1)	0.7205 0.0653	0.7197 0.0652	0.7493 0.0691	0.7016 0.0709	1.0896 0.0805	1.0910 0.0807	1.0126 0.0731	0.7379 0.0612	0.7372 0.0611	0.7659 0.0647

53

Table 5. The results of the simulation study for negative binomial processes (sample size $m = 50$).

	$\hat{C}_p$		$\hat{C}_{pc}$	$\hat{C}_{py}$	$\hat{C}_{pk}$		$\hat{C}_{pyk}$	$\hat{C}_{pm}$	
	Ansc	B-C			Ansc	B-C		Ansc	B-C
(U, L, T) = (358, 5, 85)	1.1026	1.2726	1.0357	1.0005	0.8051	0.7873	0.9989	1.0889	
(r, p) = (3, 0.03)	0.1557	0.4639	0.1732	0.0008	0.1090	0.1633	0.0013	0.1539	0.4484
(U, L, T) = (1428, 76, 462)	1.0557	1.1542	0.9503	1.0000	0.8418	0.7983	0.9987	1.0429	1.1345
(r, p) = (5, 0.01)	0.1459	0.7983	0.1544	0.0007	0.1154	0.1576	0.0015	0.1442	0.2740
(U, L, T) = (318, 5, 85)	1.0702	1.2361	0.6443	0.9986	0.7403	0.7330	0.9959	1.0569	1.2138
(r, p) = (3, 0.03)	0.1512	0.4698	0.1877	0.0020	0.1008	0.1503	0.0040	0.1494	0.4543
(U, L, T) = (1428, 26, 462)	1.4242	1.8118	2.8863	1.0012	0.8418	0.8295	0.9998	1.4069	1.7774
(r, p) = (5, 0.01)	0.1968	1.0715	2.3463	0.0010	0.1153	0.1857	0.0020	0.1945	1.0388
(U, L, T) = (328, 5, 100)	1.0786	1.2455	0.7392	0.9992	0.7572	0.7476	0.9970	1.0175	1.1667
(r, p) = (3, 0.03)	0.1524	0.4682	0.1915	0.0016	0.1029	0.1538	0.0031	0.1485	0.4169
(U, L, T) = (1428, 36, 400)	1.3148	1.5776	2.5998	1.0012	0.8418	0.8247	0.9998	1.2635	1.5175
(r, p) = (5, 0.01)	0.1817	0.7348	1.7112	0.0010	0.1154	0.1806	0.0020	0.1588	0.7202

	$\hat{C}_{pmk}$		$\hat{C}_{pTk}$	$\hat{P}_p$		$\hat{P}_{pk}$	
	Ansc	B-C		Ansc	B-C	Ansc	B-C
(U, L, T) = (358, 5, 85)	0.7948	0.7747	0.9079	1.0879	1.2471	0.7944	0.7752
(r, p) = (3, 0.03)	0.1049	0.1626	0.0709	0.1292	0.4275	0.0890	0.1541
(U, L, T) = (1428, 76, 462)	0.8311	0.7857	0.9112	1.0466	1.1409	0.8345	0.7911
(r, p) = (5, 0.01)	0.1107	0.1574	0.0675	0.1173	0.2554	0.0922	0.1465
(U, L, T) = (318, 5, 85)	0.7307	0.7212	0.9064	1.0559	1.2111	0.7304	0.7217
(r, p) = (3, 0.03)	0.0968	0.1494	0.0705	0.1254	0.4340	0.0825	0.1415
(U, L, T) = (1428, 26, 462)	0.8311	0.8165	0.9122	1.4119	1.7854	0.8345	0.8221
(r, p) = (5, 0.01)	0.1107	0.1851	0.0677	0.1583	1.0192	0.0922	0.1754
(U, L, T) = (328, 5, 100)	0.7125	0.7042	0.7522	1.0643	1.2204	0.7471	0.7360
(r, p) = (3, 0.03)	0.0862	0.1493	0.1078	0.1264	0.4322	0.0842	0.1449
(U, L, T) = (1428, 36, 400)	0.8105	0.7890	0.7837	1.3034	1.5563	0.8345	0.8173
(r, p) = (5, 0.01)	0.1131	0.1662	0.0975	0.1461	0.6945	0.0922	0.1702

Table 6. The results of the simulation study for negative binomial processes (sample size $m = 100$).

	$\hat{C}_p$		$\hat{C}_{pc}$	$\hat{C}_{py}$	$\hat{C}_{pk}$		$\hat{C}_{pyk}$	$\hat{C}_{pm}$	
	Ansc	B-C			Ansc	B-C		Ansc	B-C
(U, L, T) = (358, 5, 85)	1.0913	1.1734	1.0747	1.0007	0.7984	0.7990	0.9994	1.0833	1.1619
(r, p) = (3, 0.03)	0.1062	0.2191	0.1247	0.0004	0.0744	0.1246	0.0007	0.1067	0.2134
(U, L, T) = (1428, 76, 462)	1.0458	1.0875	0.9821	1.0000	0.8345	0.8236	0.9992	1.0381	1.0769
(r, p) = (5, 0.01)	0.1012	0.1578	0.1139	0.0004	0.0791	0.1236	0.0008	0.1013	0.1536
(U, L, T) = (318, 5, 85)	1.0592	1.1399	0.6510	0.9988	0.7342	0.7370	0.9967	1.0515	1.1287
(r, p) = (3, 0.03)	0.1031	0.2237	0.1457	0.0013	0.0688	0.1122	0.0026	0.1036	0.2178
(U, L, T) = (1428, 26, 462)	1.4108	1.5720	2.5048	1.0014	0.8345	0.8422	1.0000	1.4005	1.5552
(r, p) = (5, 0.01)	0.1365	0.5340	1.3333	0.0007	0.0791	0.1456	0.0013	0.1366	0.5186
(U, L, T) = (328, 5, 100)	1.0676	1.1486	0.7535	0.9995	0.7509	0.7534	0.9978	1.0103	1.0837
(r, p) = (3, 0.03)	0.1039	0.2225	0.1486	0.0010	0.0702	0.1155	0.0019	0.1022	0.1991
(U, L, T) = (1428, 36, 400)	1.3025	1.4132	2.3547	1.0013	0.8345	0.8398	1.0000	1.2599	1.3681
(r, p) = (5, 0.01)	0.1260	0.3809	1.1123	0.0006	0.0791	0.1422	0.0013	0.1109	0.3815

Table 6. Cont.

	$\hat{C}_{pmk}$		$\hat{C}_{pTk}$	$\hat{P}_p$		$\hat{P}_{pk}$	
	Ansc	B-C		Ansc	B-C	Ansc	B-C
(U, L, T) = (358, 5, 85) (r, p) = (3, 0.03)	0.7923 0.0724	0.7918 0.1253	0.9337 0.0516	1.0761 0.0891	1.1526 0.1997	0.7872 0.0613	0.7860 0.1185
(U, L, T) = (1428, 76, 462) (r, p) = (5, 0.01)	0.8282 0.0768	0.8161 0.1244	0.9366 0.0483	1.0373 0.0822	1.0778 0.1407	0.8277 0.0641	0.8170 0.1161
(U, L, T) = (318, 5, 85) (r, p) = (3, 0.03)	0.7286 0.0668	0.7304 0.1126	0.9320 0.0514	1.0444 0.0865	1.1196 0.2048	0.7240 0.0568	0.7250 0.1063
(U, L, T) = (1428, 26, 462) (r, p) = (5, 0.01)	0.8282 0.0768	0.8345 0.1462	0.9376 0.0486	1.3993 0.1109	1.5564 0.5101	0.8278 0.0641	0.8354 0.1388
(U, L, T) – (328, 5, 100) (r, p) = (3, 0.03)	0.7098 0.0593	0.7125 0.1134	0.7543 0.0771	1.0527 0.0872	1.1281 0.2035	0.7405 0.0580	0.7411 0.1096
(U, L, T) = (1428, 36, 400) (r, p) = (5, 0.01)	0.8080 0.0781	0.8100 0.1287	0.7825 0.0714	1.2919 0.1024	1.3996 0.3605	0.8278 0.0641	0.8331 0.1353

5. Applications

To illustrate the application of the above transformations, three examples are presented. The values of the PCIs and PPIs are computed with particular emphasis on describing Anscombe's transformation for Poisson and negative binomial data and the Freeman–Tukey's transformation for binomial data. Similarly, the indices can be calculated using the other transformations techniques. The C_{py}, C_{pyk} and C_{pTk} indices are also computed for comparison reasons. In the following examples, we present the tools and we guide practitioners on how to apply the proposed methodology. The control charts are provided in order to confirm that the process is in-control. In addition, we apply the transformation techniques to compute the PCIs and PPIs, and we discuss the results afterwards.

Example 1. We use the data on the number of nonconformities in samples of 100 printed circuit boards, which is described in more detail in Chapter 7 by Montgomery [12]. The dataset consists of 46 samples. The first 26 samples of Table 7 are from Table 7.7 of Montgomery [12] and the last 20 samples are from the Table 7.8. of Montgomery [12] The data are presented in Table 7. Moreover, it is assumed that the target value is $T = 18$ nonconformities in each sample, and the upper and lower specification limits are $U = 35$ and $L = 5$ nonconformities in each sample, respectively.

Table 7. The number of nonconformities in samples of 100 printed circuit boards.

Sample number	1	2	3	4	5	6	7	8	9	10
Nonconformities	21	24	16	12	15	5	28	20	31	25
Sample number	11	12	13	14	15	16	17	18	19	20
Nonconformities	20	24	16	19	10	17	13	22	18	**39**
Sample number	21	22	23	24	25	26	27	28	29	30
Nonconformities	30	24	16	19	17	15	16	18	12	15
Sample number	31	32	33	34	35	36	37	38	39	40
Nonconformities	24	21	28	20	25	19	18	21	16	22
Sample number	41	42	43	44	45	46				
Nonconformities	19	12	14	9	16	21				

After a retrospective data analysis, two values (in bold) in Table 7 were eliminated due to assignable causes. Fitting in the remaining 44 samples the Poisson distribution, we find that $\hat{c} = 19.0455$ is the MLE (p-value = 0.4669). In Figure 1, the c-chart is presented. Note that the control limits of the c-chart are computed with

$$UCL = \hat{c} + 3\sqrt{\hat{c}} = 19.0455 + 3\sqrt{19.0455} = 32.1378,$$

$$CL = \hat{c} = 19.0455,$$
$$LCL = \hat{c} - 3\sqrt{\hat{c}} = 19.0455 - 3\sqrt{19.0455} = 5.9532.$$

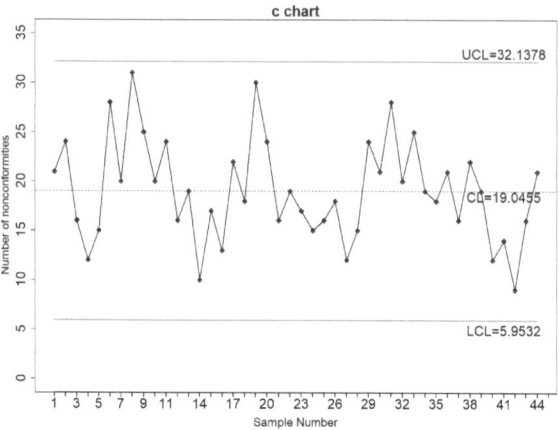

Figure 1. The c-chart for the Example 1 data.

As no point and no systematic behaviour can be seen from the c-chart, we can conclude that the process is in control. In addition, we compute all the y_i's values, for $i = 1, 2, \ldots, 44$ using Anscombe's transformation given by (2). The mean of the transformed data is $\hat{\mu}' = 8.7371$. In order to estimate the standard deviation of the transformed data, we compute their moving range. Hence, the standard deviation $\hat{\sigma}'$ is given by

$$\hat{\sigma}' = \frac{\overline{MR}}{d_2} = \frac{1.2517}{1.1284} = 1.1093$$

An $I - MR$ chart is presented in Figure 2.

The control limits for the I-chart are computed with

$$UCL = \hat{\mu}' + 3\hat{\sigma}' = 8.7371 + 3 \cdot 1.1093 = 12.065,$$
$$CL = \hat{\mu}' = 8.7371,$$
$$LCL = \hat{\mu}' - 3\hat{\sigma}' = 8.7371 - 3 \cdot 1.1093 = 5.4092,$$

and those for the MR-chart are computed with

$$UCL = D_4 \cdot \overline{MR} = 3.267 \cdot 1.2517 = 4.0893,$$
$$CL = \overline{MR} = 1.2517,$$
$$LCL = D_3 \cdot \overline{MR} = 0 \cdot 1.2517 = 0.$$

From the $I - MR$ chart in Figure 2, we confirm that the process is in control. The last step before computing the four PCIs is to transform the values of T, U and L, using transformation (2), and as a result, we have $T' = 8.5732$, $U' = 12.2270$ and $L' = 4.1833$. The values of the four PCIs are computed with

$$C_p = \frac{U' - L'}{6\hat{\sigma}'} = \frac{12.2270 - 4.1833}{6 \cdot 1.1093} = 1.2085,$$

$$C_{pk} = \frac{min\{U' - \hat{\mu}', \hat{\mu}' - L'\}}{3\hat{\sigma}'} = \frac{min\{12.2270 - 8.7371, 8.7371 - 4.1833\}}{3 \cdot 1.1093} = 1.0487,$$

$$C_{pm} = \frac{U' - L'}{6\sqrt{\hat{\sigma}'^2 + (\hat{\mu}' - T')^2}} = \frac{12.2270 - 4.1833}{6\sqrt{1.1093^2 + (8.7371 - 8.5732)^2}} = 1.1955,$$

$$C_{pmk} = \frac{min\{U' - \hat{\mu}', \hat{\mu}' - L'\}}{3\sqrt{\hat{\sigma}'^2 + (\hat{\mu}' - T')^2}} = \frac{min\{12.2270 - 8.7371, 8.7371 - 4.1833\}}{3\sqrt{1.1093^2 + (8.7371 - 8.5732)^2}} = 1.0374.$$

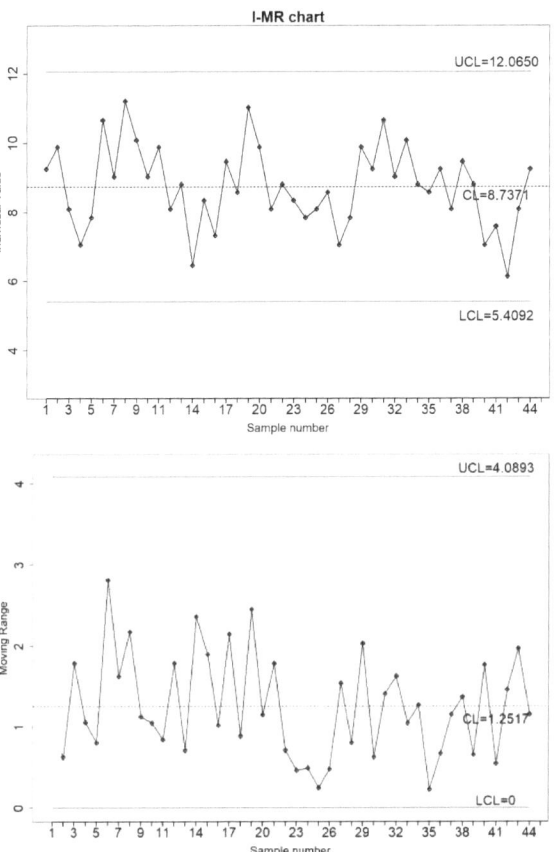

Figure 2. The $I - MR$ chart for the Example 1 data.

The long-term standard deviation of the transformed data is computed with

$$\hat{\sigma}_s' = \sqrt{\frac{\sum_{i=1}^{44}(y_i - \hat{\mu}')^2}{n-1}},$$

where y_i are the transformed data. Thus, $\hat{\sigma}_s' = 1.1732$. The values of the two PPIs are equal to

$$P_p = \frac{U' - L'}{6\hat{\sigma}_s'} = \frac{12.2270 - 4.1833}{6 \cdot 1.1732} = 1.1427,$$

$$P_{pk} = \frac{min\{U' - \hat{\mu}', \hat{\mu}' - L'\}}{3\hat{\sigma}_s'} = \frac{min\{12.2270 - 8.7371, 8.7371 - 4.1833\}}{3 \cdot 1.1732} = 0.9916.$$

Similarly, we compute the values of the indices with the other transformation techniques as well as the values of the C_{py}, C_{pyk} and C_{pTk} indices. The results are presented in Table 8. From this table, we conclude that the differences in the values of the indices applying different transformation techniques are very small. Moreover, the values of C_p, C_{pk} and C_{pmk} are very close to the corresponding values of C_{py}, C_{pyk} and C_{pTk}, while the values of P_p and P_{pk} are approximately similar to those of C_p and C_{pk}. Thus, one should consider that the process is stable. As the PCIs are higher than one, except for C_{pTk}, one should conclude that the process is capable. Consequently, the transformation techniques give similar results and one should apply them in order to make conclusions about the capability of the process.

Table 8. Indices for the Example 1 data.

Technique	C_p	C_{pk}	C_{pm}	C_{pmk}	C_{py}	C_{pyk}	C_{pTk}	P_p	P_{pk}
Anscombe's	1.2085	1.0487	1.1955	1.0374	1.0026	1.0025	0.9331	1.1427	0.9916
Freeman-Tukey's	1.2074	1.0495	1.1943	1.0382				1.1416	0.9924
Q	1.1668	1.1053	1.1500	1.0895				1.1042	1.0460

Example 2. We use the data on a number of nonconforming cans with frozen orange juice in samples of size $n = 50$ cans from an industrial process, which is also described in detail, as the previous example, in Chapter 7 by Montgomery [12]. The dataset consists of 40 samples and is presented in Table 9. It is also assumed that the target value is $T = 5$ defective bellows in each sample and that the upper and lower specification limits are $U = 13$ and $L = 1$ defective bellows per sample, respectively.

Table 9. The number of nonconforming cans in samples of size $n = 50$ cans.

Sample number	1	2	3	4	5	6	7	8	9	10
Number of nonconforming cans	8	7	5	6	4	5	2	3	4	7
Sample number	11	12	13	14	15	16	17	18	19	20
Number of nonconforming cans	6	5	5	3	7	9	6	10	4	3
Sample number	21	22	23	24	25	26	27	28	29	30
Number of nonconforming cans	5	8	11	9	7	3	5	2	1	4
Sample number	31	32	33	34	35	36	37	38	39	40
Number of nonconforming cans	5	3	7	6	4	4	6	8	5	6

Applying the binomial distribution to these data, the MLE of the parameter p is 0.109 (p-value = 0.9999).

In Figure 3, the p-chart is presented. Note that the control limits of the p-chart are computed with

$$UCL = \hat{p} + 3\sqrt{\frac{\hat{p}(1-\hat{p})}{n}} = 0.109 + 3\sqrt{\frac{0.109(1-0.109)}{50}} = 0.2412,$$

$$CL = \hat{p} = 0.109,$$

$$LCL = \hat{p} - 3\sqrt{\frac{\hat{p}(1-\hat{p})}{n}} = 0.109 - 3\sqrt{\frac{0.109(1-0.109)}{50}} = -0.0232 \Rightarrow LCL = 0.$$

From the p-chart in Figure 3, we conclude that the process is in control as no point and no systematic behaviour can be seen. In addition, we compute all the y_i's values, for $i = 1, 2, \ldots, 40$ using Freeman–Tukey's transformation, given in (7). The mean and standard deviation of the transformed data are $\hat{\mu}' = 0.6846$ and $\hat{\sigma}' = 0.1176$, respectively. An $I - MR$ chart for the transformed data is presented in Figure 4.

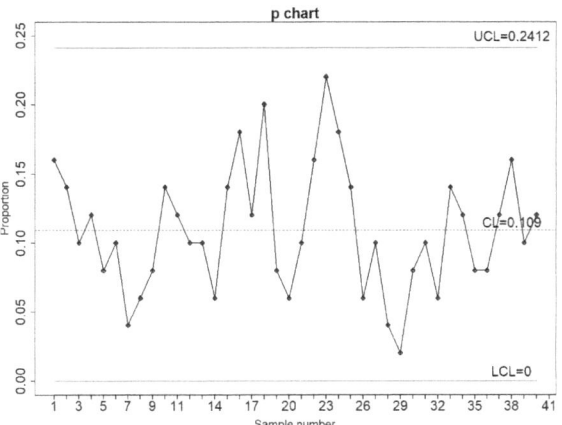

Figure 3. The p-chart for the Example 2 data.

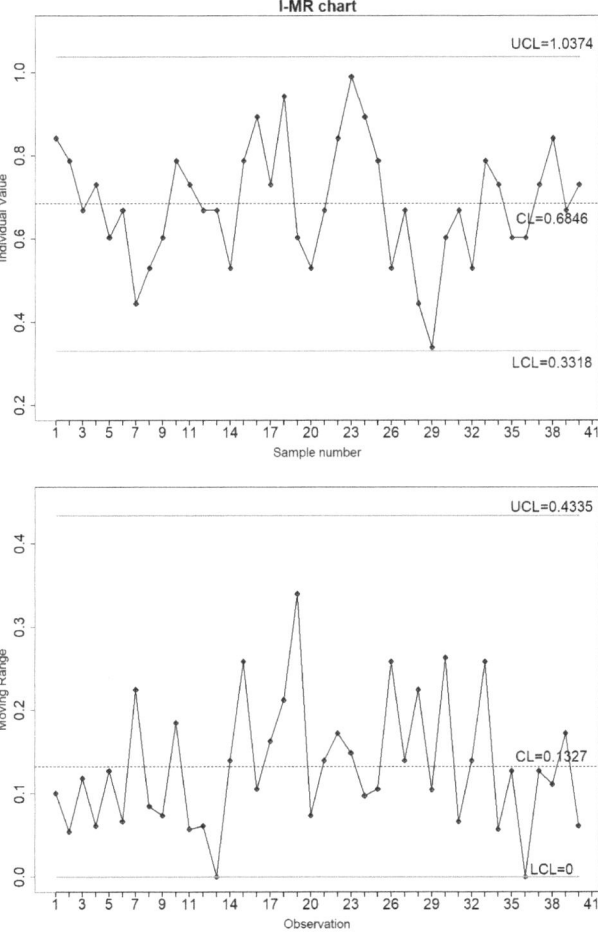

Figure 4. The $I - MR$ chart for the Example 2 data.

The control limits for the I-chart are calculated as

$$UCL = \hat{\mu}' + 3\hat{\sigma}' = 0.6846 + 3 \cdot 0.1176 = 1.0374,$$

$$CL = \hat{\mu}' = 0.6846,$$

$$LCL = \hat{\mu}' - 3\hat{\sigma}' = 0.6846 - 3 \cdot 0.1176 = 0.3318,$$

and those for the MR-chart are calculated with

$$UCL = D_4 \cdot \bar{MR} = 3.267 \cdot 0.1327 = 0.4335,$$

$$CL = \bar{MR} = 0.1327,$$

$$LCL = D_3 \cdot \bar{MR} = 0 \cdot 0.1327 = 0.$$

From the $I - MR$ chart in Figure 4, we confirm that the process is in control. The transformed values of T, U and L are $T' = 0.6686$, $U' = 1.0807$ and $L' = 0.3398$, respectively, and the values of the four PCIs are computed with

$$C_p = \frac{U' - L'}{6\hat{\sigma}'} = \frac{1.0807 - 0.3398}{6 \cdot 0.1176} = 1.0500,$$

$$C_{pk} = \frac{min\{U' - \hat{\mu}', \hat{\mu}' - L'\}}{3\hat{\sigma}'} = \frac{min\{1.0807 - 0.6846, 0.6846 - 0.3398\}}{3 \cdot 0.1176} = 0.9773,$$

$$C_{pm} = \frac{U' - L'}{6\sqrt{\hat{\sigma}'^2 + (\hat{\mu}' - T')^2}} = \frac{1.0807 - 0.3398}{6\sqrt{0.1176^2 + (0.6846 - 0.6686)^2}} = 1.0404,$$

$$C_{pmk} = \frac{min\{U' - \hat{\mu}', \hat{\mu}' - L'\}}{3\sqrt{\hat{\sigma}'^2 + (\hat{\mu}' - T')^2}} = \frac{min\{1.0807 - 0.6846, 0.6846 - 0.3398\}}{3\sqrt{0.1176^2 + (0.6846 - 0.6686)^2}} = 0.9684.$$

The long-term standard deviation of the transformed data is equal to $\hat{\sigma}_s' = 0.1415$. Thus, the values of the PPIs are

$$P_p = \frac{U' - L'}{6\hat{\sigma}_s'} = \frac{1.0807 - 0.3398}{6 \cdot 0.1415} = 0.8725,$$

$$P_{pk} = \frac{min\{U' - \hat{\mu}', \hat{\mu}' - L'\}}{3\hat{\sigma}_s'} = \frac{min\{1.0807 - 0.6846, 0.6846 - 0.3398\}}{3 \cdot 0.1415} = 0.8121.$$

The values of the indices with the other transformations techniques as well as the values of the C_{py}, C_{pyk} and C_{pTk} indices are presented in Table 10. We conclude that the differences in the values of the PCIs applying different transformation techniques are very small. The values of C_p, C_{pk} and C_{pmk} are close to the values of C_{py}, C_{pyk} and C_{pTk}, respectively. As C_{pk} and C_{pmk} and of course C_{pyk} and C_{pTk} are lower than one, one should conclude that the process is incapable. Moreover, the differences between C_p and P_p as well as C_{pk} and P_{pk} are significant. Thus, the process is unstable. Similarly to the previous example, the transformation techniques give similar results and one should apply them in order to make conclusions about the capability/performance of the process. However, one should be careful, as C_p and C_{pm} are larger than one while C_{pk}, C_{pmk}, P_p and P_{pk} are smaller than one; thus, one should evaluate all the indices to make safe conclusions about the capability/performance of the process.

Table 10. Indices for different transformation methods for the Example 2 data.

Technique	C_p	C_{pk}	C_{pm}	C_{pmk}	C_{py}	C_{pyk}	C_{pTk}	P_p	P_{pk}
Freeman-Tukey's	1.0500	0.9773	1.0404	0.9684	0.9989	0.9582	0.9350	0.8725	0.8121
Chen's	1.0493	0.9779	1.0398	0.9691				0.8719	0.8125
Q	1.0465	0.8994	1.0308	0.8859				0.8595	0.7387

Example 3. *We assume a process in which we count the total number of inspected products until $r = 5$ nonconforming items are found. It is also known that the probability p of observing a nonconforming item is equal to 0.1; the target value is $T = 45$ inspected products; and the upper and lower specification limits are $U = 140$ and $L = 5$ inspected products, respectively. The dataset can be found in Chapter 3 of Xie, Goh, and Kuralmani [42], and it is presented row by row in Table 11.*

Table 11. The number of inspected products until $r = 5$ nonconforming items are found.

71	22	88	118	27	37	47	43	39	45
30	105	33	102	49	31	15	38	18	65
61	59	30	73	39	69	34	55	29	69
99	43	38	56	38	28	16	14	106	62
61	24	48	24	48	39	58	20	46	29
46	30	39	62	77	31	43	36	19	22
45	35	20	63	43	37	45	36	68	56
90	14	73	65	50	27	23	60	27	43
36	77	28	81	50	35	67	19	47	41
24	28	28	58	36	61	31	29	62	85

The control limits of a CCC-r chart are given as the solution of the above equations

$$F(UCL_r, r, p) = \sum_{i=r}^{UCL_r} \binom{i-1}{r-1} p^r (1-p)^{i-r} = 1 - \frac{a}{2},$$

$$F(CL_r, r, p) = \sum_{i=r}^{CL_r} \binom{i-1}{r-1} p^r (1-p)^{i-r} = 0.5,$$

$$F(LCL_r, r, p) = \sum_{i=r}^{LCL_r} \binom{i-1}{r-1} p^r (1-p)^{i-r} = \frac{a}{2},$$

where α is the acceptable risk of a false alarm. In this example, the control limits of the CCC-5 chart under a standard false alarm probability level $\alpha = 0.0027$ are $UCL_5 = 134$, $CL_5 = 41$ and $LCL_5 = 10$.

In Figure 5, the CCC-5 chart is presented from which we conclude that the process is in control as no point and no systematic behaviour can be seen. In addition, we compute the y_i values, for $i = 1, 2, \ldots, 100$ using Anscombe's transformation, given in (14). The mean and standard deviation of the transformed data are $\hat{\mu}' = 3.7978$ and $\hat{\sigma}' = 0.4885$, respectively. An $I - MR$ chart for the transformed data is presented in Figure 6.

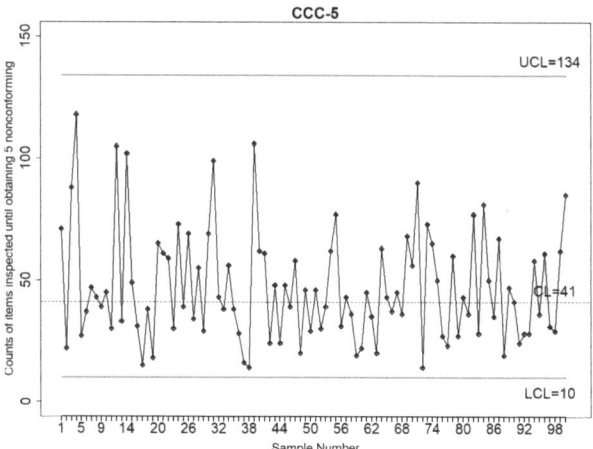

Figure 5. The CCC-5 chart for the Example 3 data.

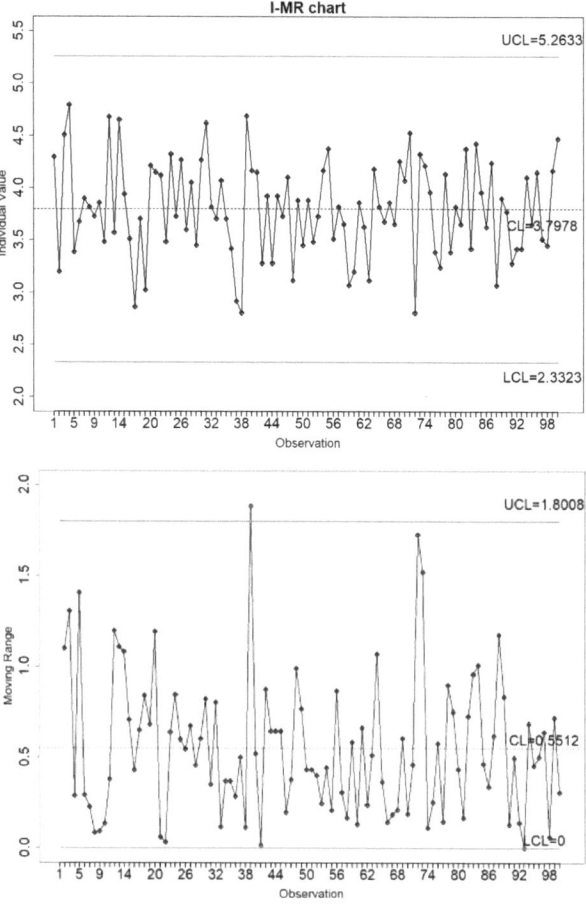

Figure 6. The $I - MR$ chart for the Example 3 data.

The control limits for the I-chart are computed with

$$UCL = \hat{\mu}' + 3\hat{\sigma}' = 3.7978 + 3 \cdot 0.4885 = 5.2633,$$

$$CL = \hat{\mu}' = 3.7978,$$

$$LCL = \hat{\mu}' - 3\hat{\sigma}' = 3.7978 - 3 \cdot 0.4885 = 2.3323,$$

and those for the MR-chart are computed with

$$UCL = D_4 \cdot \bar{MR} = 3.267 \cdot 0.5512 = 1.8008,$$

$$CL = \bar{MR} = 0.5512,$$

$$LCL = D_3 \cdot \bar{MR} - 0 \cdot 0.5512 = 0.$$

From the $I - MR$ chart in Figure 6, we note that one point, that of sample 39, is plotted over the UCL in the MR-chart. However, no assignable cause is determined, and therefore, we decide that the process is in-control. The transformed values of T, U and L are $T' = 3.8607$, $U' = 4.9593$ and $L' = 2.0149$, respectively, and the values of the four PCIs are computed with

$$C_p = \frac{U' - L'}{6\hat{\sigma}'} = \frac{4.9593 - 2.0149}{6 \cdot 0.4885} = 1.0046,$$

$$C_{pk} = \frac{\min\{U' - \hat{\mu}', \hat{\mu}' - L'\}}{3\hat{\sigma}'} = \frac{\min\{4.9593 - 3.7978, 3.7978 - 2.0149\}}{3 \cdot 0.4885} = 0.7926,$$

$$C_{pm} = \frac{U' - L'}{6\sqrt{\hat{\sigma}'^2 + (\hat{\mu}' - T')^2}} = \frac{4.9593 - 2.0149}{6\sqrt{0.4885^2 + (3.7978 - 3.8607)^2}} = 0.9963,$$

$$C_{pmk} = \frac{\min\{U' - \hat{\mu}', \hat{\mu}' - L'\}}{3\sqrt{\hat{\sigma}'^2 + (\hat{\mu}' - T')^2}} = \frac{\min\{4.9593 - 3.7978, 3.7978 - 2.0149\}}{3\sqrt{0.4885^2 + (3.7978 - 3.8607)^2}} = 0.7861.$$

The long-term standard deviation of the transformed data is equal to $\hat{\sigma}_s' = 0.4561$. Thus, the two PPIs are computed with

$$P_p = \frac{U' - L'}{6\hat{\sigma}_s'} = \frac{4.9593 - 2.0149}{6 \cdot 0.4561} = 1.0760,$$

$$P_{pk} = \frac{\min\{U' - \hat{\mu}', \hat{\mu}' - L'\}}{3\sigma_s'} = \frac{\min\{4.9593 - 3.8607, 3.8607 - 2.0149\}}{3 \cdot 0.4561} = 0.8489.$$

The values of the indices applying the two transformation methods are presented in Table 12. Moreover, the values of C_{py}, C_{pyk} and C_{pTk} are also presented in the same table. We note that the value of parameter L using the Box–Cox transformation is equal to -0.5415. Minor differences are observed in the C_p and C_{pm} values applying the two transformation methods. However, the values of the indices indicate that the process is incapable and unstable. In this example, there are small differences in the computation of the four PCIs and two PPIs. However, similar to the previous example, one should evaluate all of the indices as only C_p and P_p are larger than the one for Anscombe's transformation, while C_p, C_{pm} and P_p are larger than the one for the Q transformation.

As a general conclusion from the above three applications, one practitioner should compute all four PCIs and two PPIs in order to make safe conclusions about the capability and the performance of the process. It should also be pointed out that if a practitioner evaluates only C_p or C_{pm}, then they may wrongly conclude that the process is capable. Similarly, one should evaluate not only the P_p but also the P_{pk} index to conclude about the behaviour of the process. The results from different transformation methods are similar. Therefore, one should compute the PCIs and PPIs by applying any method. We also note that the conclusions about the capability and the performance of the processes obtained by applying the transformation techniques are the same as those obtained by computing the indices for non-normal distributions; thus, practitioners do not have to know the indices for non-normal data.

Table 12. Indices for different transformation methods for the Example 3 data.

Technique	C_p	C_{pk}	C_{pm}	C_{pmk}	C_{py}	C_{pyk}	C_{pTk}	P_p	P_{pk}
Anscombe's	1.0046	0.7926	0.9963	0.7861	1.0010	0.9994	0.8631	1.0760	0.8489
Q	1.5108	0.5903	1.4639	0.5720				1.5772	0.6162

6. Conclusions

Process capability and performance indices, named PCIs and PPIs, respectively, are practical and useful tools for continuous quality improvement. There are various methods in the literature to estimate PCIs for non-normal and continuous processes. In the case of discrete processes, new PCIs have been defined by researchers, with their use causing difficulties to many practitioners. In the present paper, we proposed various transformation techniques for data following Poisson, binomial or negative binomial distributions. The transformed data are then used to compute the well-known PCIs and PPIs. Through a simulation study, we demonstrated that the differences in the indices applying different transformation methods are very small. Comparisons with other PCIs for discrete data showed that the proposed methodology can safely be applied without any misinterpretation of the process capability. Thus, we suggest that practitioners who are not familiar with indices for non-normal data use the transformation methodology in order to obtain safe conclusions about the process capability and performance. Finally, we applied the transformation techniques in three illustrative examples for data, following different distribution.

For future research, we recommend the extension of these transformation techniques to multivariate discrete processes. Moreover, one should propose indices for data following the Conway–Maxwell Poisson distribution, which encompasses the geometric, Poisson and Bernoulli distributions as special cases, or zero-inflated distributions.

Funding: This research received no external funding

Data Availability Statement: The data that support the findings of this study are available from the respective references as mentioned in the main text.

Acknowledgments: The author thank the editors and the referees for their useful comments, which resulted in improving the quality of this article.

Conflicts of Interest: The authors declare no conflict of interest.

References

1. Juran, J.M.; *Juran's Quality Control Handbook*, 3rd ed.; McGraw-Hill: New York, NY, USA, 1974.
2. Kane, V.E. Process capability indices. *J. Qual. Technol.* **1986**, *18*, 41–52. [CrossRef]
3. Chan, L.K.; Cheng, S.W.; Spiring, F.A. A new measure of process capability: C_{pm}. *J. Qual. Technol.* **1988**, *20*, 162–175. [CrossRef]
4. Hsiang, T.C.; Taguchi, G. A tutorial on quality control and assurance—The Taguchi methods. In Proceedings of the ASA Annual Meeting, Las Vegas, NV, USA, 5–8 August 1985; p. 188.
5. Choi, B.C.; Owen, D.B. A study of a new process capability index. *Commun. Stat.-Theory Methods* **1990**, *19*, 1232–1245. [CrossRef]
6. Pearn, W.L.; Kotz, S.; Johnson, N.L. Distributional and inferential properties of process capability indices. *J. Qual. Technol.* **1992**, *24*, 216–231. [CrossRef]
7. Rodriguez, R.N. Recent developments in process capability analysis. *J. Qual. Technol.* **1992**, *24*, 176–187. [CrossRef]
8. Kotz, S.; Johnson, N.L. *Process Capability Indices*; Chapman and Hall: London, UK, 1993.
9. Kotz, S.; Johnson, N.L. Process capability indices—A review, 1992–2000. *J. Qual. Technol.* **2002**, *34*, 2–19. [CrossRef]
10. Kotz, S.; Lovelace, C.R. *Process Capability Indices in Theory and Practice*; Arnold: London, UK, 1998.

11. Pearn, W.L.; Kotz, S. *Encyclopedia and Handbook of Process Capability Indices—A Comprehensive Exposition of Quality Control Measures*; Series on Quality, Reliability and Engineering Statistics; World Scientific Publishing Co.: Singapore, 2006.
12. Montgomery, D.C. *Introduction to Statistical Quality Control*, 7th ed.; John Wiley & Sons: New York, NY, USA, 2012.
13. Clements, J.A. Process capability calculations for non-normal distributions. *Qual. Prog.* **1989**, *22*, 95–100.
14. Pearn, W.L.; Kotz, S. Application of Clements' method for calculating second and third-generation process capability indices for non-normal Pearsonian populations. *Qual. Eng.* **1994**, *7*, 139–145. [CrossRef]
15. Rivera, L.A.R.; Hubele, N.F.; Lawrence, F.P. C_{pk} index estamation using data transformation. *Comput. Ind. Eng.* **1995**, *29*, 55–58. [CrossRef]
16. Pearn, W.L.; Chen, K.S. Estimating process capability indices for non-normal Pearsonian populations. *Qual. Reliab. Eng. Int.* **1995**, *11*, 386–388. [CrossRef]
17. Castagliola, P. Evaluation of non-normal process capability indices using Burr's distributions. *Qual. Eng.* **1996**, *8*, 587–593. [CrossRef]
18. Hosseinifard, S.Z.; Abbasi, B.; Ahmad, S.; Abdollahian, M. A transformation technique to estimate the process capability index for non-normal processes. *Int. J. Adv. Manuf. Technol.* **2009**, *40*, 512–517. [CrossRef]
19. Hosseinifard, S.Z.; Abbasi, B.; Niaki, S.T.A. Process capability estimation for leukocyte filtering process in blood service: A comparison study. *IIE Trans. Healthc. Syst. Eng.* **2014**, *4*, 167–177. [CrossRef]
20. Bothe, D.R. *Measuring Process Capability: Techniques and Calculations for Quality and Manufacturing Engineers*; McGraw-Hill: New York, NY, USA, 1997.
21. Yeh, A.B.; Bhattacharya, S. A robust process capability index. *Commun. Stat. Simul. Comput.* **1998**, *27*, 565–589. [CrossRef]
22. Borges, W.; Ho, L.L. A fraction defective based capability index. *Qual. Reliab. Eng. Int.* **2001**, *17*, 447–458. [CrossRef]
23. Perakis, M.; Xekalaki, E. A process capability index that is based on the proportion of conformance. *J. Stat. Comput. Simul.* **2002**, *72*, 707–718. [CrossRef]
24. Perakis, M.; Xekalaki, E. A process capability index for discrete processes. *J. Stat. Comput. Simul.* **2005**, *75*, 175–187. [CrossRef]
25. Maiti, S.S.; Saha, M.; Nanda, A.K. On generalizing process capability indices. *Qual. Technol. Quant. Manag.* **2010**, *7*, 279–300. [CrossRef]
26. Maravelakis, P.E. Process capability indices for data following the Poisson or binomial distribution. *Qual. Technol. Quant. Manag.* **2016**, *13*, 197–206. [CrossRef]
27. Dey, S.; Saha, M. Bootstrap confidence intervals of generalized process capability index C_{pyk} using different methods of estimation. *J. Appl. Stat.* **2019**, *46*, 1843–1869. [CrossRef]
28. Pal, S.; Gauri, S.K. Measuring capability of a binomial process. *Int. J. Eng. Sci. Technol.* **2020**, *12*, 25–37. [CrossRef]
29. Gauri, S.K.; Pal, S. A Note on the Generalized Indices of Process Capability. *Int. J. Res. Ind. Eng.* **2020**, *9*, 286–303.
30. Bartlett, M.S. The square root transformation in analysis of variance. *Suppl. J. R. Stat. Soc.* **1936**, *3*, 68–78. [CrossRef]
31. Anscombe, F.J. The transformation of Poisson, binomial and negative-binomial data. *Biometrika* **1948**, *35*, 246–254. [CrossRef]
32. Freeman, M.F.; Tukey, J.W. Transformations related to the angular and the square root. *Ann. Math. Stat.* **1950**, *21*, 607–611. [CrossRef]
33. Ryan, T.P.; Schwertman, N.C. Optimal limits for attributes control charts. *J. Qual. Technol.* **1938**, *29*, 86–98. [CrossRef]
34. Quesenberry, C.P. SPC Q charts for start-up processes and short and long runs. *J. Qual. Technol.* **1991**, *23*, 213–224. [CrossRef]
35. Quesenberry, C.P. SPC Q charts for a Poisson parameter λ: Short or long runs. *J. Qual. Technol.* **1991**, *23*, 296–303. [CrossRef]
36. Quesenberry, C.P. On properties of Poisson Q charts for attributes. *J. Qual.* **1995**, *27*, 293–303.
37. Ryan, T.P. *Statistical Methods for Quality Improvement*; John Wiley & Sons: New York, NY, USA, 1989.
38. Johnson, N.L.; Kotz, S. *Discrete Distributions (Vol. 1 of the 4-Volume Set: Distributions in Statistics)*; Wiley: New York, NY, USA, 1969.
39. Chen, G. An improved p chart through simple adjustment. *J. Qual. Technol.* **1998**, *30*, 142–151. [CrossRef]
40. Quesenberry, C.P. SPC Q charts for a binomial parameter p: Short or long runs. *J. Qual. Technol.* **1991**, *23*, 239–246. [CrossRef]
41. Xie, M.; Lu, X.S.; Goh, T.N.; Chan, L.Y. A quality monitoring and decision-making scheme for automated production processes. *Int. J. Qual. Reliab. Manag.* **1999**, *16*, 148–157. [CrossRef]
42. Xie, M.; Goh, T.N.; Kuralmani, V. *Statistical Models and Control Charts for High Quality Processes*; Kluwer Academic Publishers: Boston, MA, USA, 2002.
43. Joekes, S.; Smrekar, M.; Righetti, A.F. A comparative study of two proposed CCC-r charts for high quality processes and their application to an injection molding process. *Qual. Eng.* **2016**, *28*, 467–475. [CrossRef]
44. Box, G.E.P.; Cox, D.R. An analysis of transformations. *J. R. Stat. Society. Ser. B Methodol.* **1964**, *26*, 211–252. [CrossRef]

Disclaimer/Publisher's Note: The statements, opinions and data contained in all publications are solely those of the individual author(s) and contributor(s) and not of MDPI and/or the editor(s). MDPI and/or the editor(s) disclaim responsibility for any injury to people or property resulting from any ideas, methods, instructions or products referred to in the content.

Article

New Machine-Learning Control Charts for Simultaneous Monitoring of Multivariate Normal Process Parameters with Detection and Identification

Hamed Sabahno [1,*] and Seyed Taghi Akhavan Niaki [2]

1. Department of Statistics, School of Business, Economics and Statistics, Umeå University, Umeå 901 87, Sweden
2. Department of Industrial Engineering, Sharif University of Technology, Tehran 1458889694, Iran; niaki@sharif.edu
* Correspondence: hamed.sabahno@umu.se

Abstract: Simultaneous monitoring of the process parameters in a multivariate normal process has caught researchers' attention during the last two decades. However, only statistical control charts have been developed so far for this purpose. On the other hand, machine-learning (ML) techniques have rarely been developed to be used in control charts. In this paper, three ML control charts are proposed using the concepts of artificial neural networks, support vector machines, and random forests techniques. These ML techniques are trained to obtain linear outputs, and then based on the concepts of memory-less control charts, the process is classified into in-control or out-of-control states. Two different input scenarios and two different training methods are used for the proposed ML structures. In addition, two different process control scenarios are utilized. In one, the goal is only the detection of the out-of-control situation. In the other one, the identification of the responsible variable (s)/process parameter (s) for the out-of-control signal is also an aim (detection–identification). After developing the ML control charts for each scenario, we compare them to one another, as well as to the most recently developed statistical control charts. The results show significantly better performance of the proposed ML control charts against the traditional memory-less statistical control charts in most compared cases. Finally, an illustrative example is presented to show how the proposed scheme can be implemented in a healthcare process.

Keywords: process monitoring; machine-learning techniques; simultaneous process parameters monitoring; multivariate normal process; simulation

MSC: 62P30; 68T20; 68T09; 62P10; 62H10; 62H12; 62-08

Citation: Sabahno, H.; Niaki, S.T.A. New Machine-Learning Control Charts for Simultaneous Monitoring of Multivariate Normal Process Parameters with Detection and Identification. *Mathematics* **2023**, *11*, 3566. https://doi.org/10.3390/math11163566

Academic Editors: Arne Johannssen and Nataliya Chukhrova

Received: 21 July 2023
Revised: 14 August 2023
Accepted: 15 August 2023
Published: 17 August 2023

Copyright: © 2023 by the authors. Licensee MDPI, Basel, Switzerland. This article is an open access article distributed under the terms and conditions of the Creative Commons Attribution (CC BY) license (https://creativecommons.org/licenses/by/4.0/).

1. Introduction

When there are more than two quality characteristics to be monitored in a process, multivariate control charts can be employed in the process monitoring context. In a multivariate normal process, there are two process parameters, i.e., the mean vector and the variance–covariance matrix. Simultaneous monitoring of these parameters is usually preferred over monitoring only one of them due to its better overall performance. We have two main simultaneous process parameters monitoring schemes: (i) a single-chart scheme and (ii) a double-chart scheme. In a single-chart scheme, only one chart for both process parameters is used, and in a double-chart scheme, one chart for each process parameter is employed. A single-chart scheme is usually preferred due to its simplicity, considering only one control chart should be administrated in a single-chart scheme in comparison to the double-chart scheme, in which two control charts should be administrated at the same time. Reynolds and Gyo-Young [1], Hawkins and Maboudou-Tchao [2], and Zhang and Chang [3] studied double-chart cases, and Khoo [4], Zhang et al. [5], Wang et al. [6], Sabahno et al. [7–9], Sabahno and Khoo [10], and Sabahno [11] investigated single-chart cases. Sabahno et al. [9] proposed new memory-less statistical control charts with fixed and adaptive design parameters (sample size, sampling interval, and control limits) for simultaneous monitoring of multivariate normal process parameters. They used two statistics each for monitoring one process parameter, but in the end, they combined them into one statistic.

Machine-learning techniques have been extensively used for process monitoring with control charts for different reasons, such as dimension reduction, pattern recognition, change point estimation, signal detection, identification, and fault diagnosis. Artificial neural network (ANN) is the most used ML technique in this regard. Some of the most notable research that employed ML techniques in control charts are that conducted by Chang and Ho [12], Niaki and Abbasi [13,14], Cheng and Cheng [15], Abbasi [16], Salehi et al. [17], Hosseinifard et al. [18], Weese et al. [19], Escobar and Morales-Menendez [20], Apsemidis et al. [21], Mohd Amiruddin et al. [22], Diren et al. [23], Yeganeh et al. [24], Mohammadzadeh et al. [25], and Sabahno and Amiri [26], and Yeganeh et al. [27–29].

ML structures have rarely been used to construct control charts in the literature. Niaki and Abbasi [14] developed a perceptron neural network for monitoring and classifying mean shifts in multi-attribute processes. Hosseinifard et al. [18] developed three ANN control charts for monitoring simple linear profile parameters (the intercept, the slope, and the residual variance). One of their control charts was involved in detection and identification, while the other two were only involved in detection. Mohammadzadeh et al. [25] developed an SVR (support vector regression) control chart for monitoring a logistic profile by extending Hosseinifard et al.'s [18] paper. Sabahno and Amiri [26] developed different statistical and machine-learning-based control charts with fixed and variable design parameters to monitor generalized linear regression profiles. Yeganeh et al. [27] extended Hosseinifard et al.'s [18] work for social network surveillance. Yeganeh et al. [28] proposed an ANN-based control chart to monitor binary surgical outcomes, while in another study, Yeganeh et al. [29] proposed ML-based control charts for monitoring autocorrelated profiles.

The previously mentioned papers have utilized machine-learning (ML) techniques for regression to construct control charts. While our approach in this paper is similar to theirs, by proposing a special input set for ML structures (our first input scenario), we extend them to address multivariate processes and enable simultaneous monitoring of their parameters. Moreover, while they employed a single ML structure to build their control charts, we utilize multiple machine-learning techniques and extensively compare them in different scenarios. Additionally, all the ML control charts mentioned above solely trained their ML structures using a random shift size to represent out-of-control data. In contrast, we employ an alternative method in addition to this approach.

In this paper, for the first time in the literature, the ANN, SVM (support vector machine), and RF (random forest)-based control charts are proposed for simultaneous monitoring of multivariate normal process parameters. We use two different input scenarios (in one of them, for the first time the two statistics used by Sabahno et al. [9] are utilized as the inputs) and two different control scenarios (detection and detection–identification). We also use two training methods to see which types of datasets suit each ML structure better (in one we train the ML structures with small shifts and in the other one, for the first time, we train them with both small and large mean shifts). The ML control charts are developed in the cases of two, three, and four quality characteristics, based on which their performances with one another, and with the proposed statistical control charts by Sabahno et al. [9], are compared.

This paper is structured as follows: In Section 2, machine-learning control charts are developed. In Section 3, different models are developed, and then extensive numerical analyses in each scenario under different separate and simultaneous shift sizes are conducted. In Section 4, an illustrative example is presented. In Section 5, concluding remarks and suggestions for future research are discussed.

2. Machine-Learning Control Charts

In this section, three ML control charts are proposed. After obtaining the ML structure for each control chart, which is explained later in Section 3, the upper control limit (UCL) of each control chart is obtained using the after-mentioned algorithm. First note that the design parameters in a control chart are the sample size n, the sampling interval t, and the probability of type-I error α.

The monitoring strategy for each ML control chart is as follows:

- If at sample i, the ML structure's output $\leq$ UCL, then the process is in control;
- If at sample i, the ML structure's output $>$ UCL, then the process is declared out-of-control.

The following algorithm can be used to compute the UCLs:

Step 1. Choose a value for α and n;
Step 2. Choose and train an ML structure;
Step 3. Obtain the initial value of UCL by generating and sorting 10,000 in-control samples in ascending order using the ML structure and choosing the $[10{,}000(1 - \alpha)]$th value in the range;
Step 4. Run 10,000 simulations and adjust the UCL so that the average of 10,000 run lengths (ARL) is $1/\alpha$.

The best way to estimate the UCL is to employ the above algorithm. However, it might be very time- and energy-consuming, especially if one is conducting many numerical analyses and should estimate many UCLs for different parameter settings. An easier way with almost the same accuracy is to obtain the value of the UCL by generating and sorting 10,000 in-control samples in ascending order and then choosing the $[10{,}000(1 - \alpha)]$th value in the range, repeating this 10,000 times, and finally taking the average of these 10,000 values. This proposed approach works very well in memory-less schemes in most cases; however, it is better to confirm its result using step 4 of the above algorithm.

Instead of using a statistic and developing a control chart using it, one can use supervised ML techniques. As in this research, we only need linear (continuous) outputs from the ML techniques (because, as mentioned before, we then apply a classification method to their linear outputs); the types of ML techniques that can be used are limited because not all ML techniques are capable of regression. However, fortunately, the most popular ML techniques, namely ANN, SVM, and RF, are also capable of generating linear (regression) outputs.

In what follows in the next subsections, each of the ML techniques is shortly described. It should also be noted that after the ML structures are trained, they are tested and validated by achieving the desired in-control performance of the control chart. This is explained later in Section 3.

2.1. ANN Control Chart

ANN is one of the most popular ML techniques that mimics human brain activity. It has one input layer, a different number of hidden layers, and one output layer, and each layer contains at least one node. ANNs with more than one hidden layer are called deep-learning techniques. ANNs can be used for both classification and regression. Determining the number of nodes in each layer and the number of hidden layers is very important in ANNs, which can be carried out using the trial-and-error method. For simple problems, one hidden layer usually works best. Regarding the number of nodes in the hidden layer, although there are some rules of thumb, there is no solid rule in this regard, and the number of nodes should be determined to obtain the desired performance.

Another problem in ANNs is determining the optimal values of the connection weights and node biases. There are several optimization algorithms used in ANNs for this purpose, among which are gradient descent, stochastic gradient descent, mini-batch gradient descent, Broyden–Fletcher–Goldfarb–Shanno (BFGS), momentum, Nesterov accelerated gradient, Adagrad, AdaDelta, and Adam. In this research, we use the BFGS optimization method, which is a variant of the gradient descent method. The reason for this is simply that it is the only optimization method that our selected R package ('nnet' package) uses (for more information about the computer package, refer to Section 3). The BFGS method overcomes some of the weaknesses of the gradient descent method by seeking a stationary point of the cost function; the cost function, in this case, is the mean-squared error (MSE) for regression problems and the cross-entropy (negative log-likelihood) for classification problems.

In this paper, the ANN technique is used for regression, and the trained ANN's structure's output for sample i is ANN_i. Different input scenarios, training methods, and output (control) scenarios that we use in this research for each ANN structure are explained in Section 3.

2.2. SVR Control Chart

SVM is a kernel-based method and is a powerful classification and regression technique. It can be used for classification, novelty detection (one-class classification), and regression (called SVR in this case). By ignoring errors that are smaller than a certain threshold, and creating a tube around the true output, SVM can perform regression.

SVMs usually work in two main steps: (i) transformation of the input space to a higher-dimensional feature space through a non-linear mapping function; and (ii) construction of the separating hyperplane with maximum distance from the closest points (called support vectors) of the training set. It has been shown that maximizing the margin of separation improves the generalization ability of the classifier/regressor. Training of an SVM is finding the solution for a quadratic optimization problem.

In SVMs, the calculation of dot products in a high-dimensional space can be avoided by introducing a kernel function, which allows all the necessary computations to be performed directly in the input space. The most popular kernel functions are linear, polynomial, radial basis, and sigmoid. In this research, we use an R package ('e1071' package, which is explained more in Section 3) for training and optimization, and we do not apply any optimization algorithms on our own to find the optimal values of the SVM structure's parameters. However, there are several optimization algorithms for SVMs, some of which are as follows: sequential minimization optimization (SMO), trust region Newton method (TRON), and chunking. The computer package we chose uses the SMO algorithm. The SMO algorithm solves the SVM's quadratic problem without using any numerical optimization steps. The adopted package also implements epsilon-support vector regression (ε-SVR) as the cost function for regression problems. The cost function in ε-SVR aims to minimize the deviations between the predicted values and the actual values using an ε-insensitive loss function. It also implements the hinge loss function, which penalizes misclassifications by introducing a linear error term, in classification problems. To construct a control chart, we use the SVM technique for regression; therefore, it is called SVR.

The same as the ANN case, the trained SVR structure's output at sample i is SVR_i, and the different input scenarios, training methods, and output (control) scenarios that we use in this research for each SVR structure are explained in Section 3.

2.3. RFR Control Chart

Decision trees are tree-structured classifiers/regressors with three types of nodes: (i) the root node, which is the initial node and represents the entire sample and may be split further into more nodes; (ii) the interior nodes that represent the features of a dataset that, with the branches that represent the decision rules, are connected to other nodes; and (iii) the leaf nodes, which represent the outcome. In the regression case (the case we use in this paper), they start with the root of the tree and follow splits based on variable outcomes until a leaf node is reached and a real number-type result is given. Although decision trees work best for classification, they work very well in regression cases as well. The most popular decision tree algorithms are Iterative Dichotomiser 3 (ID3), C4.5, and CART (classification and regression tree).

Random forest (RF) is an ML technique that uses ensemble learning methods for regression or classification. The ensemble learning method is a technique that combines predictions (in the case of the RF technique, by taking the average in a regression case (called RFR in the case of regression) or choosing the class with the maximum number of occurrences in a classification case) from multiple machine-learning algorithms to make a more accurate prediction than a single model (in the RF case, the single model would be a decision tree). An RF technique is powerful and accurate, and it overcomes the over-fitting issue of individual decision trees. It usually performs great on many problems, including features with non-linear relationships. The most popular RF algorithm was introduced by Leo Breiman. It builds a forest of uncorrelated trees using a CART-like procedure. The computer package we use ('randonForest' package) employs a CART algorithm and uses the mean-squared error (MSE) as the cost function for the regression problems and the Gini impurity or the cross-entropy as the cost function for the classification problems. The main problem in using RF is choosing the number of trees to be included in the model. Using more trees is not always better as it might unnecessarily and significantly increase the computational times. The best number of trees varies from problem to problem and should be determined to obtain the minimum errors, as well as the desired performance.

The same as before, the trained RFR structure's output at sample i is RFR_i.

3. Model Development and Analysis

In this section, different control–input–training scenarios are modeled for the above three ML control charts, based on which numerical analyses are conducted afterward. We, for the first time, consider several input sets, training methods, and control scenarios to see which works best for each ML structure for building a multivariate control chart for simultaneous monitoring of process parameters. As there are two input scenarios for each ML structure, two different training methods are used for each. The two training methods are (i) training the ML structures with only a certain small shift size to familiarize them with the out-of-control situations or, (ii) training them with a certain small and a certain large shift size. We have the same number of in-control and out-of-control data in both training cases. Note that in all training scenarios, training the ML structures is performed with an output of 0 representing an in-control situation and an output of 1 representing an out-of-control situation.

In the first input scenario, the two statistics employed by Sabahno et al. [9] are used. They used them in their statistical structure to develop a multivariate control chart for simultaneous monitoring of the process parameters. We, however, use these two statistics for the inputs of the ML structures. By doing so, we easily enable the ML structures to consider multivariate processes without adding complexity. These statistics are Hotelling T^2 and W.

When the in-control process parameters are known (μ_0 and Σ_0), the Hotelling T^2 statistic is evaluated for each sample i to monitor the process mean vector (μ_0) as follows:

$$T_i^2 = n(\overline{X}_i - \mu_0)'(\Sigma_0)^{-1}(\overline{X}_i - \mu_0), \tag{1}$$

where n is the sample size and $\overline{X}_i$ is the sample's mean.

Regarding the process variability (Σ_0), they used the following statistic:

$$W_i = \frac{(n-1)|\mathbf{S}_i|^{1/p}}{|\Sigma_i|^{1/p}}, \tag{2}$$

where $\mathbf{S}_i$ is the sample's variance–covariance matrix and p is the number of quality characteristics.

In the second input scenario, however, we use all the elements of the sample mean vector and variance–covariance matrix as the inputs.

For example, in the case of two quality characteristics, we have five inputs as follows: $\bar{x}_1, \bar{x}_2, s_1^2, s_2^2$, and s_{12} (covariance), where $\overline{X} = (\bar{x}_1, \bar{x}_2)$ and $\mathbf{S} = \begin{bmatrix} s_1^2 & s_{12} \\ s_{12} & s_2^2 \end{bmatrix}$.

Note that according to the above paragraphs, as the process dimension (the number of quality characteristics) increases, the number of inputs in the second input scenario also increases. However, the number of inputs in the first input scenario is always two (T^2 and W). This makes it easier and more efficient to use the first input scenario in higher dimensions.

We also consider two types of control schemes in this paper. In the first type, the goal is only the detection of the out-of-control situation, and in the other type, the goal is to identify the responsible process parameters as well. In the first one, we only have one chart/output with which we determine whether the process is in or out of control. In the second one, we have several charts/outputs to identify the responsible variables, as well as process parameters, for the out-of-control situation. In summary, two input scenarios, two training scenarios, and two control scenarios are involved.

Moreover, three different numbers of quality characteristics are considered, i.e., $p = 2$, $p = 3$, and $p = 4$. However, to reduce the paper's size, we only consider the $p = 3$ case with the first input scenario and the $p = 4$ case with the first input scenario and the first control scenario (only detection). It should also be noted that as the process dimension (the number of variables) increases, the amount of data and diversity of the dataset increases as well in both training methods. In addition, since using the second input scenario is directly affected by the number of variables, increasing the number of variables increases the number of inputs in this scenario. In addition, as the dimension increases, the number of control charts using the second control scenario increases as well, and that could increase the false-alarm rates.

The ARL (average run length) and SDRL (standard deviation of run length) are used in this paper to measure each chart's performance, but in this section, we only make comparisons based on the ARL. The ARL is the average number of runs before an out-of-control signal is detected by a control chart. A larger ARL value is preferred when the process is in control and as low as possible when out of control.

The 'svm' function of the 'e1071' package in R is employed to train the SVR structures, the 'nnet' package for the ANN structures, and the 'randomForrest' package for the RFR structures. In general, we try to use the default hyperparameters in each package, but those we had to change are explained separately in each scenario. However, one thing we had to change in all these packages is the output type, and we changed it from a classification output (the default set) to a linear (regression) output.

Note that most of the mean shifts and variation shifts in both $p = 2$ and 3 cases are chosen similarly to those of Sabahno et al. [9] to be able to compare the proposed control charts in each scenario with their proposed statistical charts. To avoid adding additional tables for the comparisons, we do not repeat their tables here and refer readers to see their work (the ARLs in their Table 4 for $p = 2$ case and in their Table 5 for $p = 3$ case are subjects of comparisons). Therefore, we only include the comparisons' results in this paper. Moreover, similar to Sabahno et al. [9], the sample size we use in this paper is 10 ($n = 10$).

3.1. Scenario a: Control Charts for Detection

In scenario a, the cases in which only signal detection is important are investigated. Since there is only one control chart involved, assuming that the probability of type-I error is equal to 0.005, the performance measure is computed as ARL = $1/\alpha$ = 200.

3.1.1. Scenario a1 (Control Type a, Input Set 1)

As mentioned before, only two inputs are considered for each ML control chart in this scenario, and they are T^2 and W statistics.

Scenario a11 (Control Type a, Input Set 1, Training Method 1)

In this section, the shift size we select for the mean shifts is 0.2, and for the variance and covariance shifts, it is 1.2 times the in-control values. We consider 250 data with only μ_1 shifted; 250 data with only μ_2 shifted; 250 data with both means shifted; 250 data with only σ_1^2 shifted; 250 data with only σ_2^2 shifted; 250 data with both σ_1^2 and σ_2^2 shifted; 250 data with only covariance shifted; 250 data with σ_1^2, σ_2^2, and covariance shifted; and 250 data with all the parameters shifted together. For the in-control dataset, the same amount of data, i.e., $9 \times 250 = 2250$ data, are included. We train each ML structure with these in and out-of-control datasets.

In this scenario, a linear kernel is used for the SVR structure and the RMSE is computed as 0.51. For the ANN structure, four nodes (twice the number of inputs) in the hidden layer are used and the trained ANN structure has an RMSE of 0.49. For the random forest package, 100 trees are used with an RMSE of 0.53. The UCLs of the ANN, SVR, and RFR control charts are computed as 0.7073, 1.016, and 0.920144, respectively. The result of this analysis, which is reported in Table 1, shows that the SVR and ANN charts perform better than the RFR chart. In general, as the mean shift increases, the SVR chart performs better, and as the variation shift increases, the ANN chart performs better. Another interesting result is that, as the shift size increases more than the values that the ML structure is trained with, the RFR chart performs worse, such that under the mean shift of size 2, it is not even able to detect the shift at all. However, this phenomenon does not happen in ANN and SVR charts.

Moreover, by comparing the results of Table 1 to Sabahno et al.'s [9] Table 4, one can see that in all cases, at least one of the ML control charts performs better than all their proposed control charts (fixed parameters and adaptive ones). Although we do not use any adaptive strategies, the proposed ML charts perform even better (much better in most cases) than their adaptive control charts.

In the case of three quality characteristics ($p = 3$), for the out-of-control dataset, we consider 150 data with only μ_1 shifted; 150 data with only μ_2 shifted; 150 data with only μ_3 shifted; 150 data with both μ_1 and μ_2 shifted; 150 data with both μ_1 and μ_3 shifted; 150 data with both μ_2 and μ_3 shifted; 150 data with only σ_1^2 shifted; 150 data with only σ_2^2 shifted; 150 data with only σ_3^2 shifted; 150 data with both σ_1^2 and σ_2^2 shifted; 150 data with both σ_1^2 and σ_3^2 shifted; 150 data with both σ_2^2 and σ_3^2 shifted; 150 data with only covariance shifted; 150 data with all three variances and covariance shifted; and 150 data with all the parameters shifted together. For the in-control dataset, we again include the same amount of data, which in this case is 2250 data. The same hyperparameters as in the case of $p = 2$ are used here as well for the ML structures. The RMSE of the ANN, SVR, and RFR structures are computed as 0.49, 0.51, and 0.53, respectively, and the computed UCLs are 0.724, 0.9955, and 0.9305, respectively. The results in Table 2 show that the RFR chart mostly performs better in higher dimensions (compared with Table 1), especially under large mean shifts, in which the deterioration in performance under shifts larger than the trained value is much less noticeable. For the ANN and SVR charts, it is kind of mediocre (in some cases they perform better, in some they perform worse); however, they perform better than the RFR chart even in higher dimensions. The other conclusions derived for the case of $p = 2$ are valid here as well.

Moreover, by comparing the results in Table 2 to Sabahno et al.'s [9] Table 5, it is evident that in most compared cases, at least one of the proposed ML control charts performs better than all their proposed control charts (fixed design parameters and adaptive ones). Only in the case of (0.5, 0.8, 0.5) mean shift, together with no/small variation shift, does at least one of their proposed charts perform a little bit better than the best of the three proposed charts in this paper.

Table 1. ARL, SDRL. Scenario a11, $p = 2$.

	ANN	SVR	RFR	ANN	SVR	RFR
(μ_1, μ_2)		$\sigma_1^2, \sigma_2^2, \sigma_{12} = 1, 1, 0.5$			$\sigma_1^2, \sigma_2^2, \sigma_{12} = 1.05, 1, 0.5$	
(0, 0)	200, 200	200, 200	200, 200	144.92, 145.2	154.96, 152.82	161.62, 158.7
(0.1, 0)	162, 160	138.66, 137.05	186.33, 189.21	122.4, 121.94	113.02, 116.26	152.91, 157.33
(0.1, 0.3)	62.19, 60.02	47.61, 48.63	140.32, 146.05	52.04, 51.61	41.22, 40.37	130.6, 125.98
(0.3, 0)	49.56, 49.55	35.99, 35.32	128.69, 125.66	40.8, 40.52	29.97, 30.13	116.94, 115.29
(0.5, 0.8)	5.12, 4.5	3.71, 3.18	60.35, 61.59	4.76, 4.3	3.46, 2.9	59.86, 58.94
(1, 1)	1.77, 1.17	1.54, 0.9	80.38, 78.19	1.83, 1.24	1.48, 0.82	74.51, 71.41
(2, 2)	1, 0	1, 0	>10,000	1, 0	1, 0	>10,000
(μ_1, μ_2)		$\sigma_1^2, \sigma_2^2, \sigma_{12} = 1.05, 1.05, 0.5$			$\sigma_1^2, \sigma_2^2, \sigma_{12} = 1.05, 1.3, 0.5$	
(0, 0)	107.32, 109.12	125.16, 121.37	146.96, 148.91	36.78, 36.65	51.55, 49.01	83.77, 81.68
(0.1, 0)	90.78, 89.31	95.82, 94.79	131.3, 129.98	32.13, 31.66	44.63, 44.43	84.89, 83.49
(0.1, 0.3)	43.42, 43.17	34.9, 34.19	113.39, 105.59	17.99, 17.23	19.29, 18.62	72.91, 73.01
(0.3, 0)	34.78, 33.34	29.68, 30.57	104.35, 99.9	16.6, 16.41	16.84, 16.49	68.34, 66.56
(0.5, 0.8)	4.38, 3.81	3.38, 2.95	56.93, 59.03	3.18, 2.56	2.87, 2.17	45.87, 46.56
(1, 1)	1.78, 1.17	1.48, 0.87	73.73, 72.28	1.57, 0.96	1.36, 0.7	64.34, 63.55
(2, 2)	1, 0	1, 0	>10,000	1, 0	1, 0	>10,000
(μ_1, μ_2)		$\sigma_1^2, \sigma_2^2, \sigma_{12} = 1.2, 1.2, 0.6$			$\sigma_1^2, \sigma_2^2, \sigma_{12} = 1.4, 1.4, 0.5$	
(0, 0)	43.95, 42.82	64.05, 62.16	90.6, 87.93	9.8, 9.55	19.51, 18.17	48.4, 45.87
(0.1, 0)	39.19, 37.37	52.2, 51.32	87.75, 90.61	9.13, 8.71	17.15, 16.7	47.2, 46.22
(0.1, 0.3)	20.18, 19.7	22.4, 22.59	71.62, 69.17	6.78, 6.28	10.3, 9.88	41.01, 39.85
(0.3, 0)	17.46, 16.44	17.58, 17.42	70.123, 69.91	5.88, 5.26	8.56, 8.3	40.36, 38.33
(0.5, 0.8)	3.37, 2.84	2.8, 2.26	47.63, 46.9	2.25, 1.65	2.29, 1.78	37.1, 34.78
(1, 1)	1.6, 0.97	1.38, 0.73	64.8, 70.47	1.35, 0.7	1.31, 0.64	58.83, 57.21
(2, 2)	1, 0	1, 0	>10,000	1, 0	1, 0	>10,000
(μ_1, μ_2)		$\sigma_1^2, \sigma_2^2, \sigma_{12} = 1.4, 1.4, 0.75$			$\sigma_1^2, \sigma_2^2, \sigma_{12} = 3, 3, 0.5$	
(0, 0)	17.5, 16.75	32.32, 31.66	61.28, 61.73	1.23, 0.53	2.79, 2.19	96.38, 94.14
(0.1, 0)	16.19, 15.61	27.22, 26.43	57.54, 56.92	1.22, 0.55	2.69, 2.12	94.52, 93.75
(0.1, 0.3)	10.5, 9.75	13.93, 13.28	53.38, 52.48	1.2, 0.5	2.46, 1.91	100.8, 94.77
(0.3, 0)	9.3, 9.08	12.07, 11.34	51.55, 49.99	1.21, 0.51	2.42, 1.85	96.16, 96.41
(0.5, 0.8)	2.75, 2.19	2.34, 2.02	40.78, 38.25	1.09, 0.33	1.51, 0.84	167.92, 165.58
(1, 1)	1.51, 0.876	1.37, 0.688	58.84, 60.1	1.04, 0.2	1.15, 0.41	325.44, 318.73
(2, 2)	1, 0	1, 0	>10,000	1, 0	1, 0	>10,000

In the case of four quality characteristics ($p = 4$), for the out-of-control dataset, we consider 125 data with only μ_1 shifted; 125 data with only μ_2 shifted; 125 data with only μ_3 shifted; 125 data with only μ_4 shifted; 125 data with μ_1 and μ_2 shifted; 125 data with μ_1 and μ_3 shifted; 125 data with μ_2 and μ_3 shifted; 125 data with μ_1 and μ_4 shifted; 125 data with μ_2 and μ_4 shifted; 125 data with μ_3 and μ_4 shifted; 125 data with μ_1, μ_2 and μ_3 shifted; 125 data with, μ_1, μ_2 and μ_4 shifted; 125 data with μ_1, μ_3 and μ_4 shifted; 125 data with μ_2, μ_3 and μ_4 shifted; 125 data with only σ_1^2 shifted; 125 data with only σ_2^2 shifted; 125 data with only σ_3^2 shifted; 125 data with only σ_4^2 shifted; 125 data with σ_1^2 and σ_2^2 shifted; 125 data with σ_1^2 and σ_3^2 shifted; 125 data with σ_2^2 and σ_3^2 shifted; 125 data with σ_1^2 and σ_4^2 shifted; 125 data with σ_2^2 and σ_4^2 shifted; 125 data with σ_3^2 and σ_4^2 shifted; 125 data with σ_1^2, σ_2^2 and σ_3^2 shifted; 125 data with σ_1^2, σ_2^2 and σ_4^2 shifted; 125 data with σ_1^2, σ_3^2 and σ_4^2 shifted; 125 data with σ_2^2, σ_3^2 and σ_4^2 shifted; 125 data with only covariance shifted; 125 data with all four variances and covariance shifted; and 125 data with all the parameters shifted together. For the in-control dataset, we again include the same amount of data, which in this case is 3875 data. The same hyperparameters as in the case of $p = 2$ are used here as well for the ML structures. The RMSE of the ANN, SVR, and RFR structures are computed as 0.49, 0.51, and 0.52, respectively, and the computed UCLs are 0.7164, 1.0439, and 0.9332, respectively.

Table 2. ARL, SDRL. Scenario a11, $p = 3$.

	ANN	SVR	RFR	ANN	SVR	RFR
(μ_1, μ_2, μ_3)	$\sigma_1^2, \sigma_2^2, \sigma_3^2, Cov = 1, 1, 1, 0.5$			$\sigma_1^2, \sigma_2^2, \sigma_3^2, Cov = 1.05, 1, 1, 0.5$		
(0, 0, 0)	200, 200	200, 200	200, 200	153.9, 156.8	158.9, 160	180.2, 179.7
(0.1, 0, 0.1)	150.8, 145.7	147.5, 147.7	181.6, 185.1	125.6, 124.1	113.3, 114.2	164.8, 170.4
(0.1, 0.3, 0.3)	53.19, 53.38	43.97, 43.62	96.25, 91.14	45.96, 44.98	37.58, 37.34	87.61, 86.93
(0.3, 0, 0)	53.66, 52.37	44.3, 44.42	95.67, 91.49	44.13, 42.38	35.45, 35.58	92.73, 93.9
(0.5, 0.8, 0.5)	5.51, 5.03	5.28, 4.65	27.19, 26.22	5.26, 4.745	4.81, 4.126	27.98, 27.18
(1, 1, 1)	1.73, 1.198	1.69, 1.06	21.56, 20.9	1.67, 1	1.65, 1.024	21.52, 19.81
(2, 2, 2)	1.02, 0.15	1, 0.03	33.14, 33.19	1.01, 0.11	1, 0	33.36, 33.13
(μ_1, μ_2, μ_3)	$\sigma_1^2, \sigma_2^2, \sigma_3^2, Cov = 1.05, 1.05, 1.05, 0.5$			$\sigma_1^2, \sigma_2^2, \sigma_3^2, Cov = 1.05, 1.3, 1.05, 0.5$		
(0, 0, 0)	94.39, 93.96	105.5, 106.1	135.2, 133.7	34.38, 34.84	43.38, 41.55	81.86, 83.72
(0.1, 0, 0.1)	79.17, 78.14	80.75, 78.77	133.9, 128.4	31.71, 30.83	36.21, 35.35	77.94, 71.35
(0.1, 0.3, 0.3)	33.21, 32.77	28.43, 28.67	75, 78.04	17.3, 16.7	15.98, 15.44	58.63, 59.42
(0.3, 0, 0)	33.71, 32.65	28.68, 27.97	83.53, 87.01	18.44, 18.62	16.73, 15.74	55.94, 54.5
(0.5, 0.8, 0.5)	4.85, 4.32	4.28, 3.8	28.38, 27.23	3.77, 3.2	3.4, 2.847	27.64, 26.99
(1, 1, 1)	1.598, 0.99	1.53, 0.89	22.56, 21.6	1.55, 0.93	1.38, 0.71	21.04, 20.5
(2, 2, 2)	1.01, 0.09	1, 0.02	27.36, 25.29	1, 0.07	1, 0.02	19.81, 19.74
(μ_1, μ_2, μ_3)	$\sigma_1^2, \sigma_2^2, \sigma_3^2, Cov = 1.2, 1.2, 1.2, 0.6$			$\sigma_1^2, \sigma_2^2, \sigma_3^2, Cov = 1.4, 1.4, 1.05, 0.5$		
(0, 0, 0)	34.11, 33.26	44.19, 43.62	79.7, 78.44	10.17, 9.71	15.38, 14.77	34.67, 35.8
(0.1, 0, 0.1)	28.58, 27.37	37.97, 38.48	73.85, 71.42	9.56, 9.17	14.02, 13.49	33.19, 31.55
(0.1, 0.3, 0.3)	16.5, 15.76	15.85, 14.85	59, 59.56	7.16, 6.69	8.18, 7.43	30.1, 29.57
(0.3, 0, 0)	16.31, 15.65	16.3, 15.62	57.71, 51.46	6.82, 6.53	7.86, 7.21	30.88, 31.9
(0.5, 0.8, 0.5)	3.72, 3.16	3.14, 2.56	27.61, 26.05	2.65, 2.07	2.44, 1.8	22.63, 21.41
(1, 1, 1)	1.51, 0.89	1.43, 0.77	21.26, 21.87	1.34, 0.68	0.68, 0.6	17.97, 18.27
(2, 2, 2)	1, 0.06	1, 0	20.33, 19.3	1, 0.02	1, 0	16.29, 15.34
(μ_1, μ_2, μ_3)	$\sigma_1^2, \sigma_2^2, \sigma_3^2, Cov = 1.4, 1.4, 1.05, 0.75$			$\sigma_1^2, \sigma_2^2, \sigma_3^2, Cov = 3, 3, 1.4, 0.5$		
(0, 0, 0)	45.86, 44.37	52.07, 50.04	104.4, 105.9	1.14, 0.41	1.92, 1.35	4.28, 3.87
(0.1, 0, 0.1)	42.02, 40.68	43.13, 42.54	99.6, 105.9	1.14, 0.41	1.87, 1.26	4.26, 3.69
(0.1, 0.3, 0.3)	20.61, 19.97	18.29, 17.97	66.77, 66.11	1.14, 0.4	1.67, 1.03	4.68, 4.35
(0.3, 0, 0)	20.79, 20.67	18.76, 18.69	67.56, 63.59	1.12, 0.36	1.77, 1.12	4.76, 4.25
(0.5, 0.8, 0.5)	3.97, 3.49	3.6, 3.04	28.77, 28.26	1.05, 0.23	1.27, 0.58	7.36, 6.835
(1, 1, 1)	1.6, 0.99	1.55, 0.9	21.1, 19.75	1.01, 0.13	1.05, 0.23	21.84, 20.87
(2, 2, 2)	1, 0.08	1, 0.02	23.68, 22.4	1, 0	1, 0	54.86, 55.44

The results in Table 3 show that the RFR chart mostly performs better again in higher dimensions (compared with both Tables 1 and 2). For the ANN and SVR charts, it is again kind of mediocre (in some cases they perform better, in some they perform worse); however, they still perform better than the RFR chart even in higher dimensions. The other conclusions derived for the case of $p = 2$ are valid here as well.

Scenario a12 (Control Type a, Input Set 1, Training Method 2)

In this scenario, we use the same input set as before, but with the second training method (trained with both small and large shifts). Here, for the out-of-control dataset, the shift size for the small mean shifts is 0.2; for the large mean shifts, it is 1; for the small variance and covariance shifts, it is 1.2 times the in-control values; and for the large variance and covariance shifts, it is 2 times the in-control values. For the small shifts, the following are considered: 150 data with only μ_1 shifted; 150 data with only μ_2 shifted; 150 data with both means shifted; 150 data with only σ_1^2 shifted; 150 data with only σ_2^2 shifted; 150 data with both variances shifted; 150 data with only covariance shifted; 150 data with both variances and covariance shifted; and 150 data with all the parameters shifted together. Similarly, for the large mean shifts, we have 150 data with only μ_1 shifted; 150 data with only μ_2 shifted; 150 data with both means shifted; 150 data with only σ_1^2 shifted; 150 data with only σ_2^2 shifted; 150 data with both variances shifted; 150 data with only covariance shifted; 150 data with both varianbces and covariance shifted; and 150 data with all the parameters shifted together. The total number of out-of-control data in this scenario is 2700; therefore, the same number of in-control data is used. For all the ML structures, the same hyperparameters as in the previous case are used. The RMSEs for the ANN, SVR, and RFR structures are 0.44, 0.53, and 0.47, respectively. The UCLs of ANN, SVR, and RFR control charts are computed as 0.8658, 0.5958, and 0.9729, respectively. The results for this case are reported in Table 4.

Table 3. ARL, SDRL. Scenario a11, $p = 4$.

	ANN	SVR	RFR	ANN	SVR	RFR
$(\mu_1, \mu_2, \mu_3, \mu_4)$	$\sigma_1^2, \sigma_2^2, \sigma_3^2, \sigma_4^2, Cov = 1, 1, 1, 1, 0.5$			$\sigma_1^2, \sigma_2^2, \sigma_3^2, \sigma_4^2, Cov = 1.05, 1, 1, 1, 0.5$		
(0, 0, 0, 0)	200, 200	200, 200	200, 200	152.8, 154.7	175.2, 177	180.2, 174.3
(0.1, 0, 0.1, 0)	167.8, 173.3	138.6, 138	169.2, 171.5	132.8, 134.3	125, 120.6	152.7, 153.2
(0.1, 0.3, 0.3, 0.1)	79.52, 80.04	45.18, 42.7	101.4, 105.8	66.18, 65.77	40.66, 40.88	92.34, 92.61
(0.3, 0, 0, 0.3)	51.29, 50.1	27.61, 26.23	77.29, 73.17	43.4, 42.92	26.25, 25.07	71.88, 68.42
(0.5, 0.8, 0.5, 0.8)	5.4, 5.05	3.41, 2.83	19.91, 19.3	5.14, 4.65	3.31, 2.79	20.03, 19.75
(1, 1, 1, 1)	1.84, 1.26	1.5, 0.86	17.81, 17.81	1.83, 1.22	1.5, 0.88	17.47, 17.37
(2, 2, 2, 2)	1, 0	1, 0.02	>10,000	1, 0	1, 0	>10,000
$(\mu_1, \mu_2, \mu_3, \mu_4)$	$\sigma_1^2, \sigma_2^2, \sigma_3^2, \sigma_4^2, Cov = 1.05, 1.05, 1.05, 1.05, 0.5$			$\sigma_1^2, \sigma_2^2, \sigma_3^2, \sigma_4^2, Cov = 1.05, 1.3, 1.05, 1.3, 0.5$		
(0, 0, 0, 0)	80.29, 78.62	103.8, 104.2	118.4, 118.3	16.56, 15.96	29.58, 27.9	42.75, 42.33
(0.1, 0, 0.1, 0)	69.13, 69.04	78.11, 79.96	108.2, 117.8	15.13, 14.66	26.21, 26.3	41.4, 42.44
(0.1, 0.3, 0.3, 0.1)	40.03, 39.38	28.81, 27.94	71.83, 64.33	11.11, 10.62	13.11, 12.92	32.93, 31.47
(0.3, 0, 0, 0.3)	28.05, 28.22	18.5, 17.23	56.97, 56.89	9.82, 9.27	9.6, 9.15	29.23, 28.9
(0.5, 0.8, 0.5, 0.8)	4.57, 4.03	3.04, 2.52	21.23, 20.74	2.97, 2.37	2.38, 1.85	18.68, 18.81
(1, 1, 1, 1)	1.75, 1.12	1.4, 0.75	19.71, 19.52	1.53, 0.92	1.32, 0.66	24.26, 25.35
(2, 2, 2, 2)	1, 0	1, 0	>10,000	1, 0	1, 0	>10,000
$(\mu_1, \mu_2, \mu_3, \mu_4)$	$\sigma_1^2, \sigma_2^2, \sigma_3^2, \sigma_4^2, Cov = 1.2, 1.2, 1.2, 1.2, 0.6$			$\sigma_1^2, \sigma_2^2, \sigma_3^2, \sigma_4^2, Cov = 1.4, 1.4, 1.05, 1.05, 0.5$		
(0, 0, 0, 0)	25.26, 24.02	47.95, 46.88	62.73, 59.53	10.38, 9.69	20.91, 20.63	28.59, 28.24
(0.1, 0, 0.1, 0)	24.18, 23.71	38.09, 38.59	54.33, 55.11	10.39, 9.58	18.75, 18.68	29.28, 27.89
(0.1, 0.3, 0.3, 0.1)	16.12, 15.35	17.27, 16.97	42.41, 41.98	7.93, 7.43	10.35, 9.55	25.98, 25.84
(0.3, 0, 0, 0.3)	13.12, 12.61	11.73, 10.91	39.12, 42.09	6.86, 6.32	7.58, 6.97	22.48, 23.2
(0.5, 0.8, 0.5, 0.8)	3.44, 2.92	2.59, 2.05	21, 20.11	2.64, 2.08	2.21, 1.66	16.46, 16.63
(1, 1, 1, 1)	1.6, 0.95	1.37, 0.71	23.52, 21.43	1.47, 0.84	1.24, 0.56	23.3, 23.02
(2, 2, 2, 2)	1, 0	1, 0	>10,000	1, 0	1, 0	>10,000
$(\mu_1, \mu_2, \mu_3, \mu_4)$	$\sigma_1^2, \sigma_2^2, \sigma_3^2, \sigma_4^2, Cov = 1.4, 1.4, 1.05, 1.05, 0.75$			$\sigma_1^2, \sigma_2^2, \sigma_3^2, \sigma_4^2, Cov = 3, 3, 1.4, 1.4, 0.5$		
(0, 0, 0, 0)	135.3, 130.2	107.6, 108.6	166.1, 162.6	1.13, 0.39	2.28, 1.65	2.69, 2.12
(0.1, 0, 0.1, 0)	117.1, 118.2	79.76, 79.14	145.8, 141.2	1.14, 0.41	2.27, 1.67	2.82, 2.31
(0.1, 0.3, 0.3, 0.1)	55.12, 54.93	29.68, 29.4	83.9, 80.92	1.11, 0.36	2.12, 1.57	2.85, 2.23
(0.3, 0, 0, 0.3)	38.91, 38.39	22.2, 22.06	68.92, 69.37	1.12, 0.36	1.87, 1.32	2.98, 2.53
(0.5, 0.8, 0.5, 0.8)	4.78, 4.327	3.34, 2.78	22.08, 21.25	1.06, 0.26	1.3, 0.65	4.34, 3.7
(1, 1, 1, 1)	1.88, 1.32	1.54, 0.93	20.19, 20.95	1.02, 0.17	1.07, 0.28	9.06, 8.13
(2, 2, 2, 2)	1, 0	1, 0.02	>10,000	1, 0	1, 0	>10,000

The results in Table 4 show that the ANN and SVR charts perform mostly better with the second training method, only in the cases of no/small variation shifts. On the contrary, this training method suits the RFR chart the most, considering it performs better than the previous training method in all the cases. In terms of which chart among the three performs better in this scenario, despite the significant improvements in the performance of the RFR chart, it still cannot outperform the ANN and SVR charts, as the ANN chart performs better in most compared cases, and in the rest (no or small variation shifts), despite the loss of performance in this training method, the SVR chart still performs better than the others. In addition, unlike the previous training method, the RFR chart does not experience a deterioration in performance as the shift size becomes larger than the trained value.

Moreover, by comparing the results of Table 4 to Sabahno et al.'s [9] Table 4, one can again see that in all the cases, at least one of the proposed ML control charts performs better (mostly much better) than all their charts.

Table 4. ARL, SDRL. Scenario a12, $p = 2$.

	ANN	SVR	RFR	ANN	SVR	RFR
(μ_1, μ_2)	$\sigma_1^2, \sigma_2^2, \sigma_{12} = 1, 1, 0.5$			$\sigma_1^2, \sigma_2^2, \sigma_{12} = 1.05, 1, 0.5$		
(0, 0)	200, 200	200, 200	200, 200	155.48, 153.15	165.01, 161.56	156.21, 152.98
(0.1, 0)	154.13, 167.78	151.92, 149.84	158.22, 155.65	125.77, 118.97	119.68, 124.59	129.18, 121.49
(0.1, 0.3)	52.94, 50.03	43.71, 43.53	56.77, 56.35	46.37, 46.44	42.12, 41.52	56.99, 55.06
(0.3, 0)	37.56, 39.89	33.85, 32.85	47.98, 44.91	34.51, 34.49	28.73, 27.4	41.32, 39.63
(0.5, 0.8)	3.78, 3.23	3.38, 2.88	4.5, 3.95	3.69, 3.19	3.28, 2.72	4.55, 4.01
(1, 1)	1.53, 0.91	1.4, 0.76	1.75, 1.13	1.5, 0.87	1.43, 0.78	1.77, 1.2
(2, 2)	1, 0	1, 0	1.09, 0.3	1, 0	1, 0	1.09, 0.32
(μ_1, μ_2)	$\sigma_1^2, \sigma_2^2, \sigma_{12} = 1.05, 1.05, 0.5$			$\sigma_1^2, \sigma_2^2, \sigma_{12} = 1.05, 1.3, 0.5$		
(0, 0)	123.73, 124.92	141.1, 142.5	120.62, 118.66	45.01, 44.81	66.29, 65.08	47.93, 48.98
(0.1, 0)	102.23, 102.65	98.54, 95.45	104.1, 104.57	37.23, 36.58	56.49, 55.84	43.89, 43.52
(0.1, 0.3)	40.99, 41.03	35.92, 36.23	46.84, 47.33	20.49, 19.86	22.62, 21.77	22.76, 22.35
(0.3, 0)	30.97, 30.54	27.23, 26.92	37, 35.01	17.84, 17.01	20.25, 19.52	21.47, 20.87
(0.5, 0.8)	3.7, 3.15	3.15, 2.56	4.28, 3.82	3.25, 2.75	2.83, 2.19	3.94, 3.32
(1, 1)	1.47, 0.83	1.41, 0.76	1.74, 1.08	1.47, 0.8	1.39, 0.75	1.77, 1.18
(2, 2)	1, 0	1, 0	1.08, 0.3	1, 0	1, 0	1.08, 0.29
(μ_1, μ_2)	$\sigma_1^2, \sigma_2^2, \sigma_{12} = 1.2, 1.2, 0.6$			$\sigma_1^2, \sigma_2^2, \sigma_{12} = 1.4, 1.4, 0.5$		
(0, 0)	51.69, 50.41	80.46, 77.9	58.56, 56.6	11.5, 11.04	29.25, 29.02	13.79, 13.32
(0.1, 0)	46.38, 47.18	62.89, 60.51	48.64, 49.39	10.32, 9.458	24.92, 25.17	13.44. 12.6
(0.1, 0.3)	23.45, 23.36	25.63, 25.32	26.47, 25.15	7.87, 7.24	14.1, 13.59	9.72, 9.48
(0.3, 0)	20.17, 19.56	21.13, 19.76	21.94, 21.43	7.12, 6.69	11.42, 11.33	8.58, 7.85
(0.5, 0.8)	3.33, 2.8	3.05, 2.51	3.87, 3.39	2.52, 1.95	2.66, 2.06	3.08, 2.42
(1, 1)	1.53, 0.91	1.41, 0.75	1.74, 1.13	1.42, 0.76	1.33, 0.67	1.66, 1.05
(2, 2)	1, 0	1, 0	1.09, 0.31	1, 0	1, 0	1.08, 0.29
(μ_1, μ_2)	$\sigma_1^2, \sigma_2^2, \sigma_{12} = 1.4, 1.4, 0.75$			$\sigma_1^2, \sigma_2^2, \sigma_{12} = 3, 3, 0.5$		
(0, 0)	21.33, 20.72	46.28, 46.25	25.54, 24.71	1.2, 0.49	4.14, 3.6	1.49, 0.87
(0.1, 0)	19.5, 19.1	39.61, 38.78	24.42, 23.76	1.17, 0.46	4.08, 3.53	1.48, 0.84
(0.1, 0.3)	12.64, 11.76	18.49, 17.03	14.88, 14.3	1.16, 0.42	3.43, 2.9	1.47, 0.84
(0.3, 0)	11.19, 10.53	15.26, 14.42	13.72, 13.02	1.16, 0.45	3.36, 2.82	1.51, 0.89
(0.5, 0.8)	2.96, 2.36	2.79, 2.18	3.43, 2.97	1.08, 0.3	1.85, 1.2	1.27, 0.59
(1, 1)	1.49, 0.86	1.43, 0.76	1.77, 1.18	1.03, 0.18	1.31, 0.66	1.13, 0.39
(2, 2)	1, 0	1, 0	1.1, 0.32	1, 0	1, 0.3	1.02, 0.15

In the case of $p = 3$, for the out-of-control dataset, the following are considered: 125 data with μ_1 shifted 0.2 and 125 data shifted 1; 125 data with μ_2 shifted 0.2 and 125 data shifted 1; 125 data with only μ_3 shifted 0.2 and 125 data shifted 1; 125 data with μ_1 and μ_2 shifted 0.2 and 125 data shifted 1; 125 data with μ_1 and μ_3 shifted 0.2 and 125 data shifted 1; 125 data with μ_2 and μ_3 shifted 0.2 and 125 data shifted 1; 125 data with σ_1^2 shifted 1.2 times and 125 data shifted 2 times; 125 data with σ_2^2 shifted 1.2 times and 125 data shifted 2 times; 125 data with σ_3^2 shifted 1.2 times and 125 data shifted 2 times; 125 data with σ_1^2 and σ_2^2 shifted 1.2 times and 125 data shifted 2 times; 125 data with σ_1^2 and σ_3^2 shifted 1.2 times and 125 data shifted 2 times; 125 data with σ_2^2 and σ_3^2 shifted 1.2 times and 125 data shifted 2 times; 125 data with only covariance shifted 1.2 times and 125 data shifted 2 times; 125 data with all three variances and covariance shifted 1.2 times and 125 data shifted 2 times; 125 data with all the parameters shifted together with small shift sizes; and 125 data shifted together with large shift sizes. For the in-control dataset, we include the same amount of data, which in this scenario is 3750 data. By keeping the previous hyperparameters, the RMSE of the ANN, SVR, and RFR structures are 0.44, 0.52, and 0.5, respectively. The UCLs are computed as 0.8635, 0.5781, and 0.9783, respectively. The results are presented in Table 5.

Table 5. ARL, SDRL. Scenario a12, $p = 3$.

	ANN	SVR	RFR	ANN	SVR	RFR
(μ_1, μ_2, μ_3)	\multicolumn{3}{c}{$\sigma_1^2, \sigma_2^2, \sigma_3^2, Cov = 1, 1, 1, 0.5$}			$\sigma_1^2, \sigma_2^2, \sigma_3^2, Cov = 1.05, 1, 1, 0.5$		
(0, 0, 0)	200, 200	200, 200	200, 200	162.9, 170.9	167.6, 163	177.2, 186
(0.1, 0, 0.1)	148.9, 145.8	141.3, 136.4	145.8, 148.7	123.2, 125.3	119.5, 118.7	132.2, 131.8
(0.1, 0.3, 0.3)	44.9, 45.35	38.54, 37.57	52.27, 52.87	41.05, 38.39	34.83, 34.83	44, 44.2
(0.3, 0, 0)	44.83, 44.09	36.98, 35.26	55.52, 53.86	38.93, 37.94	33.9, 32.46	43.5, 43.37
(0.5, 0.8, 0.5)	4.55, 3.9	3.93, 3.21	5.74, 4.92	4.44, 4.04	3.9, 3.55	5.53, 4.97
(1, 1, 1)	1.47, 0.81	1.42, 0.79	1.69, 1	1.51, 0.87	1.4, 0.73	1.78, 1.17
(2, 2, 2)	1, 0	1, 0	1.05, 0.23	1, 0	1, 0	1.06, 0.25
(μ_1, μ_2, μ_3)	\multicolumn{3}{c}{$\sigma_1^2, \sigma_2^2, \sigma_3^2, Cov = 1.05, 1.05, 1.05, 0.5$}			$\sigma_1^2, \sigma_2^2, \sigma_3^2, Cov = 1.05, 1.3, 1.05, 0.5$		
(0, 0, 0)	109.1, 110.3	117.8, 117.4	114.2, 120.1	46.17, 45.71	65.58, 67.24	48.03, 46.07
(0.1, 0, 0.1)	87.09, 88.44	93.44, 91.19	90.04, 94.33	39.46, 39.27	51.23, 49.16	40.36, 41.61
(0.1, 0.3, 0.3)	33.59, 32.73	30.85, 31.25	37.79, 36.54	19.72, 19.87	20.69, 21.13	21.6, 21.77
(0.3, 0, 0)	33.12, 33.06	29.23, 29.36	36.43, 35.96	20.39, 19.7	21.64, 21.35	21.92, 20.55
(0.5, 0.8, 0.5)	4.37, 3.84	3.81, 3.2	5.17, 4.93	3.76, 3.27	3.49, 2.82	4.42, 3.87
(1, 1, 1)	1.49, 0.88	1.35, 0.7	1.76, 1.16	1.47, 0.87	1.38, 0.71	1.7, 1.11
(2, 2, 2)	1, 0	1, 0	1.06, 0.24	1, 0	1, 0	1.07, 0.29
(μ_1, μ_2, μ_3)	\multicolumn{3}{c}{$\sigma_1^2, \sigma_2^2, \sigma_3^2, Cov = 1.2, 1.2, 1.2, 0.6$}			$\sigma_1^2, \sigma_2^2, \sigma_3^2, Cov = 1.4, 1.4, 1.05, 0.5$		
(0, 0, 0)	42.13, 41.58	68.06, 67.12	44.54, 42.65	12.74, 12.42	25.15, 24.26	13.02, 12.54
(0.1, 0, 0.1)	37.37, 35.82	54.46, 55.72	39, 39.25	12.14, 12.15	22.95, 22.95	12.32, 12.16
(0.1, 0.3, 0.3)	20, 19.04	20.41, 20.32	20.24, 19.76	8.43, 7.74	13.01, 12.13	9.54, 8.17
(0.3, 0, 0)	19.19, 19.2	20.66, 19.71	20.83, 20.14	8.34, 7.62	12.11, 11.96	8.66, 8.44
(0.5, 0.8, 0.5)	3.83, 3.29	3.32, 2.82	4.42, 4.03	3, 2.4	2.9, 2.38	3.36, 2.76
(1, 1, 1)	1.49, 0.9	1.38, 0.73	1.69, 1.1	1.37, 0.73	1.32, 0.63	1.52, 0.88
(2, 2, 2)	1, 0	1, 0	1.06, 0.25	1, 0	1, 0	1.06, 0.27
(μ_1, μ_2, μ_3)	\multicolumn{3}{c}{$\sigma_1^2, \sigma_2^2, \sigma_3^2, Cov = 1.4, 1.4, 1.05, 0.75$}			$\sigma_1^2, \sigma_2^2, \sigma_3^2, Cov = 3, 3, 1.4, 0.5$		
(0, 0, 0)	54.68, 54.76	69.61, 72.43	56.6, 55.83	1.2, 0.48	3.72, 3.25	1.36, 0.72
(0.1, 0, 0.1)	48.04, 48.68	52.62, 50.82	49.07, 49.57	1.19, 0.48	3.53, 3.1	1.36, 0.72
(0.1, 0.3, 0.3)	21.29, 21.31	20.82, 20.21	24.63, 23.86	1.17, 0.45	3.02, 2.5	1.3, 0.62
(0.3, 0, 0)	22.63, 22.02	19.92, 19.11	24.3, 24.31	1.19, 0.47	3.06, 2.42	1.34, 0.7
(0.5, 0.8, 0.5)	3.84, 3.32	3.39, 2.85	4.59, 4.16	1.1, 0.34	1.82, 1.23	1.25, 0.55
(1, 1, 1)	1.53, 0.89	1.47, 0.84	1.74, 1.2	1.03, 0.19	1.21, 0.52	1.08, 0.28
(2, 2, 2)	1, 0	1, 0, 02	1.08, 0.3	1, 0	1, 0	1.01, 0.11

This table shows that again under no/small variation shifts, the SVR chart mostly performs better than the other two, but as the variation shift increases, the other two begin to perform better than the SVR chart, with the ANN chart having the best performance. By comparing Tables 2 and 5 (the first and second training methods), the conclusions driven for the $p = 2$ case (comparing Tables 1 and 4) can be driven for the $p = 3$ case as well. By comparing the cases of $p = 3$ and $p = 2$ (Tables 4 and 5), we realize that the conclusions are similar to those of Scenario a11.

Moreover, by comparing the results of Table 5 to Sabahno et al.'s [9] Table 5, we can see that with the second training method, at least one of the ML control charts performs better than all their control charts, in all the shift cases.

In the case of four quality characteristics ($p = 4$), for the out-of-control dataset, we consider 100 data with only μ_1 shifted 0.2 and 100 data shifted 1; 100 data with only μ_2 shifted 0.2 and 100 data shifted 1; 100 data with only μ_3 shifted 0.2 and 100 data shifted 1; 100 data with only μ_4 shifted 0.2 and 100 data shifted 1; 100 data with μ_1 and μ_2 shifted 0.2 and 100 data shifted 1; 100 data with μ_1 and μ_3 shifted 0.2 and 100 data shifted 1; 100 data with μ_2 and μ_3 shifted 0.2 and 100 data shifted 1; 100 data with μ_1 and μ_4 shifted 0.2 and

100 data shifted 1; 100 data with μ_2 and μ_4 shifted 0.2 and 100 data shifted 1; 100 data with μ_3 and μ_4 shifted 0.2 and 100 data shifted 1; 100 data with μ_1, μ_2 and μ_3 shifted 0.2 and 100 data shifted 1; 100 data with μ_1, μ_2 and μ_4 shifted 0.2 and 100 data shifted 1; 100 data with μ_1, μ_3 and μ_4 shifted 0.2 and 100 data shifted 1; 100 data μ_2, μ_3 and μ_4 shifted 0.2 and 100 data shifted 1; 100 data with only σ_1^2 shifted 1.2 times and 100 data shifted 2 times; 100 data with only σ_2^2 shifted 1.2 times and 100 data shifted 2 times; 100 data with only σ_3^2 shifted 1.2 times and 100 data shifted 2 times; 100 data with only σ_4^2 shifted 1.2 times and 100 data shifted 2 times; 100 data with σ_1^2 and σ_2^2 shifted 1.2 times and 100 data shifted 2 times; 100 data with σ_1^2 and σ_3^2 shifted 1.2 times and 100 data shifted 2 times; 100 data with σ_2^2 and σ_3^2 shifted 1.2 times and 100 data shifted 2 times; 100 data with σ_1^2 and σ_4^2 shifted 1.2 times and 100 data shifted 2 times; 100 data with σ_2^2 and σ_4^2 shifted 1.2 times and 100 data shifted 2 times; 100 data with σ_3^2 and σ_4^2 shifted 1.2 times and 100 data shifted 2 times; 100 data with σ_1^2, σ_2^2 and σ_3^2 shifted 1.2 times and 100 data shifted 2 times; 100 data with σ_1^2, σ_2^2 and σ_4^2 shifted 1.2 times and 100 data shifted 2 times; 100 data with σ_1^2, σ_3^2 and σ_4^2 shifted 1.2 times and 100 data shifted 2 times; 100 data with σ_2^2, σ_3^2 and σ_4^2 shifted 1.2 times and 100 data shifted 2 times; 100 data with only covariance shifted 1.2 times and 100 data shifted 2 times; 100 data with all four variances and the covariance shifted 1.2 times and 100 data shifted 2 times; and 100 data with all the parameters shifted together with small shift sizes and 100 data shifted together with large shift sizes. For the in-control dataset, we include the same amount of data, which in this scenario is 6200 data.

By keeping the previous hyperparameters, the RMSE of the ANN, SVR, and RFR structures are 0.43, 0.53, and 0.46, respectively. The UCLs are computed as 0.856, 0.5108, and 0.9735, respectively. The results are presented in Table 6.

This table shows that again under no/small variation shifts, the SVR chart mostly performs better than the other two, but as the variation shift increases, the other two begin to perform better than the SVR chart, with the ANN chart having the best performance. By comparing Tables 3 and 6 (first and second training methods), the conclusions driven for the $p = 2$ case (comparing Tables 1 and 4) can be driven for the $p = 4$ case as well. The results in Table 6 show that all the charts mostly perform better in higher dimensions when the second training method is used.

3.1.2. Scenario a2 (Control Type a, Input Set 2)

As mentioned before, different inputs for the ML structures are considered in this scenario. The sample's mean vector and variance–covariance matrix are used and in the case of $p = 2$, there are going to be five inputs (individual variances, means, and covariance) as described in the following subsections.

Scenario a21 (Control Type a, Input Set 2, Training Method 1)

In-control and out-of-control datasets are the same as in Scenario a11. In this scenario, again, a linear kernel is used for the SVR structure, and the RMSE is computed as 0.51. For the ANN control chart, 10 nodes (twice the number of inputs) are used in the hidden layer and the trained ANN structure has an RMSE of 0.48. For the random forest structure, again 100 trees are used with an RMSE of 0.5. The UCLs of the ANN, SVR, and RFR control charts are computed as 0.8746, 1.0631, and 0.8143, respectively. The results of this analysis are reported in Table 7. Based on these results, the SVR chart performs better than the other two under all shift cases. By comparing their performance against the previous input scenario (comparing Tables 1 and 7), we realize that the ANN chart mostly performs worse when the second input scenario is used. The situation is kind of mediocre with the SVR chart, and it seems that the input method overall does not affect its performance. The RFR chart on the other hand, especially under moderate and large mean shifts, mostly performs better in this input scenario, and the deterioration in performance as the shift size becomes larger than the trained value is much less when this input set is used.

Table 6. ARL, SDRL. Scenario a12, $p = 4$.

	ANN	SVR	RFR	ANN	SVR	RFR
$(\mu_1, \mu_2, \mu_3, \mu_4)$	$\sigma_1^2, \sigma_2^2, \sigma_3^2, \sigma_4^2, Cov = 1, 1, 1, 1, 0.5$			$\sigma_1^2, \sigma_2^2, \sigma_3^2, \sigma_4^2, Cov = 1.05, 1, 1, 1, 0.5$		
(0, 0, 0, 0)	200, 200	200, 200	200, 200	163.9, 158.3	174.04, 169.22	160.2, 170.4
(0.1, 0, 0.1, 0)	148, 149.2	133.58, 131.92	152.2, 159.5	127.9, 126.5	121.35, 122.85	127.7, 126.7
(0.1, 0.3, 0.3, 0.1)	52.81, 51.7	44.38, 43.31	58.32, 56.33	46.43, 44.73	37.04, 36.51	54.4, 54.9
(0.3, 0, 0, 0.3)	34.3, 34.65	25.74, 25.18	40.84, 40.07	29.85, 29.68	23.79, 23.6	35.56, 36.57
(0.5, 0.8, 0.5, 0.8)	3.72, 3.2	3.21, 2.67	4.79, 4.23	3.73, 3.19	3.15, 2.57	4.47, 3.93
(1, 1, 1, 1)	1.58, 0.98	1.44, 0.78	1.83, 1.19	1.53, 0.89	1.4, 0.74	1.79, 1.203
(2, 2, 2, 2)	1, 0	1, 0	1.02, 0.15	1, 0	1, 0	1.02, 0.16
$(\mu_1, \mu_2, \mu_3, \mu_4)$	$\sigma_1^2, \sigma_2^2, \sigma_3^2, \sigma_4^2, Cov = 1.05, 1.05, 1.05, 1.05, 0.5$			$\sigma_1^2, \sigma_2^2, \sigma_3^2, \sigma_4^2, Cov = 1.05, 1.3, 1.05, 1.3, 0.5$		
(0, 0, 0, 0)	90.44, 90.41	113.47, 112.57	92.41, 100.2	19.97, 19.34	38.03, 37.17	22.14, 21.16
(0.1, 0, 0.1, 0)	73.3, 72.54	80.82, 80.86	75.77, 81.53	18.39, 18.98	31.29, 30.39	20.76, 20.77
(0.1, 0.3, 0.3, 0.1)	32.89, 34.2	29.7, 30.17	35.04, 36.03	11.77, 11.75	15.89, 15.8	13.84, 12.81
(0.3, 0, 0, 0.3)	21.97, 21.22	18.94, 18.27	26.83, 27.45	9.25, 8.63	10.53, 9.91	10.68, 10.29
(0.5, 0.8, 0.5, 0.8)	3.3, 2.62	2.9, 2.38	4.18, 3.6	2.51, 1.92	2.43, 1.88	3.2, 2.61
(1, 1, 1, 1)	1.47, 0.8	1.37, 0.72	1.8, 1.17	1.36, 0.7	1.31, 0.62	1.65, 1.05
(2, 2, 2, 2)	1, 0	1, 0	1.02, 0.17	1, 0	1, 0	1.03, 0.19
$(\mu_1, \mu_2, \mu_3, \mu_4)$	$\sigma_1^2, \sigma_2^2, \sigma_3^2, \sigma_4^2, Cov = 1.2, 1.2, 1.2, 1.2, 0.6$			$\sigma_1^2, \sigma_2^2, \sigma_3^2, \sigma_4^2, Cov = 1.4, 1.4, 1.05, 1.05, 0.5$		
(0, 0, 0, 0)	31.71, 31.05	55.85, 54.67	36.09, 33.88	13.98, 13.43	27.48, 27.52	15.11, 15.02
(0.1, 0, 0.1, 0)	29.45, 28.48	45.9, 47.48	28.89, 28.02	12.06, 11.21	22.92, 22.49	14.42, 13.81
(0.1, 0.3, 0.3, 0.1)	16.28, 15.78	19.25, 18.42	18.4, 17.29	8.66, 8.11	12.33, 12.17	10.21, 10.01
(0.3, 0, 0, 0.3)	11.75, 11.49	13.26, 12.59	14.95, 14.45	6.87, 6.63	9.14, 8.6	8.51, 7.64
(0.5, 0.8, 0.5, 0.8)	2.75, 2.21	2.59, 2.08	3.56, 3.1	2.28, 1.69	2.3, 1.69	2.97, 2.32
(1, 1, 1, 1)	1.46, 0.81	1.36, 0.69	1.68, 1.06	1.31, 0.65	1.3, 0.61	1.54, 0.88
(2, 2, 2, 2)	1, 0	1, 0	1.04, 0.21	1, 0	1, 0	1.05, 0.24
$(\mu_1, \mu_2, \mu_3, \mu_4)$	$\sigma_1^2, \sigma_2^2, \sigma_3^2, \sigma_4^2, Cov = 1.4, 1.4, 1.05, 1.05, 0.75$			$\sigma_1^2, \sigma_2^2, \sigma_3^2, \sigma_4^2, Cov = 3, 3, 1.4, 1.4, 0.5$		
(0, 0, 0, 0)	111.4, 112.1	96.52, 96.41	120.3, 118.1	1.15, 0.41	2.99, 2.39	1.3, 0.63
(0.1, 0, 0.1, 0)	92.3, 90.43	79.73, 80.99	106.6, 99.79	1.18, 0.46	2.95, 2.49	1.3, 0.63
(0.1, 0.3, 0.3, 0.1)	35.14, 33.45	28.81, 28.11	41.53, 40.01	1.16, 0.42	2.64, 2.01	1.27, 0.57
(0.3, 0, 0, 0.3)	25.72, 25.52	21.05, 20.43	30.23, 28.25	1.17, 0.45	2.37, 1.84	1.26, 0.59
(0.5, 0.8, 0.5, 0.8)	3.55, 3.06	3.11, 2.68	4.1, 3.57	1.06, 0.26	1.44, 0.80	1.17, 0.44
(1, 1, 1, 1)	1.59, 0.95	1.5, 0.84	1.86, 1.25	1.02, 0.16	1.12, 0.37	1.07, 0.27
(2, 2, 2, 2)	1, 0.02	1, 0	1.03, 0.17	1, 0	1, 0	1.03, 0.17

Moreover, by comparing the results in Table 7 to Sabahno et al.'s [9] Table 4, it can be seen that in most cases, at least one of the proposed ML control charts performs better (mostly much better) than all their proposed control charts. Only in the case of (0.5, 0.8) mean shift, together with no/small variation shift, at least one of their proposed charts performs a little bit better than the best of ours.

Scenario a22 (Control Type a, Input Set 2, Training Method 2)

In-control and out-of-control datasets are the same as in Scenario a12. The same as before, a linear kernel is used for the SVR structure, and the RMSE is computed as 0.48. For the ANN control chart, 10 nodes (twice the number of inputs) in the hidden layer are used and the trained ANN structure has an RMSE of 0.42. For the random forest structure, 100 trees are used again with an RMSE of 0.45. The UCLs of the ANN, SVR, and RFR control charts are computed as 0.8846, 0.8136, and 0.894, respectively.

Table 7. ARL, SDRL. Scenario a21, $p = 2$.

(μ_1, μ_2)	ANN	SVR	RFR	ANN	SVR	RFR
	$\sigma_1^2, \sigma_2^2, \sigma_{12} = 1, 1, 0.5$			$\sigma_1^2, \sigma_2^2, \sigma_{12} = 1.05, 1, 0.5$		
(0, 0)	200, 200	200, 200	200, 200	186.2, 187.6	142, 138.4	176.7, 177.4
(0.1, 0)	231.6, 222.2	162.3, 167.9	181.6, 174.8	212.4, 221.1	124.5, 121.7	181.7, 179.7
(0.1, 0.3)	39.04, 40.21	49.88, 49.47	50.44, 49.63	35.28, 34.55	40.77, 41.27	50.68, 48.63
(0.3, 0)	274.8, 282.8	116.7, 117.2	165.7, 171.1	225.2, 225.2	86.65, 84.26	168, 170.8
(0.5, 0.8)	5.17, 4.66	6.08, 5.38	12.31, 12.28	5, 4.38	5.52, 4.93	12.26, 11.75
(1, 1)	2.28, 1.74	2.39, 1.76	10.02, 8.69	2.17, 1.63	2.29, 1.69	10.06, 9.615
(2, 2)	1, 0.09	1, 0.08	11.04, 10.03	1, 0.08	1, 0.05	11.28, 10.79
(μ_1, μ_2)	$\sigma_1^2, \sigma_2^2, \sigma_{12} = 1.05, 1.05, 0.5$			$\sigma_1^2, \sigma_2^2, \sigma_{12} = 1.05, 1.3, 0.5$		
(0, 0)	150, 150.2	112.4, 111.3	178.2, 170.9	56.75, 53.19	35.34, 34.7	135.6, 133.3
(0.1, 0)	167.8, 165.7	91.67, 88.75	172.5, 167.6	63.89, 60	29.74, 30.07	108.6, 110.8
(0.1, 0.3)	31.64, 31.06	32.33, 32.36	46.03, 47.7	17.87, 17.33	13.91, 12.76	47.6, 45.41
(0.3, 0)	181.1, 185.6	66.24, 66.01	145, 139.8	70.16, 68.29	24.4, 23.79	98.41, 93.14
(0.5, 0.8)	4.89, 4.46	4.72, 4.41	12, 12.29	3.99, 3.49	3.11, 2.55	12.31, 11.62
(1, 1)	2.16, 1.6	2.18, 1.61	10.49, 9.53	2.01, 1.386	1.77, 1.172	11.4, 11.41
(2, 2)	1, 0.07	1, 0.05	11.06, 10.5	1, 0.08	1, 0.03	14.01, 13.45
(μ_1, μ_2)	$\sigma_1^2, \sigma_2^2, \sigma_{12} = 1.2, 1.2, 0.6$			$\sigma_1^2, \sigma_2^2, \sigma_{12} = 1.4, 1.4, 0.5$		
(0, 0)	82.78, 81.39	40.88, 41.09	112.5, 112.3	21.28, 20.96	9.24, 8.37	91.04, 88.54
(0.1, 0)	84.68, 82.83	35.39, 34.28	102, 101.2	22.17, 22.54	8.59, 7.91	86.13, 84.34
(0.1, 0.3)	22.31, 22.49	15.23, 14.85	35.87, 33.51	8.81, 8.45	5.02, 4.51	41.04, 42.17
(0.3, 0)	85.71, 88.17	27.79, 27.52	102, 106.5	21.26, 21.9	7.29, 6.61	74.15, 69.47
(0.5, 0.8)	3.84, 3.17	3.24, 2.67	9.72, 9.28	2.65, 2.13	1.91, 1.35	12.47, 12.18
(1, 1)	1.88, 1.29	1.79, 1.21	9.06, 8.25	1.59, 0.95	1.34, 0.66	12.14, 12.33
(2, 2)	1.01, 0.09	1, 0.06	10.77, 10.35	1, 0.06	1, 0.03	17.15, 16.48
(μ_1, μ_2)	$\sigma_1^2, \sigma_2^2, \sigma_{12} = 1.4, 1.4, 0.75$			$\sigma_1^2, \sigma_2^2, \sigma_{12} = 3, 3, 0.5$		
(0, 0)	42.26, 39.87	16.45, 15.6	75.73, 71.09	1.91, 1.3	1.2, 0.48	105.9, 108.5
(0.1, 0)	41.82, 39.94	15.16, 14.77	68.19, 70.49	1.98, 1.4	1.22, 0.53	92.67, 90.9
(0.1, 0.3)	13.78, 14.03	7.58, 7.1	26.01, 24.69	1.49, 0.86	1.13, 0.38	57.07, 56.54
(0.3, 0)	40.9, 40.55	12.26, 11.34	65.37, 64.64	1.97, 1.39	1.17, 0.45	84.41, 78.99
(0.5, 0.8)	3.02, 2.45	2.39, 1.84	8.55, 8.2	1.17, 0.47	1.03, 0.19	23.51, 23.63
(1, 1)	1.67, 1.04	1.54, 0.9	7.81, 7.57	1.11, 0.35	1.01, 0.12	22.48, 21.63
(2, 2)	1, 0.08	1, 0.07	10.28, 9.77	1, 0.08	1, 0	59.71, 60.83

According to the results of this analysis (reported in Table 8), the ANN chart is the worst-performing in all the compared cases. Under zero variation shift, the RFR chart performs better than the SVR chart. Other than that, under small variation shifts, together with small mean shifts, the SVR chart performs better, and together with moderate and large mean shifts, the RFR chart performs better. In large variation shift cases, the SVR chart performs better. By comparing the first and second training cases (Tables 7 and 8), it is evident that the ANN and RFR charts perform better with the second training method being applied, in all the shift cases. On the other hand, under no or small variation shifts, the SVR chart also performs better with the second training method, but under larger variation shifts, the chart performance is rather the same no matter which training method is used. In addition, no deterioration in performance as the shift size becomes larger than the trained value is noticeable in any of the charts in this scenario. By comparing Tables 4 and 8 (different input sets), we realize that all the charts mostly perform better with the second input scenario, with the RFR chart having the least improved cases.

Table 8. ARL, SDRL. Scenario a22, $p = 2$.

	ANN	SVR	RFR	ANN	SVR	RFR
(μ_1, μ_2)		$\sigma_1^2, \sigma_2^2, \sigma_{12} = 1, 1, 0.5$			$\sigma_1^2, \sigma_2^2, \sigma_{12} = 1.05, 1, 0.5$	
(0, 0)	200, 200	200, 200	200, 200	123.77, 120.89	112.98, 110.77	146.04, 137.69
(0.1, 0)	175.12, 169.66	155.3, 149.36	122.8, 117.12	107.61, 104.76	87.61, 86.69	100.7, 94.8
(0.1, 0.3)	55.5, 53.37	49.26, 48.86	37.66, 35.35	40.88, 40.46	32.22, 32.18	30.83, 29.54
(0.3, 0)	68.36, 69.04	93.47, 91.08	33.89, 32.93	49.6, 46.63	57.54, 59.13	30.19, 30.05
(0.5, 0.8)	4.73, 4.28	5.52, 5.12	3.47, 2.97	4.07, 3.59	4.19, 3.63	3.27, 2.64
(1, 1)	1.94, 1.35	2.1, 1.51	1.39, 0.72	1.77, 1.16	1.87, 1.26	1.36, 0.73
(2, 2)	1, 0.03	1, 0.03	1, 0.04	1, 0.05	1, 0.04	1, 0.05
(μ_1, μ_2)		$\sigma_1^2, \sigma_2^2, \sigma_{12} = 1.05, 1.05, 0.5$			$\sigma_1^2, \sigma_2^2, \sigma_{12} = 1.05, 1.3, 0.5$	
(0, 0)	119.7, 117.43	113.66, 111.26	145.8, 143.26	36.98, 36.29	35.36, 34.15	67.92, 71.98
(0.1, 0)	109.92, 110.25	91.58, 89.65	94.87, 95.08	32.44, 31.9	28.9, 29.51	57.81, 57.1
(0.1, 0.3)	39.17, 38.3	31.75, 30.94	32.75, 32.57	16.98, 16.67	13.64, 12.93	19.89, 19.42
(0.3, 0)	46.38, 46.47	57.45, 56.99	30.82, 28.69	21.45, 21.07	22.37, 21.61	26.26, 25.91
(0.5, 0.8)	4.2, 3.63	4.33, 3.76	3.36, 2.79	3.36, 2.75	3.03, 2.5	3.07, 2.52
(1, 1)	1.79, 1.18	1.9, 1.32	1.32, 0.68	1.63, 1.03	1.62, 0.98	1.36, 0.69
(2, 2)	1, 0.03	1, 0.02	1, 0.07	1, 0.04	1, 0.03	1, 0.07
(μ_1, μ_2)		$\sigma_1^2, \sigma_2^2, \sigma_{12} = 1.2, 1.2, 0.6$			$\sigma_1^2, \sigma_2^2, \sigma_{12} = 1.4, 1.4, 0.5$	
(0, 0)	49.03, 47.37	43.5, 43.13	70.52, 70.58	8.71, 8.04	9.21, 8.72	26.31, 26.03
(0.1, 0)	46.69, 46.07	35.71, 37.08	50.66, 50.53	7.93, 7.31	8.41, 7.83	22.27, 21.87
(0.1, 0.3)	21.07, 20.83	15.34, 14.99	20.19, 19.52	5.9, 5.29	5.08, 4.4	10.86, 10.34
(0.3, 0)	26.63, 25.92	24.93, 24.83	22.78, 23.16	6.36, 6.04	6.86, 6.31	12.83, 12.98
(0.5, 0.8)	3.61, 3.05	3.18, 2.74	2.86, 2.32	2.23, 1.56	1.89, 1.27	2.41, 1.76
(1, 1)	1.66, 1.05	1.64, 1	1.39, 0.75	1.39, 0.76	1.28, 0.59	1.3, 0.66
(2, 2)	1, 0.05	1, 0.05	1, 0.03	1.01, 0.12	1, 0	1, 0, 07
(μ_1, μ_2)		$\sigma_1^2, \sigma_2^2, \sigma_{12} = 1.4, 1.4, 0.75$			$\sigma_1^2, \sigma_2^2, \sigma_{12} = 3, 3, 0.5$	
(0, 0)	21.16, 20.8	17.8, 17.78	29.82, 28.27	1.26, 0.57	1.26, 0.58	2.01, 1.5
(0.1, 0)	19.72, 19.11	15.32, 14.3	25.04, 24.31	1.24, 0.53	1.2, 0.48	1.96, 1.32
(0.1, 0.3)	11.45, 10.87	7.954, 7.319	11.55, 10.3	1.26, 0.58	1.14, 0.39	1.67, 1.09
(0.3, 0)	14.49, 13.59	11.25, 10.4	13.41, 12.79	1.2, 0.47	1.18, 0.46	1.85, 1.26
(0.5, 0.8)	2.93, 2.43	2.26, 1.67	2.57, 2	1.23, 0.54	1.03, 0.2	1.26, 0.57
(1, 1)	1.59, 0.97	1.42, 0.76	1.34, 0.7	1.22, 0.54	1.01, 0.12	1.1, 0.35
(2, 2)	1, 0.09	1, 0.07	1, 0.04	1.345, 0.67	1, 0	1, 0, 05

Moreover, by comparing the results in Table 8 to Sabahno et al.'s [9] Table 4, we can again see that in all the cases, at least one of the proposed ML control charts performs better (mostly much better) than all their control charts. Therefore, when the second input scenario is used, if the second training method is utilized, the proposed scheme performs better under all shift sizes and types, unlike the first training method (Scenario a21).

3.2. Scenario b: Control Charts for Detection and Identification

In scenario b, several ML structures are involved, and consequently, so are several control charts in each control scheme. Each process parameter has its output (control chart), and if the corresponding chart signals, it means that this parameter (and consequently the variable associated with it) has shifted. However, as the number of quality characteristics increases, the number of control charts (outputs) also increases in this scenario. This might increase the false-alarm rates. As such, this scenario should be used with more caution, especially in larger dimensions. In addition, even before conducting any numerical analysis (next subsections), overall, worse performance compared to scenario a is expected in this scenario because more than one control chart is being monitored together. Having said that, the advantage of this scenario is not its performance, but helping one to identify the responsible variable parameter by allowing a little bit of performance to be sacrificed.

In the case of $p = 2$, since we have five process parameters, namely μ_1, μ_2, σ_1^2, σ_2^2 and σ_{12}, five ML structures, each with its own control chart, are therefore required. These charts are monitored together and obtaining a signal from either one of them means that the process is out of control. Before we start developing control charts for each scenario, we should mention that it was more difficult to choose suitable hyperparameters for this scenario to obtain the desired ARL performance, especially in the case of $p = 3$. Therefore, we tried many combinations of hyperparameter values for the ML structures to obtain the desired overall performance. The desired performance, in this case, is computed (assuming the independency of control charts as well as equality of their type-I error probability) as follows: $\alpha_{overall} = 1 - (1 - \alpha)^m$, where m is the number of control charts, and ARL = $\frac{1}{\alpha_{overall}} = \frac{1}{0.005} = 200$. Consequently, for the case of $p = 2$, in which a maximum of five control charts is required ($m = 5$), by using the above formulae, we have $\alpha = 0.001$. This means that each control chart's performance is $\frac{1}{0.001} = 1000$, but they all together should have a performance of $\frac{1}{0.005} = 200$. Similarly, for the case of $p = 3$, individual αs are computed as 0.000716, which results in ARL = $\frac{1}{0.000716} = 1396.6$. Note that, for simplicity, we assume that all the covariances are equal in the case of $p = 3$. Therefore, we only have one control chart for monitoring the covariance, making a total of seven control charts ($m = 7$, i.e., three to monitor the means, three to monitor the variances, and one to monitor the covariance).

3.2.1. Scenario b1 (Control Type b, Input Set 1)

Similar to Scenario a1, the inputs, in this case, are T^2 and W statistics. Again, two different training methods are applied on each control chart (one trained with only small shifts and the other one trained with small and large shifts).

Scenario b11 (Control Type b, Input Set 1, Training Method 1)

In this scenario, the ML structures are trained with only small shift sizes. First, we generate 1000 in-control data that are going to be used in all control charts. Second, we generate 1000 out-of-control data with 0.2 shifts in μ_1 and use it for the control chart that monitors μ_1; 1000 out-of-control data with 0.2 shifts in μ_2 and use it for the control chart that monitors μ_2; 1000 out-of-control data with shifts of 1.2 times the in-control σ_1^2 and use it for the control chart that monitors σ_1^2; 1000 out-of-control data with shifts of 1.2 times the in-control σ_2^2 and use it for the control chart that monitors σ_2^2; and 1000 out-of-control data with shifts of 1.2 times the in-control σ_{12} and use it for the control chart that monitors σ_{12}. Now that we have five datasets, we need to train five different ML structures. For the ANN scheme, for monitoring μ_1, μ_2, σ_1^2, σ_2^2 and σ_{12}, we use 5, 5, 5, 5, and 4 nodes in the hidden layers, respectively, to obtain an overall performance of 200. The RMSE of these ANN structures, respectively, are 0.49, 0.49, 0.48, 0.48, and 0.49. Finally, the UCLs are, respectively, computed as 1.0292, 1.1628, 0.8846, 1.048, and 0.8393.

For the SVR scheme, the major difference was that the linear kernel did not work for all the structures in this scenario, and we had to try the radial kernel as well (other kernel types did not work at all). The kernel types we used for μ_1, μ_2, σ_1^2, σ_2^2 and σ_{12} control charts are, respectively, radial, radial, radial, radial, and linear. The RMSEs and UCLs are as follows. The RMSEs are 0.52, 0.52, 0.52, 0.53, and 0.51, and the UCLs are 1.1174, 1.1204, 1.1089, 1.0972, and 1.283. Regarding the RFR scheme, 100 trees still worked well for each structure in this scenario. The RMSEs and UCLs for the RFR charts are computed as follows. The RMSEs are 0.53, 0.54, 0.52, 0.52, and 0.52, and the UCLs are 0.9576, 0.9409, 0.9597, 0.9681, and 0.9433.

According to the reported results in Table 9, all the control charts mostly experience a deterioration in performance compared to when only one control chart is used (Table 1). However, the RFR scheme mostly performs better in this control case when the mean shift is large. The worst deterioration in performance can be seen in the SVR scheme. It even experiences a deterioration in performance under shifts larger than the trained value in this scenario (unlike Table 1). The RFR scheme, however, experiences that effect only when the variation shift size is very large in this scenario. Out of these three, the ANN is the best-performing scheme. Its performance is even very close to that of being reported in Table 1 (used only for detection).

Table 9. ARL, SDRL. Scenario b11 (with identification), $p = 2$.

	ANN	SVR	RFR	ANN	SVR	RFR
(μ_1, μ_2)		$\sigma_1^2, \sigma_2^2, \sigma_{12} = 1, 1, 0.5$			$\sigma_1^2, \sigma_2^2, \sigma_{12} = 1.05, 1, 0.5$	
(0, 0)	200, 200	200, 200	200, 200	149.42, 151.52	201.87, 203.46	181.8, 177.7
(0.1, 0)	160.4, 155.4	173.9, 176.6	186.7, 191	124, 120	184.7, 191.4	180.1, 189
(0.1, 0.3)	62.01, 58.38	110.02, 110.37	155.2, 158.5	52.15, 50.89	109.1, 107.4	143, 151.4
(0.3, 0)	48.04, 46.43	93.51, 90	136.3, 138.5	40.94, 40.87	97.53, 97.58	139.4, 143.1
(0.5, 0.8)	4.75, 4.3	36.58, 36.75	49.6, 49.63	4.55, 4.16	38.09, 38.81	51.89, 50.52
(1, 1)	1.79, 1.15	51.36, 49.96	29.7, 30.33	1.77, 1.19	47.34, 46.34	31.93, 32.45
(2, 2)	1, 0.03	>10,000	19.38, 18.91	1, 0.03	>10,000	20.84, 20.32
(μ_1, μ_2)		$\sigma_1^2, \sigma_2^2, \sigma_{12} = 1.05, 1.05, 0.5$			$\sigma_1^2, \sigma_2^2, \sigma_{12} = 1.05, 1.3, 0.5$	
(0, 0)	118.52, 119.79	202.07, 207.02	180.8, 175.1	39.63, 39.97	168.7, 169	131.2, 131.3
(0.1, 0)	97.77, 99.47	172.9, 172.1	176.5, 178.1	36.89, 36.02	161.9, 158.9	130.8, 130.4
(0.1, 0.3)	43.95, 44.88	108.1, 106.8	138.2, 133	20.56, 20.27	102.3, 103.9	113.9, 116.9
(0.3, 0)	35.95, 35.47	95.47, 93.45	134.1, 128.2	19.35, 19.4	88.08, 86.51	122.5, 117.7
(0.5, 0.8)	4.422, 3.934	38.92, 37.7	56.43, 55.79	3.69, 3.19	39.36, 39.55	74.89, 70.93
(1, 1)	1.76, 1.16	47.73, 44.82	34.79, 32.97	1.77, 1.16	47.39, 47.61	52.86, 50.55
(2, 2)	1, 0.02	>10,000	22.8, 22.2	1, 0.05	>10,000	35.4, 34.89
(μ_1, μ_2)		$\sigma_1^2, \sigma_2^2, \sigma_{12} = 1.2, 1.2, 0.6$			$\sigma_1^2, \sigma_2^2, \sigma_{12} = 1.4, 1.4, 0.5$	
(0, 0)	47.96, 47.6	186.1, 183.4	132.9, 133.4	11.57, 11.02	146.5, 145.7	94.65, 99.79
(0.1, 0)	41.51, 40.24	157.3, 157.5	144.8, 156.5	11.2, 10.7	135.7, 139.9	97.2, 96.26
(0.1, 0.3)	23.57, 22.57	104.1, 106.2	126.2, 117.9	8.69, 8.15	103.9, 105.3	94.03, 94.87
(0.3, 0)	20.93, 20.87	88.51, 86.74	122.6, 121.7	7.41, 6.69	87.51, 88.02	96.93, 95.25
(0.5, 0.8)	3.79, 3.23	38.36, 38.36	69.34, 69.38	3, 2.4	43.3, 44.68	106.1, 105.8
(1, 1)	1.83, 1.24	48.31, 49.39	49.93, 52.86	1.65, 1.08	56.18, 54.86	109.8, 109.8
(2, 2)	1, 0.04	>10,000	33.44, 32.79	1, 0.07	>10,000	91.38, 85.05
(μ_1, μ_2)		$\sigma_1^2, \sigma_2^2, \sigma_{12} = 1.4, 1.4, 0.75$			$\sigma_1^2, \sigma_2^2, \sigma_{12} = 3, 3, 0.5$	
(0, 0)	19.79, 18.66	160.9, 161.4	108, 107.1	1.27, 0.6	725.8, 726.7	222.9, 221.5
(0.1, 0)	19.09, 18.97	141.7, 132.8	103.9, 106.7	1.27, 0.6	700.6, 692.5	206.9, 200.7
(0.1, 0.3)	13.15, 12.39	99.46, 98.47	104.4, 106	1.26, 0.58	735.2, 706.8	217.9, 228.2
(0.3, 0)	11.99, 10.93	88.22, 83.6	108.71, 104.8	1.28, 0.58	709.2, 712.3	221.7, 215.4
(0.5, 0.8)	3.46, 3.03	41.47, 41.09	94.66, 93.9	1.13, 0.38	643.3, 611.6	436.7, 432.3
(1, 1)	1.81, 1.18	53.56, 55.79	69.6, 75.21	1.08, 0.28	812.3, 578	1434, 1027
(2, 2)	1, 0.08	>10,000	59.42, 58.75	1, 0.02	>10,000	>10,000

Moreover, by comparing the results of Table 9 to Sabahno et al.'s [9] Table 4, one can still see that in most cases (except mostly in the case of (0.5, 0.8) mean shift together with no/small variation shift), at least one of the proposed ML control charts performs better than all their proposed control charts, even though their control charts are only designed for detection. In the case of three quality characteristics, the construction of the dataset is the same as the case of $p = 2$, except here we add the following: 1000 out-of-control data with 0.2 shifts in μ_3 and use it for the control chart that monitors μ_3 and 1000 out-of-control data with shifts 1.2 times the in-control σ_3^2 and use it for the control chart that monitors σ_3^2. For the ANN scheme, for monitoring μ_1, μ_2, μ_3, σ_1^2, σ_2^2, σ_3^2 and the covariance, we use 5, 5, 5, 5, 5, 5, and 4 nodes in the hidden layers, respectively, with RMSEs of 0.49, 0.49, 0.49, 0.49, 0.49, and 0.48, and UCLs of 0.9169, 1.2852, 0.9164, 0.822, 0.8787, 0.8193, and 0.7305. For the SVR scheme, the used kernels, respectively, are radial, radial, radial, radial, radial, radial, and linear. The RMSEs are 0.51, 0.51, 0.51, 0.52, 0.52, 0.52, and 0.52, and the UCLs are 1.1506, 1.0763, 1.0906, 1.0908, 1.1556, 1.0914, and 1.6437. Regarding the RFR scheme, the least numbers of trees that we had to use for each structure (to obtain the overall ARL of 200), respectively, 100, 100, 100, 500, 100, 300, and 100. The RMSEs are 0.53, 0.53, 0.53,

0.53, 0.52, 0.53, and 0.52, and the UCLs are 0.9485, 0.9585, 0.9474, 0.9577, 0.9647, 0.9647, and 0.9677. The results of this analysis are reported in Table 10.

Table 10. ARL, SDRL. Scenario b11 (with identification), $p = 3$.

	ANN	SVR	RFR	ANN	SVR	RFR
(μ_1, μ_2, μ_3)	\multicolumn{3}{c}{$\sigma_1^2, \sigma_2^2, \sigma_3^2, Cov = 1, 1, 1, 0.5$}			$\sigma_1^2, \sigma_2^2, \sigma_3^2, Cov = 1.05, 1, 1, 0.5$	
(0, 0, 0)	200, 200	200, 200	200, 200	173.2, 170	200, 197	180.7, 171.2
(0.1, 0, 0.1)	162.7, 158.5	191.7, 180.4	193.2, 182.2	140.1, 139.8	184.4, 186.3	169.1, 178.3
(0.1, 0.3, 0.3)	65.37, 63.74	112.4, 113.3	140.2, 136.4	62.19, 58.74	104.3, 106.4	137.1, 134.9
(0.3, 0, 0)	70, 68.85	112.1, 108.8	145.3, 147.1	62.41, 62.24	105.7, 102.8	136.1, 129.4
(0.5, 0.8, 0.5)	8.79, 8.26	39.13, 40.1	67.1, 65.66	8.88, 8.04	39.58, 38.35	64.91, 66.47
(1, 1, 1)	2.45, 1.88	41.89, 39.86	85.13, 79.04	2.41, 1.79	40.68, 41.82	81.15, 81.58
(2, 2, 2)	1, 0	>10,000	>10,000	1, 0	>10,000	>10,000
(μ_1, μ_2, μ_3)	\multicolumn{3}{c}{$\sigma_1^2, \sigma_2^2, \sigma_3^2, Cov = 1.05, 1.05, 1.05, 0.5$}			$\sigma_1^2, \sigma_2^2, \sigma_3^2, Cov = 1.05, 1.3, 1.05, 0.5$	
(0, 0, 0)	119.6, 120.1	166.9, 173.5	185.3, 174.1	58.1, 56.94	117, 121.3	122, 123.7
(0.1, 0, 0.1)	111.3, 112	161.5, 155.2	152, 144.9	53.26, 53.94	113.3, 114.5	114.6, 112.5
(0.1, 0.3, 0.3)	54.7, 54.6	98.81, 95.89	113.1, 106.3	34.07, 34.35	79.56, 82.71	98.41, 96.06
(0.3, 0, 0)	53.08, 51.83	96.81, 100.2	117.3, 121.1	32.76, 32.4	83.34, 78.16	106.1, 108
(0.5, 0.8, 0.5)	8.64, 7.86	39.58, 37.68	64.28, 62.58	6.88, 6.14	41.4, 40	70.57, 68.66
(1, 1, 1)	2.43, 1.86	41.77, 42.75	89.83, 86.32	2.27, 1.63	47.5, 45.62	99.91, 96.18
(2, 2, 2)	1, 0.04	>10,000	>10,000	1, 0.03	>10,000	>10,000
(μ_1, μ_2, μ_3)	\multicolumn{3}{c}{$\sigma_1^2, \sigma_2^2, \sigma_3^2, Cov = 1.2, 1.2, 1.2, 0.6$}			$\sigma_1^2, \sigma_2^2, \sigma_3^2, Cov = 1.4, 1.4, 1.05, 0.5$	
(0, 0, 0)	54.77, 54.89	115.1, 112	126.4, 129.6	17.75, 16.96	71.26, 68.07	74.38, 82.69
(0.1, 0, 0.1)	51.34, 48.68	108.7, 106.1	113.1, 105.3	17.79, 17.88	70.62, 71.74	79.58, 83.47
(0.1, 0.3, 0.3)	32.2, 32.72	81.05, 76.15	90.85, 86.53	13.11, 13.1	59.12, 58.77	71.82, 68.78
(0.3, 0, 0)	31.45, 30.61	77.98, 76.26	94.03, 86.58	12.84, 12.08	61.83, 58.73	80.5, 82.38
(0.5, 0.8, 0.5)	6.52, 5.72	40.93, 40.51	70.07, 71.17	4.77, 3.92	50.43, 49.86	76.7, 73.9
(1, 1, 1)	2.21, 1.58	47.82, 46.98	105.9, 98.5	1.96, 1.33	60.61, 61.3	138.3, 146.4
(2, 2, 2)	1, 0	>10,000	>10,000	1, 0.03	>10,000	>10,000
(μ_1, μ_2, μ_3)	\multicolumn{3}{c}{$\sigma_1^2, \sigma_2^2, \sigma_3^2, Cov = 1.4, 1.4, 1.05, 0.75$}			$\sigma_1^2, \sigma_2^2, \sigma_3^2, Cov = 3, 3, 1.4, 0.5$	
(0, 0, 0)	72.1, 76.37	130.7, 130	135.9, 136.5	1.24, 0.58	132.5, 133.2	66.75, 65.52
(0.1, 0, 0.1)	66.25, 67.53	129.3, 128.7	126.2, 122.1	1.25, 0.57	140.8, 132.4	72.06, 71.05
(0.1, 0.3, 0.3)	38.7, 39.57	90.35, 95.11	101.9, 105.1	1.22, 0.52	174.8, 182.1	80.7, 78.22
(0.3, 0, 0)	35.49, 36.39	84.25, 83.67	103.4, 106.2	1.2, 0.47	141.8, 138.7	72.15, 73.2
(0.5, 0.8, 0.5)	7.22, 6.9	43.91, 44.9	72.57, 70.33	1.16, 0.45	320.2, 319.3	167.3, 171.3
(1, 1, 1)	2.24, 1.62	43.59, 42.46	106.5, 100.2	1.06, 0.26	953.8, 947.7	944.5, 964.4
(2, 2, 2)	1, 0.05	>10,000	>10,000	1, 0	>10,000	>10,000

According to the results in Table 10, the ANN chart performs better than the others. By comparing Tables 2 and 10, one can conclude that when the identification is a goal as well, all the schemes perform worse. Also, both the SVR and RFR schemes experience significant deterioration in performance as the shift size becomes larger than the trained value in this scenario. By comparing Tables 9 and 10 ($p = 2$ and $p = 3$ cases), we realize that the ANN scheme mostly performs worse in higher dimensions and the SVR scheme mostly performs better (except under no variation shift). Regarding the RFR scheme, only under small mean shifts and very large variation shifts, it mostly performs better in higher dimensions.

Moreover, by comparing the results in Table 10 to Sabahno et al.'s [9] Table 5, we can see that the number of cases in which at least one of our control charts performs better than all their charts is almost equal to the cases that at least one of their proposed charts performs better than ours. Once again, their charts are designed for detection, and in this scenario, we are looking for both detection and identification; therefore, the performance deterioration is normal.

Scenario b12 (Control Type b, Input Set 1, Training Method 2)

In this scenario, the ML structures are trained with both small and large shift sizes. Here again, 1000 in-control data are generated to be used in all control charts. Regarding the out-of-control dataset, we generate 500 out-of-control data with 0.2 shifts and 500 out-of-control data with 1 shift in μ_1 and use it for the control chart that monitors μ_1; 500 out-of-control data with 0.2 shifts in μ_2 and 500 out-of-control data with 1 shift in μ_2 and use it for the control chart that monitors μ_2; 500 out-of-control data with shifts 1.2 times that of the in-control σ_1^2 and 500 out-of-control data with shifts 2 times that of the in-control σ_1^2 and use it for the control chart that monitors σ_1^2; 500 out-of-control data with shifts 1.2 times that of the in-control σ_2^2 and 500 out-of-control data with shifts 2 times that of the in-control σ_2^2 and use it for the control chart that monitors σ_2^2; and finally, 500 out-of-control data with shifts 1.2 times that of the in-control σ_{12} and 500 out-of-control data with shifts 2 times that of the in-control σ_{12} and use it for the control chart that monitors σ_{12}. As five datasets are considered, five different ML structures for each scheme are required.

For the ANN scheme, for μ_1, μ_2, σ_1^2, σ_2^2 and σ_{12}, we use five nodes in all the hidden layers. The RMSE of these ANN structures, respectively, are 0.42, 0.42, 0.46, 0.46, and 0.43. Finally, the UCLs are computed as 1.1086, 0.9762, 0.9519, 0.985, and 1.007. For the SVR scheme, we use the radial kernel for all the structures. The RMSEs and UCLs are as follows. The RMSEs are 0.45, 0.44, 0.49, 0.49, and 0.46, and the UCLs are 1.0793, 1.1018, 1.0266, 1.0194, and 1.0103. Regarding the RFR scheme, 100 trees worked well for each structure. The RMSEs and UCLs for the RFR charts are as follows. The RMSEs are 0.45, 0.45, 0.49, 0.49, and 0.45, and the UCLs are 0.9998, 0.9998, 0.9937, 0.9961, and 0.999.

The results in Table 11 show that in most cases, the ANN scheme performs better than the other charts. However, under no variation shift, and small variation shifts when the mean shift is also small, the RFR scheme performs better. By comparing the results of this table with its equivalent scenario when only detection was the goal (comparing Tables 4 and 11), one can see the performance deterioration in all the schemes when identification is added to the goal as well. However, again same as in the previous case (Scenario b11), deterioration is at its lowest for the ANN chart. In addition, by comparing the results of two training methods (Scenario b11/Table 9 and Scenario b12/Table 11), one can see that the SVR and RFR schemes benefit from the second training method, with the RFR benefiting the most so that we do not see any deterioration in performance under shifts larger than the trained value. The ANN scheme, however, experiences a slight deterioration in performance in most cases.

Moreover, by comparing the results in Table 11 to Sabahno et al.'s [9] Table 4, we realize that although more cases are added to the ones in which at least one of their proposed control charts performs better, we can still see that the cases in which at least one of the proposed ML schemes performs better than all their proposed control charts are more, even though their control charts are only designed for detection.

In the case of three quality characteristics, the construction of the dataset is similar to the $p = 2$ case, with the only difference being that since two more process parameters (μ_3 and σ_3^2)/charts are added in the case of $p = 3$, 500 out-of-control data with 0.2 shifts and 500 out-of-control data with 1 shift in μ_3 are generated to be used for the control chart that monitors μ_3, and 500 out-of-control data with shifts 1.2 times that of the in-control σ_3^2 and 500 out-of-control data with shifts 2 times that of the in-control σ_3^2 are considered to be used for the control chart that monitors σ_3^2.

For the ANN control charts to monitor μ_1, μ_2, μ_3, σ_1^2, σ_2^2, σ_3^2, and the covariance, five nodes are utilized in each hidden layer, with RMSEs of 0.38, 0.38, 0.38, 0.45, 0.44, 0.44, and 0.27 as well as UCLs of 0.913, 0.9082, 0.96518, 0.9581, 0.9339, 0.9797, and 0.9951. For the SVR control charts, the used kernels are all radial. The RMSEs and UCLs are as follows. The RMSEs are 0.39, 0.39, 0.39, 0.48, 0.48, 0.48, and 0.29, and the UCLs are 1.0722, 1.0839, 1.1073, 1.0892, 1.0128, 1.0823, and 1.0281. Regarding the RFR control charts, the least numbers of trees that we had to use for each structure (to obtain the overall ARL of 200) are 100, 100, 100, 500, 300, 300, and 100. The RMSEs are 0.4, 0.4, 0.4, 0.45, 0.46, 0.46, and 0.29, and the UCLs are 0.7026, 0.7052, 0.7254, 0.3872, 0.4453, 0.4077, and 0.8023. Note that for the RFR structures in this scenario, for the first time, we had to activate a feature that the 'randomForest' package offers, and it is called 'corr.bias', which performs bias correction for the regression model. Without that feature being activated, we were not able to obtain the desired performance, no matter how many trees we tried.

Table 11. ARL, SDRL. Scenario b12 (with identification), $p = 2$.

	ANN	SVR	RFR	ANN	SVR	RFR
(μ_1, μ_2)	$\sigma_1^2, \sigma_2^2, \sigma_{12} = 1, 1, 0.5$			$\sigma_1^2, \sigma_2^2, \sigma_{12} = 1.05, 1, 0.5$		
(0, 0)	200, 200	200, 200	200, 200	167.4, 170.5	193.6, 186.8	172.1, 177.2
(0.1, 0)	171.1, 171.3	181.2, 203.5	166.5, 145.6	139, 139.6	167.5, 170.9	157.8, 164
(0.1, 0.3)	79.9, 78.78	99.75, 105.6	63.04, 65.82	69.15, 70.55	83.67, 75.27	70.87, 68.67
(0.3, 0)	62.84, 59.85	75.31, 76.62	50.5, 52.81	53.81, 54.14	75.63, 74.51	44.89, 40.81
(0.5, 0.8)	5.51, 5.04	30.7, 35.05	5.29, 4.51	5.57, 5.1	24.71, 23	5.65, 4.97
(1, 1)	1.92, 1.27	31.48, 30.6	2.01, 1.36	1.89, 1.24	28.24, 25.56	2.01, 1.29
(2, 2)	1, 0.03	>10,000	1.22, 0.5414	1, 0.03	>10,000	1.26, 0.6
(μ_1, μ_2)	$\sigma_1^2, \sigma_2^2, \sigma_{12} = 1.05, 1.05, 0.5$			$\sigma_1^2, \sigma_2^2, \sigma_{12} = 1.05, 1.3, 0.5$		
(0, 0)	133.9, 131.2	157.5, 144.2	127.9, 123.6	52.48, 51.18	109.5, 115	55.23, 53.85
(0.1, 0)	116.3, 115.4	144.6, 145.5	119.3, 116.9	47.29, 46.04	102.5, 108.9	54.11, 51.69
(0.1, 0.3)	54.95, 57.15	83.44, 89.2	56.89, 56.27	26.13, 26.2	66.23, 71.56	29, 01, 30.95
(0.3, 0)	45.51, 45	74.85, 71.91	44.08, 41.1	23.33, 22.87	56.56, 56.35	28.85, 28.8
(0.5, 0.8)	5.32, 4.73	23.63, 21.63	4.93, 4.2	4.29, 3.84	18.89, 20	5.05, 4.82
(1, 1)	1.83, 1.22	27.98, 25.33	1.98, 1.28	1.73, 1.09	22.17, 19.72	2.07, 1.56
(2, 2)	1, 0	>10,000	1.26, 0.51	1, 0	>10,000	1.34, 0.71
(μ_1, μ_2)	$\sigma_1^2, \sigma_2^2, \sigma_{12} = 1.2, 1.2, 0.6$			$\sigma_1^2, \sigma_2^2, \sigma_{12} = 1.4, 1.4, 0.5$		
(0, 0)	60.34, 59.04	124.5, 132.4	75.91, 77.46	14.94, 14.66	58.04, 54.73	25.81, 25.53
(0.1, 0)	55.62, 53.89	107.8, 98.68	65.75, 63.07	14.1, 14.42	53.29, 48.72	21.36, 20.77
(0.1, 0.3)	31.88, 31.27	62.41, 56.23	34.39, 33.71	10.38, 9.79	36.14, 35.83	15.69, 15.49
(0.3, 0)	26.52, 25.58	55.24, 51.44	33.01, 31.47	9.14, 8.97	34.89, 34.84	14.04, 13.73
(0.5, 0.8)	4.41, 3.95	21.47, 22.07	5.49, 4.335	3.03, 2.45	15.77, 14.66	4.94, 4.5
(1, 1)	1.76, 1.18	24.05, 24.97	2.04, 1.35	1.55, 0.92	21.01, 21.06	2.38, 1.64
(2, 2)	1, 0.03	>10,000	1.29, 0.6	1, 0	3684.07, 2428.36	1.61, 0.98
(μ_1, μ_2)	$\sigma_1^2, \sigma_2^2, \sigma_{12} = 1.4, 1.4, 0.75$			$\sigma_1^2, \sigma_2^2, \sigma_{12} = 3, 3, 0.5$		
(0, 0)	27.24, 26.27	79.79, 76.59	35.43, 35.04	1.5, 0.85	64.87, 66.76	14.6, 14.12
(0.1, 0)	24.84, 23.87	70.3, 71.4	33.03, 32.2	1.49, 0.85	63.42, 63.09	14.04, 13.35
(0.1, 0.3)	16.94, 16.28	45.1, 47.75	21.66, 21.23	1.44, 0.8	62.29, 61.14	12.93, 11.99
(0.3, 0)	14.64, 13.86	42.77, 41.06	19.98, 19.13	1.46, 0.81	60.07, 59.96	12.91, 12.09
(0.5, 0.8)	3.5, 2.93	19.98, 19.23	4.85, 4.39	1.34, 0.64	55.18, 50.77	11.38, 10.41
(1, 1)	1.71, 1.12	22.8, 21.87	2.25, 1.657	1.16, 0.45	79.59, 74.07	11.11, 10.37
(2, 2)	1, 0.03	4761.09, 5179.76	1.4, 0.73	1, 0.04	4256.1, 4516.53	11.47, 10.73

The result of this analysis is reported in Table 12. This table shows that in this case, the RFR scheme performs better than the other two in all the shift cases. By comparing this case with the first training method (Tables 10 and 12), we realize that the RFR scheme performs better with the second training method, and on the contrary, the ANN and SVR schemes mostly perform worse. Comparing the $p = 3$ and $p = 2$ cases (Tables 11 and 12), the RFR scheme mostly performs better in higher dimensions. On the other hand, the ANN and SVR charts mostly perform worse in higher dimensions (however, under very large mean shifts, the SVR scheme performs better). The highest deterioration in performance belongs to the ANN scheme. By comparing two process control scenarios (Tables 5 and 12), one can see that all the schemes experience deterioration in performance, more so for the SVR scheme, as unlike the previous control case, it experiences a significant deterioration in performance as the shift size becomes larger than the trained value.

Table 12. ARL, SDRL. Scenario b12 (with identification), $p = 3$.

	ANN	SVR	RFR	ANN	SVR	RFR
(μ_1, μ_2, μ_3)	\multicolumn{3}{c}{$\sigma_1^2, \sigma_2^2, \sigma_3^2, Cov = 1, 1, 1, 0.5$}			$\sigma_1^2, \sigma_2^2, \sigma_3^2, Cov = 1.05, 1, 1, 0.5$		
(0, 0, 0)	200, 200	200, 200	200, 200	231.6, 237.3	222.7, 234.9	168.8, 170.3
(0.1, 0, 0.1)	190.1, 196.3	191.7, 190.8	160.7, 155	216.5, 214.1	212.96, 212.47	147.2, 157.1
(0.1, 0.3, 0.3)	145.1, 157.2	143.6, 145.7	53.17, 54.18	171, 171.5	153.5, 157.2	47.92, 49.33
(0.3, 0, 0)	155.2, 150.9	143.7, 148.2	52.73, 53.01	152.1, 143.3	157.2, 155.2	49.32, 51.13
(0.5, 0.8, 0.5)	18.42, 17.86	72.88, 72.84	6.11, 5.824	18.61, 18.54	80.7, 75.61	5.48, 4.97
(1, 1, 1)	3.56, 2.95	58.57, 58.1	1.97, 1.44	3.58, 3.03	61.34, 61.48	1.87, 1.27
(2, 2, 2)	1.11, 0.37	1494, 1397	1.2, 0.5	1.11, 0.34	1504, 1629	1.2, 0.4738
(μ_1, μ_2, μ_3)	\multicolumn{3}{c}{$\sigma_1^2, \sigma_2^2, \sigma_3^2, Cov = 1.05, 1.05, 1.05, 0.5$}			$\sigma_1^2, \sigma_2^2, \sigma_3^2, Cov = 1.05, 1.3, 1.05, 0.5$		
(0, 0, 0)	272, 274.4	277.6, 294.1	127.3, 127.9	389.4, 380.8	350.1, 358.3	57.92, 54.64
(0.1, 0, 0.1)	275.7, 287.7	268.9, 261.7	99.14, 94.62	361.7, 363.8	335.5, 331.6	54.21, 53.41
(0.1, 0.3, 0.3)	176.8, 173	184.2, 192.1	39.81, 39.34	191.4, 205.4	250.8, 270	27.95, 26.8
(0.3, 0, 0)	171.5, 174.2	172.6, 163.8	41.3, 39.46	224.7, 198.1	263.7, 257.4	30.73, 29.67
(0.5, 0.8, 0.5)	19.84, 19.12	88.31, 84.7	5.93, 4.911	23.47, 22.21	103.5, 104	5.5, 5.17
(1, 1, 1)	4.18, 3.65	62.5, 60.63	1.94, 1.35	4.89, 4.44	76.9, 73.98	1.93, 1.48
(2, 2, 2)	1.2, 0.5	1453.11, 1538	1.18, 0.45	1.32, 0.61	1120.58, 1064.46	1.28, 0.59
(μ_1, μ_2, μ_3)	\multicolumn{3}{c}{$\sigma_1^2, \sigma_2^2, \sigma_3^2, Cov = 1.2, 1.2, 1.2, 0.6$}			$\sigma_1^2, \sigma_2^2, \sigma_3^2, Cov = 1.4, 1.4, 1.05, 0.5$		
(0, 0, 0)	427.05, 391.7	382.5, 392.9	62.2, 68.76	522.8, 501.9	602.7, 5, 583	22.73, 20.64
(0.1, 0, 0.1)	400.4, 414.5	333, 328.5	54.46, 54.61	426.9, 392.3	496.7, 505.2	21.92, 21.14
(0.1, 0.3, 0.3)	215.4, 216.1	228.3, 233.4	27.86, 26.09	219.6, 217.7	354.7, 347.3	15.05, 14.52
(0.3, 0, 0)	234.6, 226.5	263.7, 278.7	27.52, 26.72	173.5, 167	384.7, 382.9	14.34, 13.9
(0.5, 0.8, 0.5)	22.8, 22.6	116.9, 126.7	5.2, 4.68	29.7, 28.48	139, 149.3	4.61, 4.04
(1, 1, 1)	5.22, 5.02	66.91, 65.02	2.02, 1.35	8.44, 7.63	68.62, 67.18	2.21, 1.58
(2, 2, 2)	1.33, 0.65	932.7, 871.2	1.28, 0.66	1.73, 1.12	974.3, 982.5	1.32, 0.6323
(μ_1, μ_2, μ_3)	\multicolumn{3}{c}{$\sigma_1^2, \sigma_2^2, \sigma_3^2, Cov$ 1.4, 1.4, 1.05, 0.75}			$\sigma_1^2, \sigma_2^2, \sigma_3^2, Cov$ 3, 3, 1.4, 0.5		
(0, 0, 0)	286.3, 291.4	285.6, 255.2	68.35, 64.36	63.12, 60.09	458.3, 432.8	2.47, 1.84
(0.1, 0, 0.1)	286.4, 284	271.5, 280	64.79, 74.13	63.15, 61.07	427.3, 411.2	2.5, 2.01
(0.1, 0.3, 0.3)	159.1, 150.6	189.4, 188.1	30.58, 30.93	55.69, 53.68	344, 340.6	2.51, 2.01
(0.3, 0, 0)	158.1, 165	219.4, 192.7	30.31, 31.45	50.76, 48.21	337.1, 351.6	2.45, 1.94
(0.5, 0.8, 0.5)	18.46, 16.9	106.8, 106.7	5.12, 4.67	29.64, 29.25	203.3, 191.9	1.97, 1.41
(1, 1, 1)	4.3, 4.01	70.44, 77.17	2.05, 1.33	19.25, 18.14	108, 110.8	1.48, 0.85
(2, 2, 2)	1.23, 0.51	803.2, 715.9	1.2, 0.49	13.82, 12.84	1979.95, 2028.43	1.13, 0.36

Moreover, by comparing the results in Table 12 to Sabahno et al.'s [9] Table 5, we can see that in the cases of large mean shifts and/or large variation shifts, at least one of their proposed charts performs better than all our proposed charts. Otherwise, at least one of our charts performs better than all their charts.

3.2.2. Scenario b2 (Control Type b, Input Set 2)

The individual elements of each sample's mean vector and variance–covariance matrix are considered as the inputs in this scenario.

Scenario b21 (Control Type b, Input Set 2, Training Method 1)

The first training method is used in this scenario. The construction of in and out-of-control datasets are the same as in Scenario b11. For the ANN scheme, 11 nodes in each hidden layer are used. The obtained RMSEs are 0.45, 0.43, 0.46, 0.46, and 0.45. Also, the computed UCLs are 0.9644, 1.1054, 1.3556, 1.1599, and 1.177. For the SVR scheme, we use linear kernels in all the structures in this scenario. The RMSEs are 0.49, 0.47, 0.5, 0.51, and 0.5. The UCLs are 1.2462, 1.2957, 1.3706, 1.3492, and 1.1669. For the RFR scheme, we use 100 trees for each structure. The RMSEs are 0.49. 0.48, 0.49, 0.49, and 0.5. The UCLs are 0.8489, 0.9042, 0.88, 0.8534, and 0.8459. The result of this scenario is reported in Table 13.

Table 13. ARL, SDRL. Scenario b21 (with identification), $p = 2$.

	ANN	SVR	RFR	ANN	SVR	RFR
(μ_1, μ_2)		$\sigma_1^2, \sigma_2^2, \sigma_{12} = 1, 1, 0.5$			$\sigma_1^2, \sigma_2^2, \sigma_{12} = 1.05, 1, 0.5$	
(0, 0)	200, 200	200, 200	200, 200	176.7, 177.2	141.8, 144.9	173.2, 178.9
(0.1, 0)	179.64, 185.25	165.69, 162.43	180.9, 184.9	160.4, 164.7	118.7, 121.7	172.2, 171.5
(0.1, 0.3)	105.6, 109.4	55.46, 54.82	124.8, 125.5	111.2, 112.3	51.25, 50.59	129.1, 127.6
(0.3, 0)	141, 140.7	40.08, 38.88	153.3, 148.8	126, 123.9	32.26, 31.24	143, 142.8
(0.5, 0.8)	52.1, 51.68	9.35, 8.8	117.3, 118.2	52.31, 52.34	8.49, 7.61	118.2, 111.6
(1, 1)	57.59, 58.31	5.07, 4.43	403.6, 369.5	52.29, 50.81	4.74, 4.28	382.2, 413.6
(2, 2)	48.33, 47.02	1.03, 0.19	>10,000	43.19, 41.41	1.04, 0.22	>10,000
(μ_1, μ_2)		$\sigma_1^2, \sigma_2^2, \sigma_{12} = 1.05, 1.05, 0.5$			$\sigma_1^2, \sigma_2^2, \sigma_{12} = 1.05, 1.3, 0.5$	
(0, 0)	169.6, 167.2	111.5, 114	160.3, 155.6	99.24, 98.47	30.91, 31.46	111.5, 107.7
(0.1, 0)	144.2, 142.4	98.1, 100.21	157, 157.8	85.06, 84.01	33.06, 31.67	116, 125
(0.1, 0.3)	117.3, 116.6	42.04, 39.82	121.8, 125.3	123.7, 127.8	16.58, 17.14	109.6, 106.4
(0.3, 0)	106.5, 110.8	31.94, 29.97	145.3, 143.2	56.9, 56.08	20.03, 18.95	117, 119.1
(0.5, 0.8)	54.98, 51.53	7.92, 7.35	125.4, 126.1	73.16, 72.22	6.06, 5.25	134, 136.4
(1, 1)	50.73, 49.55	4.52, 3.93	370.3, 353.2	55.04, 56.33	3.83, 3.26	318.4, 337.9
(2, 2)	41.48, 43.25	1.03, 0.19	>10,000	45.47, 43.74	1.04, 0.21	>10,000
(μ_1, μ_2)		$\sigma_1^2, \sigma_2^2, \sigma_{12} = 1.2, 1.2, 0.6$			$\sigma_1^2, \sigma_2^2, \sigma_{12} = 1.4, 1.4, 0.5$	
(0, 0)	138.6, 134.6	54.05, 54.47	132, 131.3	38.51, 36.3	8.93, 9.49	73.89, 75.73
(0.1, 0)	127.4, 131	50.91, 49.72	125.2, 117.8	32.91, 32.83	8.87, 8.62	72.35, 75.25
(0.1, 0.3)	122, 118.4	25.85, 24.53	108.3, 104.1	54.57, 52.75	6.55, 6.16	83.54, 83.87
(0.3, 0)	83.93, 82.46	21.56, 19.51	131.3, 130	24.62, 24.27	6.96, 6.24	82.86, 79.47
(0.5, 0.8)	60.66, 62.22	7.25, 6.57	123.7, 128.5	46.98, 45.42	3.75, 3.27	151.9, 146.2
(1, 1)	52.75, 54.07	3.99, 3.56	342.6, 346.2	32.56, 31.69	2.57, 2.02	364.4, 311.5
(2, 2)	45.81, 48.53	1.05, 0.24	>10,000	35.74, 34.2	1.03, 0.2	>10,000
(μ_1, μ_2)		$\sigma_1^2, \sigma_2^2, \sigma_{12} = 1.4, 1.4, 0.75$			$\sigma_1^2, \sigma_2^2, \sigma_{12} = 3, 3, 0.5$	
(0, 0)	112.3, 112.9	24.84, 23.7	118.8, 120.6	2.53, 2.05	1.16, 0.43	58.39, 58.91
(0.1, 0)	90.95, 95.76	24.12, 23.53	114.7, 114.9	2.53, 1.98	1.13, 0.38	62.33, 63.42
(0.1, 0.3)	121.1, 119.5	15.19, 15.28	95.5, 89.7	2.8, 2.16	1.15, 0.43	72.95, 70.41
(0.3, 0)	67.03, 68.44	13.98, 13.48	122.5, 120.3	2.55, 1.97	1.13, 0.36	66.19, 65.73
(0.5, 0.8)	68.94. 66.51	5.32, 4.85	113.6, 110.5	3.36, 2.78	1.11, 0.36	224.6, 232.5
(1, 1)	55.07, 52.25	3.34, 2.84	260.3, 259.9	3.44, 2.74	1.07, 0.29	679.6, 665.7
(2, 2)	54.17, 54.49	1.05, 0.23	>10,000	6.23, 5.54	1, 0.06	>10,000

It is clear from the results in Table 13 that in all cases, the SVR scheme performs better than the other two in this scenario. By comparing Tables 7 and 13 (to its equivalent scenario with only detection as the goal), we realize that the ANN scheme mostly performs worse in this scenario. The RFR scheme mostly performs worse as well, but the deterioration is more severe, especially as the mean shifts increase. The SVR scheme mostly performs worse too, but the deterioration in performance in most cases is not that noticeable compared to the ANN and RFR schemes. By comparing Tables 9 and 13 (two different input methods, but the same training methods), we can see that the ANN scheme performs significantly worse with the second input set. On the contrary, the SVR scheme performs significantly better with this input set so that it does not experience any deterioration in performance under large shifts. Regarding the RFR scheme, it performs better under small mean shifts, but as the mean shift increases, its performance becomes significantly worse.

Moreover, by comparing the results in Table 13 to Sabahno et al.'s [9] Table 4, one can see that the situation is improved compared to the previous scenario, and in most cases (except in the case of (0.5, 0.8) mean shifts), at least one of the proposed ML control charts performs better than all their proposed control charts again, even though their control charts are only designed for detection.

Scenario b22 (Control Type b, Input Set 2, Training Method 2)

The construction of in-control and out-of-control datasets is the same as in Scenario b12. For the ANN scheme, 10 nodes in each hidden layer are used. The obtained RMSEs are 0.38, 0.36, 0.41, 0.42, and 0.38. The UCLs are computed as 0.9637, 1.2445, 1.7237, 1.034, and 1.061. For the SVR scheme, again linear kernels are used in all the structures, with the RMSEs equal to 0.41, 0.4, 0.47, 0.47, and 0.43 and the UCLs computed as 0.8145, 0.8271, 0.9899, 0.9878, and 0.9457. Regarding the RFR scheme, 100 trees for each ML structure are used. The obtained RMSEs are 0.42, 0.4, 0.45, 0.45, and 0.43. The computed UCLs are 0.9401, 0.9528, 0.9458, 0.9284, and 0.894. The results of this analysis are reported in Table 14.

Table 14. ARL, SDRL. Scenario b22 (with identification), $p = 2$.

	ANN	SVR	RFR	ANN	SVR	RFR
(μ_1, μ_2)		$\sigma_1^2, \sigma_2^2, \sigma_{12} = 1, 1, 0.5$			$\sigma_1^2, \sigma_2^2, \sigma_{12} = 1.05, 1, 0.5$	
(0, 0)	200, 200	200, 200	200, 200	178.1, 179.7	148.2, 152	153.3, 149.7
(0.1, 0)	153.9, 151.9	155.4, 152	148, 142	132.5, 132	117.4, 121.1	129.3, 133.3
(0.1, 0.3)	108.7, 107.5	54.17, 56.74	69.72, 70.86	118.4, 120.6	44.3, 44.38	65.13, 63.27
(0.3, 0)	44.82, 42.93	40.26, 39.92	47.45, 47.45	38.06, 36.72	32.54, 32.8	44.68, 47.01
(0.5, 0.8)	28.11, 29.32	8.69, 7.84	23.04, 21.48	30.53, 29.86	8.05, 7.17	22.21, 21.46
(1, 1)	16.27, 15.87	4.88, 4.32	60.94, 56.68	16.11, 14.79	4.57, 4.01	54.22, 54.52
(2, 2)	8.55, 8.31	1.02, 0.15	>10,000	9.41, 8.35	1.03, 0.18	>10,000
(μ_1, μ_2)		$\sigma_1^2, \sigma_2^2, \sigma_{12} = 1.05, 1.05, 0.5$			$\sigma_1^2, \sigma_2^2, \sigma_{12} = 1.05, 1.3, 0.5$	
(0, 0)	160.7, 165.5	117.5, 119.7	140.7, 146.2	68.12, 68.09	31.45, 31.49	65.31, 66.4
(0.1, 0)	115.2, 113.9	97.86, 100.5	110, 116.6	54.93, 53.79	30.67, 30.35	65.18, 64
(0.1, 0.3)	103.9, 106.1	36.53, 35.09	63.48, 61.96	65.09, 65.87	16.59, 15.88	44.09, 42.66
(0.3, 0)	35.67, 35.27	32.35, 31.18	42.7, 39.96	26.15, 26.13	20.27, 19.6	34.88, 33.61
(0.5, 0.8)	26.22, 26	7.21, 6.7	21.78, 21.91	16.03, 15.26	5.81, 5.3	22.63, 22.88
(1, 1)	14.59, 14.39	4.46, 3.8	54.27, 54.18	8.78, 7.95	3.76, 3.33	41.85, 42.24
(2, 2)	9.35, 8.75	1.04, 0.2	>10,000	8.85, 8.08	1.04, 0.21	>10,000
(μ_1, μ_2)		$\sigma_1^2, \sigma_2^2, \sigma_{12} = 1.2, 1.2, 0.6$			$\sigma_1^2, \sigma_2^2, \sigma_{12} = 1.4, 1.4, 0.5$	
(0, 0)	121.5, 126.7	55.63, 54.19	94.1, 97.06	26.39, 25.6	8.66, 8.15	28.33, 27.67
(0.1, 0)	84.17, 84.25	49.78, 50.41	75.83, 76.77	22.12, 21.59	8.36, 7.67	27.04, 25.97
(0.1, 0.3)	93.44, 92.86	24.24, 23.77	55.67, 56.03	33.73, 34.4	6.8, 6.24	26.5, 26.05
(0.3, 0)	28.67, 28.06	22.83, 22.92	34.83, 34.89	12.12, 11.71	7.4, 7.06	21.43, 20.75
(0.5, 0.8)	21.62, 20.51	6.47, 5.76	22.83, 22.38	15.73, 14.94	3.42, 2.81	21.8, 20.96
(1, 1)	11.13, 10.06	4, 3.5	48.59, 44.41	7.24, 6.82	2.66, 2.1	40.76, 38.22
(2, 2)	7.87, 7.28	1.04, 0.2	>10,000	8.43, 7.65	1.02, 0.16	>10,000
(μ_1, μ_2)		$\sigma_1^2, \sigma_2^2, \sigma_{12} = 1.4, 1.4, 0.75$			$\sigma_1^2, \sigma_2^2, \sigma_{12} = 3, 3, 0.5$	
(0, 0)	79.65, 74.89	26.41, 74.89	26.41, 26.34	2.99, 2.39	1.13, 0.37	12.99, 12.41
(0.1, 0)	55.06, 55.22	26.12, 25.51	64.21, 59.84	2.74, 2.13	1.15, 0.41	13.86, 13.97
(0.1, 0.3)	68.62, 71.7	15.83, 15.82	46.4, 44.21	3.19, 2.55	1.12, 0.35	14.38, 13.47
(0.3, 0)	22.16, 21.4	14.23, 13.54	33.38, 31.51	2.46, 1.87	1.13, 0.4	16.09, 15.11
(0.5, 0.8)	15.89, 15.82	5.39, 4.8	25.09, 24.34	2.82, 2.27	1.09, 0.31	23.9, 23.52
(1, 1)	8.03, 7.9	3.44, 2.83	51.26, 51.36	2.35, 1.74	1.08, 0.3	61.49, 58.96
(2, 2)	5.75, 5.25	1.03, 0.19	>10,000	1.85, 1.3	1, 0.08	>10,000

According to the results in Table 14, in most cases, the SVR scheme performs better than the other two. By comparing Table 14 to Table 13 (two different training methods), we realize that the ANN and RFR schemes perform better with the second training method. On the other hand, on average, the performance of the SVR scheme is very similar in the two training methods. By comparing Table 14 to its equivalent in scenario a (which is Table 8), we can conclude that all the schemes mostly perform worse when the identification is a goal and when the second training method is used. Having said that, the RFR scheme is the most affected one, while the SVR scheme is the least affected. By comparing Table 14 to Table 11 (different input sets), we realize that the ANN scheme mostly performs worse with this input set. The RFR scheme mostly performs worse too, but it even experiences a significant deterioration in performance as the shifts become larger than the training value in this input set. The SVR scheme performs better with this input method such that there is no deterioration in performance anymore as the shift size becomes larger than the trained value (unlike the RFR scheme).

Moreover, by comparing the results in Table 14 to Sabahno et al.'s [9] Table 4, one can see that similar to the previous scenario, except in the case of (0.5, 0.8) mean shift, at least one of the proposed ML control charts performs better than all their proposed control charts.

4. An Illustrative Example

A real case originally discussed by Hawkins and Maboudou-Tchao [2] regarding a healthcare process for monitoring blood pressure and heart rate is used in this section to illustrate the application of the proposed ML schemes. The main indicators, in this case, are heart attack and stroke. The quality characteristics are x_1 = systolic blood pressure, x_2 = diastolic blood pressure, and x_3 = heart rate. They follow a multivariate normal distribution with the following in-control parameter values.

$$\mu_0 = (126.61, 77.48, 80.95)' \text{ and } \Sigma_0 = \begin{bmatrix} 15.04 & 8.66 & 10.51 \\ 8.66 & 5.83 & 5.56 \\ 10.51 & 5.56 & 15.17 \end{bmatrix}.$$

To identify the quality characteristic and the process parameter responsible for the chart signal, the second control scenario is used for this practical case. Based on the results of the numerical analyses section, the first training method for the ANN and SVR schemes and the second training method for the RFR scheme (because the results showed that the RFR scheme performs better with the second training method) are utilized. The mean shift size used for training is $0.2 \times \sigma$ (note that in the numerical analyses section, since all the standard deviations were equal to 1, we simply used $0.2 \times 1 = 0.2$) for small shifts, and similarly, it is $1 \times \sigma$ for large shifts. The shifted variances used for training are as they were in Section 3 (because in both cases, they are multiplied by a coefficient). In addition, we assume that detection of the covariance shift is not a priority for the quality system; therefore, we only consider six control charts for monitoring μ_1, μ_2, μ_3, σ_1^2, σ_2^2, and σ_3^2. The UCLs are computed using the proposed algorithm in Section 2 with $\alpha = 0.005$ and $n = 10$. Also, the same R packages as in the simulations study section are used in this section. Similarly, the only changes we carried out in those packages' default settings were changing the output type to regression and changing the number of trees in the RFR scheme, the number of nodes in the hidden layer in the ANN scheme, and the kernel type in the SVR scheme.

For the SVR scheme, radial kernels are used in all the structures. The RMSEs are 0.47, 0.47, 0.51, 0.44, 0.46, and 0.52. The UCLs 1.0714, 1.0792, 1.0735, 1.0744, 1.0033, and 1.0932. For the ANN scheme, five nodes are used in the hidden layers. The RMSEs are 0.45, 0.45, 0.49, 0.42, 0.44, and 0.49. The UCLs are 0.9934, 1.7028, 1.0451, 0.9771, 0.9736, and 0.8285. For the RFR scheme, each structure's number of trees is as follows: 100, 100, 100, 300, 100, and 500, respectively. The bias correction feature is only turned on for the last structure. The RMSEs are 0.4, 0.4, 0.43, 0.38, 0.41, and 0.47. The UCLs are 0.9858, 0.98, 0.9999, 0.9986, 0.9993, and 0.9914.

To see how each control chart would react in the case of an out-of-control situation, we make an artificial shift to the process and take ten consecutive samples from the process. To do so, we shift the third mean by 1.1 ($\mu_3 = 80.95 + 1.1$) and the third variance by 1.4 ($\sigma_3^2 = 1.4 \times 15.17$). It should be noted that we use the detection plus identification scenario; therefore, we have six control charts for each of the proposed ML schemes, and if any of those six charts signal in each scheme, we call the process out of control.

The results of ten consecutive random samplings from the process for each control scheme, i.e., SVR, ANN, and RFR, are reported in Tables 15–17, respectively. Note that, since there are six control charts involved in each scheme (making it a total of 18 control charts), to reduce the paper size, they are presented in tables.

Table 15. The SVM scheme for the illustrative example.

Sample		1	2	3	4	5	6	7	8	9	10
input	T^2	6.5128	15.7031	1.3442	1.3567	3.2905	1.2372	6.0249	10.778	4.049	14.8453
	W	7.2728	8.1192	6.4587	9.7029	5.818	6.3745	9.3089	7.5685	14.8656	16.7297
SVR1	output	0.8457	0.8835	0.1029	0.1004	0.0977	0.1025	0.9014	0.7506	0.1469	0.9076
	UCL	1.0714	1.0714	1.0714	1.0714	1.0714	1.0714	1.0714	1.0714	1.0714	1.0714
	status	In-control	In-control	In-control	In-control	In-control	In-control	In-control	In-control	In-control	In-control
SVR2	output	0.4584	0.411	0.0593	0.1037	0.0547	0.0609	0.9269	0.9038	0.4274	0.8055
	UCL	1.0792	1.0792	1.0792	1.0792	1.0792	1.0792	1.0792	1.0792	1.0792	1.0792
	status	In-control	In-control	In-control	In-control	In-control	In-control	In-control	In-control	In-control	In-control
SVR3	output	0.8544	0.2529	0.1407	0.2313	0.1673	0.137	0.8628	0.8551	0.1161	0.7355
	UCL	1.0735	1.0735	1.0735	1.0735	1.0735	1.0735	1.0735	1.0735	1.0735	1.0735
	status	In-control	In-control	In-control	In-control	In-control	In-control	In-control	In-control	In-control	In-control
SVR4	output	0.9032	0.9224	0.8991	0.0927	0.8993	0.8989	0.9874	0.9166	0.1002	0.4982
	UCL	1.0744	1.0744	1.0744	1.0744	1.0744	1.0744	1.0744	1.0744	1.0744	1.0744
	status	In-control	In-control	In-control	In-control	In-control	In-control	In-control	In-control	In-control	In-control
SVR5	output	0.9005	0.8972	0.9018	0.1064	0.9019	0.9049	0.9948	0.9173	0.0978	0.502
	UCL	1.0033	1.0033	1.0033	1.0033	1.0033	1.0033	1.0033	1.0033	1.0033	1.0033
	status	In-control	In-control	In-control	In-control	In-control	In-control	In-control	In-control	In-control	In-control
SVR6	output	0.8993	0.7554	0.8613	0.8612	0.3868	0.8588	0.9125	0.8946	0.0479	0.4052
	UCL	1.0932	1.0932	1.0932	1.0932	1.0932	1.0932	1.0932	1.0932	1.0932	1.0932
	status	In-control	In-control	In-control	In-control	In-control	In-control	In-control	In-control	In-control	In-control

Table 16. The ANN scheme for the illustrative example.

Sample			1	2	3	4	5	6	7	8	9	10
input	T^2		6.5128	15.7031	1.3442	1.3567	3.2905	1.2372	6.0249	10.778	4.049	14.8453
	W		7.2728	8.1192	6.4587	9.7029	5.818	6.3745	9.3089	7.5685	14.8656	16.7297
ANN1	output		0.5011	0.8582	0.2728	0.1333	0.1906	0.2616	0.4963	0.7294	0.3312	0.8917
	UCL		0.9934	0.9934	0.9934	0.9934	0.9934	0.9934	0.9934	0.9934	0.9934	0.9934
	status		In-control	In-control	In-control	In-control	In-control	In-control	In-control	In-control	In-control	In-control
ANN2	output		0.4603	0.8571	0.0971	0.0984	0.298	0.0894	0.4286	0.7975	0.3115	0.778
	UCL		1.7028	1.7028	1.7028	1.7028	1.7028	1.7028	1.7028	1.7028	1.7028	1.7028
	status		In-control	In-control	In-control	In-control	In-control	In-control	In-control	In-control	In-control	In-control
ANN3	output		0.5155	0.9792	0.1965	0.3162	0.2041	0.1703	0.4309	0.7775	0.3236	**1.1416**
	UCL		1.0451	1.0451	1.0451	1.0451	1.0451	1.0451	1.0451	1.0451	1.0451	**1.0451**
	status		In-control	In-control	In-control	In-control	In-control	In-control	In-control	In-control	In-control	**Out-of-control**
ANN4	output		0.7431	0.9343	0.6765	0.2614	0.8417	0.6917	0.4771	0.8653	0.0965	0.2469
	UCL		0.9771	0.9771	0.9771	0.9771	0.9771	0.9771	0.9771	0.9771	0.9771	0.9771
	status		In-control	In-control	In-control	In-control	In-control	In-control	In-control	In-control	In-control	In-control
ANN5	output		0.7848	0.9268	0.668	0.3124	0.8712	0.683	0.4604	0.8637	0.1478	0.535
	UCL		0.9736	0.9736	0.9736	0.9736	0.9736	0.9736	0.9736	0.9736	0.9736	0.9736
	status		In-control	In-control	In-control	In-control	In-control	In-control	In-control	In-control	In-control	In-control
ANN6	output		0.7737	**0.8694**	0.793	0.4344	0.7838	0.7943	0.6009	**0.8315**	0.3094	0.4433
	UCL		0.8285	**0.8285**	0.8285	0.8285	0.8285	0.8285	0.8285	**0.8285**	0.8285	0.8285
	status		In-control	**Out-of-control**	In-control	In-control	In-control	In-control	In-control	**Out-of-control**	In-control	In-control

Table 17. The RFR scheme for the illustrative example.

Sample		1	2	3	4	5	6	7	8	9	10
input	T^2	6.5128	15.7031	1.3442	1.3567	3.2905	1.2372	6.0249	10.778	4.049	14.8453
	W	7.2728	8.1192	6.4587	9.7029	5.818	6.3745	9.3089	7.5685	14.8656	16.7297
RFR1	output	0.1858	0.7008	0.0095	0.0045	0.412	0.02083	0.7746	0.9075	0.4241	**1**
	UCL	0.9858	0.9858	0.9858	0.9858	0.9858	0.9858	0.9858	0.9858	0.9858	**0.9858**
	status	In-control	In-control	In-control	In-control	In-control	In-control	In-control	In-control	In-control	**Out-of-control**
RFR2	output	0.1595	0.639	0.0875	0.0506	0.3076	0.1833	0.4221	0.7321	0.6366	0.9625
	UCL	0.98	0.98	0.98	0.98	0.98	0.98	0.98	0.98	0.98	0.98
	status	In-control	In-control	In-control	In-control	In-control	In-control	In-control	In-control	In-control	In-control
RFR3	output	0.2488	0.9231	0.1631	0.3348	0.1533	0.1613	0.8306	0.5681	0.126	**1**
	UCL	0.9999	0.9999	0.9999	0.9999	0.9999	0.9999	0.9999	0.9999	0.9999	**0.9999**
	status	In-control	In-control	In-control	In-control	In-control	In-control	In-control	In-control	In-control	**Out-of-control**
RFR4	output	0.7712	0.8366	0.7156	0.2556	0.7266	0.5866	0.7521	0.8556	0.4362	0.5173
	UCL	0.9986	0.9986	0.9986	0.9986	0.9986	0.9986	0.9986	0.9986	0.9986	0.9986
	status	In-control	In-control	In-control	In-control	In-control	In-control	In-control	In-control	In-control	In-control
RFR5	output	0.4531	0.8206	0.775	0.4805	0.948	0.7651	0.903	0.8585	0.4093	0.5771
	UCL	0.9993	0.9993	0.9993	0.9993	0.9993	0.9993	0.9993	0.9993	0.9993	0.9993
	status	In-control	In-control	In-control	In-control	In-control	In-control	In-control	In-control	In-control	In-control
RFR6	output	0.3192	0.4266	0.2557	0.2768	0.3157	0.379	0.3957	0.3162	0.1386	0.1974
	UCL	0.9914	0.9914	0.9914	0.9914	0.9914	0.9914	0.9914	0.9914	0.9914	0.9914
	status	In-control	In-control	In-control	In-control	In-control	In-control	In-control	In-control	In-control	In-control

Regarding the SVM scheme (Table 15), none of its control charts signal during the first ten consecutive samplings. Regarding the ANN scheme (Table 16), the ANN6 chart, which is responsible for the detection of σ_3^2 shifts, signals at samples no. 2 and no. 8, and the ANN3 chart, which is responsible for the detection of μ_3 shifts, signals at sample no. 10. Regarding the RFR scheme (Table 17), its RFR1 and RFR3 charts, which are responsible for the detection of μ_1 and μ_3 shifts, respectively, both signal at sample no. 10. Clearly in the case of RFR1, we received a false alarm from the RFR scheme, considering there is no shift in the first mean.

Note that, this was only a simple example to show how each proposed control scheme can be implemented in practice, and based on only 10 samples, no comparisons can be made. Performance comparisons were the purpose of the previous section.

5. Concluding Remarks

This paper proposed new control charts for simultaneous monitoring of multivariate normal process mean vectors and the variance–covariance matrix. For the first time, machine-learning techniques were used for this purpose. Three used ML techniques are ANN, SVM, and RF. We received linear outputs from these ML structures and then applied control chart rules to decide whether the process is in control or out of control. Two different input sets and two different training methods were employed for the proposed ML structures. In the first input set, two statistics (one representing the process mean vector and the other for the process variability) were employed, and for the second input set, we used each process parameter for each quality characteristic separately as inputs. In the first training method, we only trained the ML structures with a small shift size, and in the other method, a small and a large shift size were considered. We also used two different process control scenarios. In the first scenario, the only goal was the detection of the out-of-control situation, regardless of which variable and process parameter is responsible for it. In the second scenario, on the other hand, other than detection, identifying which variable (s)/process parameter (s) is responsible for the signal was also a goal, which involved several control charts to be monitored together.

For each of these control–input–training scenarios, the ML structures were trained, and control charts were developed. Numerical analyses were performed for the cases of two, three, and four quality characteristics. The results, in general, showed that depending on which control-input-training scenario is used, as well as the number of variables, each of these ML control charts performed better in some cases, and there is no absolute winner among them. However, considering how its decision-making procedure works based on dividing tree branches, the RFR scheme tended to mostly perform better when there were more inputs, more diverse training, and more quality characteristics (higher dimension). However, this did not happen in all cases (except for the diverse training part), and most importantly, even if its performance was improved by all these diversities, it did mean that it should perform better than the ANN and SVR charts (which actually in most cases it did not). It was also concluded that when identification was also a goal, the charts performed worse. However, this deterioration in performance was mostly at its lowest for one ML scheme (it differed based on the scenario).

We also compared the proposed ML charts with some recently developed multivariate statistical control charts with fixed and adaptive chart parameters (designed only for detection). For the case of $p = 2$, the results showed that in the detection-only scenario and with the first input set, at least one of the proposed ML charts performed better than all their proposed charts in all the shift cases, even though our proposed schemes are all fixed parameters. With the second input set together with the first training method, our proposed charts performed better in most cases, and together with the second training method, they performed better in all cases. Regarding the detection–identification scenario, our proposed ML charts still performed better in more cases, even though their charts have only been designed for detection and that usually by default means better performance. For the case of $p = 3$, the results showed that in the detection-only scenario, at least one of the ML charts performed better than all their proposed charts in most shift cases, and with the second training method, they performed better in all the shift cases. In the detection–identification scenario, one can say that the cases in which at least one of our proposed schemes performed better than all their charts and vice versa were almost the same. However, keep in mind that, unlike their charts, our proposed charts are also capable of identification. Lastly, an illustrative example based on a healthcare-related practical case was presented to show how the proposed schemes can be implemented in practice.

Highlighting the primary focus of this paper, our investigation centered the utilization of diverse machine-learning techniques in constructing control charts, effectively substituting traditional statistical methods. This exploration involved rigorous testing of different input sets and training methodologies to surpass the performance of statistical control charts. While our study concentrated on specific control charts, along with a limited selection of input sets and training approaches, it presents an opportunity for further exploration of a wide range of control charts, as well as diverse input sets and training methods, thus broadening the horizons of research in this field.

For future developments, one might be interested in trying different input sets, training methods, and even output sets for ML structures. Adding adaptive features to the proposed control charts would also be a major improvement. Since ML control charts have rarely been developed, developing them for many other applications and comparing them to traditional statistical control charts might also be interesting. In particular, since all the developed ML control charts are so far memory-less, developing memory-type ML control charts and comparing them to memory-type statistical control charts might also be interesting for some researchers. In addition, how to train the ML structures in the case of unknown distributions is another challenge that might be worth investigating by some other researchers.

Author Contributions: Conceptualization, H.S.; methodology, H.S. and S.T.A.N.; software, H.S.; validation, H.S.; formal analysis, H.S.; investigation, H.S. and S.T.A.N.; resources, H.S..; data curation, H.S.; writing—original draft preparation, H.S.; writing—review and editing, S.T.A.N.; visualization, H.S.; supervision, S.T.A.N.; project administration, H.S. All authors have read and agreed to the published version of the manuscript.

Funding: This research received no external funding.

Institutional Review Board Statement: Not applicable.

Informed Consent Statement: Not applicable.

Data Availability Statement: All data are available in the manuscript.

Acknowledgments: The authors thank the journal's editorial board and appreciate the esteemed reviewers for their constructive comments, which led to significant improvements in the quality of the paper.

Conflicts of Interest: The authors declare no conflict of interest.

References

1. Reynolds, M.R., Jr.; Gyo-Young, C. Multivariate control charts for monitoring the mean vector and covariance matrix. *J. Qual. Technol.* **2006**, *38*, 230–253.
2. Hawkins, D.M.; Maboudou-Tchao, E.M. Multivariate exponentially weighted moving covariance matrix. *Technometrics* **2008**, *50*, 155–166. [CrossRef]
3. Zhang, G.; Chang, S.I. Multivariate EWMA control charts using individual observations for process mean and variance monitoring and diagnosis. *Int. J. Prod. Res.* **2008**, *46*, 6855–6881. [CrossRef]
4. Khoo, M.B.C. A new bivariate control chart to monitor the multivariate process mean and variance simultaneously. *Qual. Eng.* **2004**, *17*, 109–118. [CrossRef]
5. Zhang, J.; Li, Z.; Wang, Z. A multivariate control chart for simultaneously monitoring process mean and variability. *Comput. Stat. Data Anal.* **2010**, *54*, 2244–2252. [CrossRef]
6. Wang, K.; Yeh, A.B.; Li, B. Simultaneous monitoring of process mean vector and covariance matrix via penalized likelihood estimation. *Comput. Stat. Data Anal.* **2014**, *78*, 206–217. [CrossRef]
7. Sabahno, H.; Castagliola, P.; Amiri, A. A variable parameters multivariate control chart for simultaneous monitoring of the process mean and variability with measurement errors. *Qual. Reliab. Eng. Int.* **2020**, *36*, 1161–1196. [CrossRef]

8. Sabahno, H.; Castagliola, P.; Amiri, A. An adaptive variable-parameters scheme for the simultaneous monitoring of the mean and variability of an autocorrelated multivariate normal process. *J. Stat. Comput. Simul.* **2020**, *90*, 1430–1465. [CrossRef]
9. Sabahno, H.; Amiri, A.; Castagliola, P. A new adaptive control chart for the simultaneous monitoring of the mean and variability of multivariate normal processes. *Comput. Ind. Eng.* **2021**, *151*, 106524. [CrossRef]
10. Sabahno, H.; Khoo, M.B.C. A multivariate adaptive control chart for simultaneously monitoring of the process parameters. *Commun. Stat. Simul. Comput.* **2022**, 1–19. [CrossRef]
11. Sabahno, H. An adaptive max-type multivariate control chart by considering measurement errors and autocorrelation. *J. Stat. Comput. Simul.* **2023**, 1–26. [CrossRef]
12. Chang, S.I.; Ho, E.S. A two-stage network approach for process variance change detection and classification. *Int. J. Prod. Res.* **1999**, *37*, 1581–1599. [CrossRef]
13. Niaki, S.T.A.; Abbasi, B. Fault diagnosis in multivariate control charts using artificial neural networks. *Qual. Reliab. Eng. Int.* **2005**, *21*, 825–840. [CrossRef]
14. Niaki, S.T.A.; Abbasi, B. Detection and classification mean-shifts in multiattribute processes by artificial neural networks. *Int. J. Prod. Res.* **2008**, *46*, 2945–2963. [CrossRef]
15. Cheng, C.-S.; Cheng, H.-P. Identifying the source of variance shifts in the multivariate process using neural networks and support vector machines. *Expert Syst. Appl.* **2008**, *35*, 198–206. [CrossRef]
16. Abbasi, B. A neural network applied to estimate process capability of nonnormal processes. *Expert Syst. Appl.* **2009**, *36*, 3093–3100. [CrossRef]
17. Salehi, M.; Bahreininejad, A.; Nakhai, I. On-line analysis of out-of-control signals in multivariate manufacturing processes using a hybrid learning-based model. *Neurocomputing* **2011**, *74*, 2083–2095. [CrossRef]
18. Hosseinifard, S.Z.; Abdollahian, M.; Zeephongsekul, P. Application of artificial neural networks in linear profile monitoring. *Expert Syst. Appl.* **2011**, *38*, 4920–4928. [CrossRef]
19. Weese, M.; Martinez, W.; Megahed, F.M.; Jones-Farmer, L.A. Statistical learning methods applied to process monitoring: An overview and perspective. *J. Qual. Technol.* **2016**, *48*, 4–24. [CrossRef]
20. Escobar, C.A.; Morales-Menendez, R. Machine learning techniques for quality control in high conformance manufacturing environment. *Adv. Mech. Eng.* **2018**, *10*, 1–16. [CrossRef]
21. Apsemidis, A.; Psarakis, S.; Moguerza, J.M. A review of machine learning kernel methods in statistical process monitoring. *Comput. Ind. Eng.* **2020**, *142*, 106376. [CrossRef]
22. Mohd Amiruddin, A.A.A.; Zabiri, H.; Taqvi, S.A.A.; Dendena Tufa, L. Neural network applications in fault diagnosis and detection: An overview of implementations in engineering-related systems. *Neural Comput. Appl.* **2020**, *32*, 447–472. [CrossRef]
23. Demircioglu Diren, D.; Boran, S.; Cil, I. Integration of machine learning techniques and control charts in multivariate processes. *Sci. Iran.* **2020**, *27*, 3233–3241. [CrossRef]
24. Yeganeh, A.; Pourpanah, F.; Shadman, A. An ANN-based ensemble model for change point estimation in control charts. *Appl. Soft Comput.* **2021**, *110*, 107604. [CrossRef]
25. Mohammadzadeh, M.; Yeganeh, A.; Shadman, A. Monitoring logistic profiles using variable sample interval approach. *Comput. Ind. Eng.* **2021**, *158*, 107438. [CrossRef]
26. Sabahno, H.; Amiri, A. New statistical and machine learning based control charts with variable parameters for monitoring generalized linear model profiles. *Comput. Ind. Eng.* 2023; *submitted*.
27. Yeganeh, A.; Chukhrova, N.; Johannssen, A.; Fotuhi, H. A network surveillance approach using machine learning based control charts. *Expert Syst. Appl.* **2023**, *219*, 119660. [CrossRef]
28. Yeganeh, A.; Shadman, A.; Shongwe, S.C.; Abbasi, S.A. Employing evolutionary artificial neural network in risk-adjusted monitoring of surgical performance. *Neural Comput. Appl.* **2023**, *35*, 10677–10693. [CrossRef]
29. Yeganeh, A.; Johannssen, A.; Chukhrova, N.; Abbasi, S.A.; Pourpanah, F. Employing machine learning techniques in monitoring autocorrelated profiles. *Neural Comput. Appl.* **2023**, *35*, 16321–16340. [CrossRef]

Disclaimer/Publisher's Note: The statements, opinions and data contained in all publications are solely those of the individual author(s) and contributor(s) and not of MDPI and/or the editor(s). MDPI and/or the editor(s) disclaim responsibility for any injury to people or property resulting from any ideas, methods, instructions or products referred to in the content.

Article

Data-Driven Surveillance of Internet Usage Using a Polynomial Profile Monitoring Scheme

Unarine Netshiozwi [1], Ali Yeganeh [1], Sandile Charles Shongwe [1,*] and Ahmad Hakimi [2]

[1] Department of Mathematical Statistics and Actuarial Science, Faculty of Natural and Agricultural Sciences, University of the Free State, Bloemfontein 9301, South Africa; 2016157592@ufs4life.ac.za (U.N.); yeganeh.ali1369@gmail.com (A.Y.)
[2] Department of Industrial Engineering, Faculty of Engineering, University of Kurdistan, Sanandaj 0098, Iran; a.hakimi@uok.ac.ir
* Correspondence: shongwesc@ufs.ac.za

Abstract: Control charts, which are one of the major tools in the Statistical Process Control (SPC) domain, are used to monitor a process over time and improve the final quality of a product through variation reduction and defect prevention. As a novel development of control charts, referred to as profile monitoring, the study variable is not defined as a quality characteristic; it is a functional relationship between some explanatory and response variables which are monitored in such a way that the major aim is to check the stability of this model (profile) over time. Most of the previous works in the area of profile monitoring have focused on the development of different theories and assumptions, but very little attention has been paid to the practical application in real-life scenarios in this field of study. To address this knowledge gap, this paper proposes a monitoring framework based on the idea of profile monitoring as a data-driven method to monitor the internet usage of a telecom company. By definition of a polynomial model between the hours of each day and the internet usage within each hour, we propose a framework with three monitoring goals: (i) detection of unnatural patterns, (ii) identifying the impact of policies such as providing discounts and, (iii) investigation of general social behaviour variations in the internet usage. The results shows that shifts of different magnitudes can occur in each goal. With the aim of different charting statistics such as Hoteling T^2 and MEWMA, the proposed framework can be properly implemented as a monitoring scheme under different shift magnitudes. The results indicate that the MEWMA scheme can perform well in small shifts and has faster detection ability as compared to the Hoteling T^2 scheme.

Keywords: control chart; internet usage monitoring; profile monitoring; statistical process control (SPC); telecom company

MSC: 03-08

1. Introduction

Statistical Process Control (SPC) is a novel idea with the aim of high-quality process monitoring and control. It has been utilised frequently in industrial processes since the 1920s [1] to monitor numeric or attribute quality characteristics and it includes seven major techniques, the most prominent of which is the control chart. To conduct a process monitoring procedure, two phases (namely, Phases I and II) are defined for each control chart in such a way that Phase I aims to estimate the process parameters from the historical data (off-line monitoring), and Phase II aims to adequately detect unnatural conditions, denoted by Out-of-Control (OC), from the stable or normal situations, denoted by In-Control (IC) as soon as possible. According to Montgomery [2], an IC process has only process-inherent random variability while a process that can be affected by assignable (non-random) causes is referred to as an OC situation. So, it could be said that the major aim of

Phase II monitoring is about the detection of assignable causes from the process-inherent random ones.

To monitor a process in SPC with control charts, two approaches are usually performed. The first (conventional) one is to consider a univariate or multivariate distribution for a single or multiple quality characteristics. For the second one, the process quality is formulated and monitored by a functional relationship between a response variable and one or more explanatory variables, which is denoted by *profile monitoring* [3]. It could be said that the major goal of profile monitoring is to check the stability of a specified IC relationship (or profile) over time [4].

In recent decades, several studies have focused on the development of novel control charts in the area of profile monitoring either in Phase I or II. As a few examples of the pioneer works, one can refer to Kang and Albin [5], Kim, et al. [6] and Woodall, et al. [7] in which some basic control charts were developed for the linear profiles whose shape is established by a simple straight line. Although linear profiles are the simplest and the most frequent case [8], other profile shapes may be needed in the complicated process. To name just a few, these are ordinal [9], binary logistic [10,11], Poisson [12], nonlinear [13], complex relationship [14,15] and so forth. In addition to these types, polynomial profiles have been considered in several studies which are in line with the aim of this study.

The polynomial profiles are built in the form of $y = b_0 + b_1 x + b_2 x^2 + \ldots + b_k x^k$ and were first discussed by Kazemzadeh, et al. [16]. They proposed three control charts for Phase I analysis based on the F-test, Hotelling's T^2 (hereafter T^2) and Likelihood Ratio Test (LRT) statistics. Amiri, et al. [17] considered a quadratic (polynomial with order 2) profile type in monitoring the relationship between the torque produced by an engine and the engine speed in revolutions per minute as a case study in the automotive industry. In addition, they discussed the observed challenges of monitoring polynomial profiles in Phase I and II. Zhang, et al. [18] proposed two other modified methods for a quadratic form of polynomial profiles when the IC model was linear. They concluded that T^2 and LRT control charts yielded the same performance. In a similar problem, i.e., changing the linear IC form to a quadratic polynomial profile, Zhang, et al. [19] proposed a new Exponential Weighted Moving Average (EWMA) charting statistic in combination with a score test approach. The simulation results indicated that their proposed approach was robust against the variations of the profiles' shape. Furthermore, a cumulative sum (CUSUM) approach was proposed by Zhang, et al. [20] in similar problem adjustments. Despite the three previous works which were limited to speculation about the quadratic polynomial profiles, Yao, et al. [21] considered the general form of polynomial profiles and compared four common control charts for this situation.

Considering previous research in the area of profile monitoring, it could be observed that the emphasis of most studies has been on the development of new control charts not only in the polynomial model but also in all other profile types, while little attention has been paid to the practical application of the profile monitoring idea. To the best of the authors' knowledge, there is no research about the real application of polynomial profiles, but some sporadic works could be found in other profile types. For instance, to count a few: monitoring of highway safety [22], calibration system [23] and chemical gas sensors [24] with a general linear model, plasma etch process in semiconductor manufacturing with nonlinear profiles [25], social networks and healthcare with a logistic model [26–28], engine procedure in the automotive industry with autocorrelated profiles [17] and customer satisfaction with a nonparametric approach [29]. Furthermore, it is noteworthy to mention that there are several practical applications of control charts for monitoring quality characteristics; see, for example, [30–33].

Similar to the above research, the aim of this study is to propose a practical application of control charts with the idea of profile monitoring in the field of internet usage monitoring. From the (telecommunication) internet companies' point of view, the trend of total internet usage in some specific platforms (e.g., social networks or applications) during different hours in each day is crucial. Identifying different trends and behaviours in internet usage can reveal unnatural conditions and situations as well as lead to new decision and policy making such as offering discounts and markups, changing service tariff and costs, and so forth. Although a few studies investigated internet usage from other practical aspects [34–36], there is no monitoring scheme that has been extended yet in this area.

Thus, as the major contribution of this study, a novel monitoring scheme based on the idea of profiles is proposed to monitor the internet usage of a well-known anonymous internet company. By the aim of the proposed method, one can monitor the internet usage in each day and identify the unnatural trends and patterns. The problem definition, assumption, formulation and adjustments are completely described in the next sections. Previous studies usually investigated two common phases, Phase I and Phase II, in the practical application. Recently, Zwetsloot, et al. [37] suggested a more comprehensive SPC framework for practical application with four phases. Based on their idea and as the second major contribution, both phases (i.e., I and II) and some other steps are considered in this paper as a comprehensive framework for implementation of control charts in the area of this problem. It includes different steps such as model estimation, verification, Phase I analysis, Phase II signalling and signal interpretation. Although some other frameworks have been extended for other aims in the SPC domain [17,38–40], the latter are not compatible for the provided real-life data used in this paper. The proposed framework could be applied in real application of not only polynomial profiles, but also other profile types and other SPC monitoring problems.

To sum up, the contributions of this paper can be listed as follows:

(i) A practical application of profile monitoring in the area of internet usage monitoring.
(ii) Development of a novel SPC framework for detection of unnatural trends and patterns in some specific platforms (e.g., social networks or applications).
(iii) Consideration of both SPC common phases and some other steps including model estimation, verification, Phase I analysis, Phase II signalling and signal interpretation in the proposed framework.
(iv) Definition of charting statistics based on the polynomial profile model.

The remainder of the article is organized as follows. Section 2 proposes the problem definition and formulation. Section 3 presents a unified framework based on the SPC approach to solve the problem. The monitoring procedure based on the proposed framework is described in Section 4 and finally, Section 5 concludes the article and provides some direction for future research.

2. Problem Definition

In some companies such as online shops, hypermarkets, manufacturing plants and so forth, the trend (pattern) of purchase orders, consumption of a specific product or some other quality characteristics at different hours within a day can have several important consequences for the performance of the company. This type of information is very helpful in such companies to arrange the products and personnel, investigate the sales program, provide some hourly discounts and so forth. On the other hand, a change in the pattern of a specific day from the previous days could be caused by some important assignable causes, which may lead to unfavourable events. For example, if it is expected that the purchase orders will decrease with a nearly constant rate in all hours of holidays or weekends, but instead some occurrence of a sharp decrease or increase in a specific hour is observed, then it is sometimes necessary to discover the reason behind that unexpected event.

The above idea is also important in internet telecommunication (telecom) companies which provide data/broadband and internet services. As such, these companies can obtain some beneficial results from monitoring internet usage during different hours of the day. From the latter scenario, the following three major aims are investigated in this paper:

(i) First, unnatural patterns, network failure and some other abnormal situations in which the total trend of a day is affected and varied are detected by this procedure.
(ii) Identifying the impact of some policies such as providing discounts or markups for specific hours is considered as the second aim.

(iii) Some general social behaviour variations, which could also be considered as customer requirements, are detected by this idea. In this case, some specific products are subject to change and it is important to identify whether this change can affect the overall trend or not.

To reach the above aims, process monitoring is implemented, which usually refers to surveillance of an industrial process or manufacturing lines over time to achieve a high level of final product quality, by defining control limits for the desired quality characteristics and producing signals in the unfavourable situations at every step of the manufacturing process [2]. The SPC techniques, specifically control charts, provide several beneficial tools and techniques to discover unusual variability and prevent the occurrence of these again. Hence, it is necessary to define and utilize a proper control chart in the area of this problem. The simple and conventional approach is to provide some univariate or multivariate control charts to monitor the process at each hour of the day. However, it is a more complex scheme in this area, as a large number of charting statistics and control limits need to be defined [41]. To avoid these challenges, some techniques and tools are required to aggregate all the day's results and monitor the process entirely. For this aim, the profile monitoring idea which monitors a relationship between some explanatory and response variables over time is suggested in this paper. More details about how profile monitoring deals with this situation can be found in Zou, et al. [42] and Yeganeh, et al. [43].

Hence, a well-known anonymous telecom company provided some real-life data to be studied as per the above aims using the idea of profile monitoring. The data includes the internet usage (in terms of GigaBytes (GB)) of more frequently active subscribers in two major social networks, i.e., Instagram and WhatsApp, in each hour of the days between 1 January 2021 and 30 April 2022. So, the dataset of each social network has 485 rows (days) and 24 columns (hours). For better understanding, Figure 1 depicts the profile of internet usage in the first ten days of 2021 for (a) Instagram and (b) WhatsApp. Each line (colour) represents a specific day. As shown, there seem to be low usage of internet between 3:00 and 8:00, while the rate is at the highest volume during 18:00 to 23:00 in all ten days.

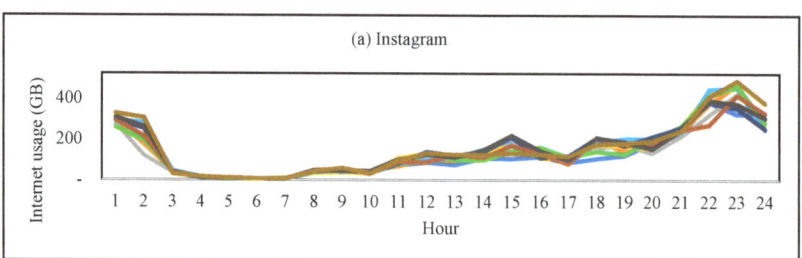

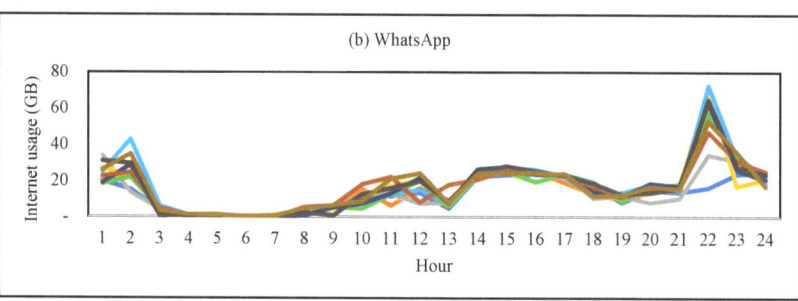

Figure 1. The internet usage of more frequently active subscribers for (**a**) Instagram and (**b**) WhatsApp in the first 10 days of 2021 (each line is an indicator of a specific day—10 days in total).

Considering Figure 1, the polynomial profile models are suitable for the internet usage in each day. For this aim, we followed the approach of Wang, Li, Ma, Song and Wang [29] in which the number of months were assumed as the explanatory variables in such a way that the hours of the day establish the independent variables. Hence, the responses are defined as the usage of internet in a specific hour. For the jth day of the monitoring procedure, the polynomial profile model is as given in Equation (1).

$$y_{ij} = a_0 + a_1 x_i^1 + a_2 x_i^2 + \ldots + a_k x_i^k + \varepsilon_{ij},$$
$$\varepsilon_{ij} \sim N(0, \sigma^2),$$
$$i = 1, 2, \ldots, n,$$
$$j = 1, 2, \ldots.$$
(1)

As the explanatory variables are hours, they are fixed (constant) in all the profiles and it could be written that $x_i = 1, 2, \ldots, 24$. In other words, it is not necessary to use a time index (here j) for the explanatory variable [44]. Furthermore, it is assumed that the error terms are completely independent, that is, there is no between and within autocorrelated error terms. At the end of the jth day, there are 24 values ($n = 24$) for the internet usage at each hour so that the parameters of the model in Equation (1) are estimated by the Ordinary Least Square (OLS) algorithm as shown in Equation (2) [16].

$$[\hat{a}_{0j}, \hat{a}_{1j}, \ldots, \hat{a}_{kj}] = (X'X)^{-1} X' Y_j,$$

$$X = \begin{pmatrix} 1 & 1 & 1 & \ldots & 1 \\ 1 & 2 & 2^2 & \ldots & 2^k \\ 1 & 3 & 3^2 & \ldots & 3^k \\ \vdots & \vdots & \vdots & \vdots & \vdots \\ 1 & 24 & 24^2 & \ldots & 24^k \end{pmatrix},$$
(2)

$$Y_j = [y_{1j}, y_{2j}, \ldots, y_{24j}]'.$$

In Equation (2), the fixed explanatory variable is a $24 \times (k+1)$ matrix and the response variable is a 24×1 vector. Furthermore, the error variance is estimated by the Mean Squared Error (MSE) of the residuals as follows [42]:

$$\hat{\sigma}_j^2 = \frac{\sum_{i=1}^{n} (y_{ij} - (\hat{a}_{0j} + \hat{a}_{1j} x_i^1 + \hat{a}_{2j} x_i^2 + \ldots + \hat{a}_{kj} x_i^k))^2}{n - k - 1}.$$
(3)

To solve this problem, the proper polynomial model is first assigned on the data to reach the IC model and then, the stability of this model is checked over time by the idea of profile monitoring and control chart theory. In fact, the IC regression model parameters in Equation (1), denoted by 0 index hereafter, including $a_{00}, a_{10}, \ldots, a_{k0}$ and σ_0^2, are monitored over time and triggering a signal on the jth day is equivalent to changing the IC model to OC, or the occurrence of a new pattern in internet usage on the jth day. Considering this point, the response variables do not directly monitor the process and it does not matter in this approach whether the internet usage is located in a specific range in different hours, but it is necessary that the IC polynomial predefined model has a constant form at the end of each day over time. Therefore, when the internet usage at a specific hour changes from the previous day's pattern, and if the usage in other hours also changes according to a predefined IC polynomial model in the way that the IC model remains constant, the process is still IC; otherwise, an OC signal is triggered by this approach.

3. The Proposed Framework

As a general SPC approach, previous historical data are used to establish a control chart in Phase I and estimate the IC model. The information of m previous days, with the most common range being between 20 and 30, is also considered as the Phase I (reference) data in this study (it is usually denoted by the number of subgroups in Montgomery [2], hence we can say there are m subgroups). The value of m is considered as a hyper parameter of the model and could be obtained based on expert judgment or some other criteria. On the other hand, it is necessary to verify the polynomial model assumptions before Phases I and II analyses. Except the work presented by Amiri, Jensen and Kazemzadeh [17] and Zwetsloot, Jones-Farmer and Woodall [37], there is no other research on model verification in the profile monitoring area. Five steps are defined in the proposed approach to reach a suitable decision about the process condition; these are model estimation, model verification, Phase I analysis, Phase II analysis and signal interpretation. These steps could be applied not only in polynomial profiles but also in other profile types. The details of the five proposed steps are given in the following subsections.

3.1. Model Estimation

In the first step, the polynomial model is fitted on each subgroup and the coefficients as well as the residuals of m profiles are estimated considering a specific value of k. Hence, the residual and coefficient matrices have the form $m \times n$ and $m \times (k + 1)$, respectively.

3.2. Model Verification

The error terms of the regression model in Equation (1) contain some principal assumptions; those are as follows:

(i) Following normal distribution;
(ii) Equality of variances;
(iii) Lack of any autocorrelation.

The importance of these assumptions is due to the validation of OLS estimators under the normal distribution and independence of error terms. So, some requirements are necessary to ensure proper design principals under the prepared real data. The following checklist and solutions are followed in this framework:

- The residuals of m profiles are aggregated (to have an $m \times n$ matrix) and the normality assumption is checked by the well-known Kolmogorov–Smirnov test. To modify the non-normal data, which is equal to the acceptance of an alternative hypothesis, Montgomery [2] suggested the use of transformation on the raw data. Following this idea, the responses (internet usage values) are transformed as $y^* = y^s$, where ($0 < s < 1$). For different values of s (from greater than 0 to lower than 1), the polynomial model is fitted on the data, and the normality of residuals is investigated to reach the best value of s based on the model's accuracies.
- The equality of variance between the samples is checked by the Levene test computed by performing analysis of variance (ANOVA) on the squared deviations of the residual values from other samples means. It is expected to reach large p-values for the equality of variances, whereas transformation on response values could also be modified this way.
- The existence of autocorrelation may occur between or within the profiles. These two situations were completely investigated by Noorossana, et al. [45] and Soleimani, et al. [46], in which the autocorrelation coefficients were denoted by ϕ and ρ, respectively. Furthermore, sometimes both of them (between and within correlation) may happen at the same time; this scenario has been discussed by Ahmadi, et al. [47]. Note that the observations of different lags are possible when studying the autocorrelation effect. So, a broad range of situations should be checked, and it is worth mentioning that this is a very time-consuming procedure. Considering the above references, it is suggested that the first order autocorrelation (i.e., lag = 1) is the most probable condition and thus, we only consider it in this paper. On the other hand, there is no general approach or hypothesis test for checking the autocorrelation in all the profiles simultaneously in the current Phase I literature.
- In line with this aim, the first order autocorrelation effect is checked on the rows and columns of the $m \times n$ obtained residual matrix to reach the m-variate vector $[\phi_1, \phi_2, \ldots, \phi_m]$ and the n-variate vector $[\rho_1, \rho_2, \ldots, \rho_n]$. Then, their average could be the indicator of existence of autocorrelation $\varnothing$ between and within profiles, which is computed

as $\bar{\varnothing} = \frac{\sum_{a=1}^{m} \varnothing_a}{m}$ and $\bar{\rho} = \frac{\sum_{a=1}^{n} \rho_a}{n}$. Based on the results of Noorossana, Amiri and Soleimani [45] and Soleimani, Noorossana and Amiri [46], the effect of autocorrelation can be disregarded in case of absolute values lower than 0.3 values for $\bar{\varnothing}$ and $\bar{\rho}$. When there is autocorrelation in the model, Amiri, Jensen and Kazemzadeh [17] suggested to use a Linear Mixed Model (LMM). It generates some new coefficients for the model and the effect of autocorrelation could be reduced or removed. Another possible approach is to employ the proposed statistics by Noorossana, Amiri and Soleimani [45], Soleimani, Noorossana and Amiri [46] and Ahmadi, Yeganeh and Shadman [47] for Phase I analysis, depending on the type of autocorrelation.

The best polynomial model is obtained by checking the order of different models, i.e., the determination of k in Equation (1), and yields to the satisfaction of the above three conditions. It is naturally an iterative procedure which starts from $k = 2$ and is terminated when the best value of k is reached. Note that the value of k is identical for all the m profiles. At the end of this step, there is a matrix of coefficients with dimension $m \times (k + 1)$.

3.3. Phase I Analysis

As mentioned, the principal aims of Phase I analysis are to check the stability of the process, estimate the parameters, remove the outliers, ensure the distribution of the data and check the false alarm rate. Kang and Albin [5] and Yeh, et al. [48] suggested the T^2 control chart for Phase I profile monitoring analysis due to the existence of large shifts. Furthermore, a proper power and detection ability were reported for this approach in Kazemzadeh, Noorossana and Amiri [16]. Different extensions of the T^2 control chart have been proposed based on different estimations of the variance–covariance matrix, one of which has been extended by Williams, et al. [49]. This approach is also applied here by defining the charting statistic for the hth profile as outlined in Equation (4).

$$\hat{\beta}_h = [\hat{a}_{0h}, \hat{a}_{1h}, \ldots, \hat{a}_{kh}] = (X'X)^{-1}X'Y_h,$$
$$Y_h = [y_{1h}, y_{2h}, \ldots, y_{24h}]',$$
$$T_h^2 = (\hat{\beta}_h - \bar{\beta})'S^{-1}(\hat{\beta}_h - \bar{\beta}), \quad (4)$$
$$\bar{\beta} = \frac{1}{m}\sum_{h=1}^{m}\hat{\beta}_h,$$
$$h = 1, 2, \ldots, m.$$

Note that in this paper, the index of each profile is indicated by h in Phase I, and by j in Phase II. In Equation (4), the estimation of the variance–covariance matrix is computed as shown in Equation (5).

$$S = \frac{\hat{V}_h'\hat{V}_h}{2(m-1)},$$
$$\hat{V}_h = \begin{pmatrix} \hat{v}_1' \\ \hat{v}_2' \\ \ldots \\ \hat{v}_{(m-1)}' \end{pmatrix}, \quad (5)$$
$$\hat{v}_h' = \hat{\beta}_{h+1} - \hat{\beta}_h.$$

To obtain the control limit, the exact distribution of T_h^2 is needed; while it may not be known, some approximations have been proposed. For example, Williams, et al. [50] showed that $\chi^2_{(k+1),\alpha}$ could be considered as a proper control limit (α is determined by the user specifications). By this approach, all the m charting statistics are compared with $\chi^2_{(k+1),\alpha}$ and the profiles with charting statistic greater than $\chi^2_{(k+1),\alpha}$, i.e., $T_h^2 > \chi^2_{(k+1),\alpha}$ are removed and the procedure is iterated until there are no OC profiles. At the end of the Phase I analysis, the IC process parameters including $a_{00}, a_{10}, \ldots, a_{k0}, \phi_0, \rho_0, \sigma_0$ and the variance–covariance matrix (S_0) are estimated based on the averaging of the remaining profiles. Note that the 0 index indicates the IC condition.

3.4. Phase II Analysis

As mentioned in the literature review, three general categories were extended in polynomial profiles entailing (i) T^2-, (ii) LRT- and (iii) EWMA-based methods. Our simulations revealed similar performance between the LRT and EWMA approaches (the results can be given to the interested readers), so it is neglected here. Two other groups were chosen for the Phase II monitoring aim. As the shift in magnitude in Phase II is usually lower than that of Phase I, the T^2 statistic proposed by Williams, Woodall and Birch [49] is not efficient. Yao, Li, He and Zhang [21] extended a modified version of the T^2 control chart for Phase II analysis in polynomial profiles. In this approach, the variance–covariance matrix (S) is modified based on the idea of Kang and Albin [5]. To compute the charting statistics in the jth profile in Phase II, the parameters are first estimated using the OLS approach ($\hat{\beta}_j$) and then the charting statistic (T_j^2) are computed as shown in Equation (6).

$$\begin{aligned}
\hat{\beta}_j &= [\hat{a}_{0j}, \hat{a}_{1j}, \ldots, \hat{a}_{kj}] = (X'X)^{-1}X'Y_j, \\
\beta_0 &= [a_{00}, a_{10}, \ldots, a_{k0}], \\
Y_j &= [y_{1j}, y_{2j}, \ldots, y_{24j}]', \\
T_j^2 &= (\hat{\beta}_j - \beta_0)' S_0^{-1} (\hat{\beta}_j - \beta_0), \\
j &= 1, 2, \ldots, \\
S_0 &= \sigma_0^2 X'X^{-1}.
\end{aligned} \quad (6)$$

In Equation (6), δ_0 is the IC standard deviation of errors which is estimated in Phase I analysis using Equation (3). The OC signal is triggered when T_j^2 is greater than UCL_{T^2}. As the distribution of T_j^2 is not known in Phase II and the proposed control limit by Williams, Woodall, Birch and Sullivan [50] is suitable for Phase I, UCL_{T^2} is computed by a Monte Carlo simulation. For this aim, it is common to define a new criterion denoted by Average Run Length (ARL), which means, on average, the number of required samples to reach a signal. When the simulation studies are performed using IC profiles, then it is denoted as ARL_0. It is expected to be reached as a target by the adjustment of UCL_{T^2} [51]. To better clarify, the following Algorithm 1 is suggested to obtain UCL_{T^2}. Following previous works such as those of Kang and Albin [5] and Yeganeh, et al. [52], in this paper, we also set ARL_0 at 200.

Algorithm 1. Procedure of computations of UCL_{T^2} based on the desired ARL_0

Maxiteration = 10,000; Set UCL_{T^2} to an approximate value;
while ARL_0 is not equal to desired ARL_0
  Let ARL = [];
  for t = 1:Maxiteration **do**
    j = 0;
    while (charting statistic < UCL_{T^2})
      Generate random IC profile;
      Compute charting statistic;
      j = j + 1;
    end while
    Store the signalling time (j) in ARL;
  end for; % ARL would be a 1 × 10,000 vector after this procedure
  ARL_0 = average of (ARL)
  if (ARL_0 < desired ARL_0) increase UCL_{T^2}; **else** decrease UCL_{T^2}; **end if**
end while

However, the T^2 control chart is not solely sufficient for Phase II analysis due to its weakness in detection of small shifts. Several authors such discussed this issue and suggested the use of EWMA control charts [5,6]. To remedy this challenge, the Multivariate EWMA (MEWMA) approach, proposed by Zou, Tsung and Wang [42], is also applied in

Phase II monitoring of polynomial profiles. The comprehensive details of this scheme were discussed in Yeganeh, Shadman and Amiri [52]. Similar to UCL_{T^2}, UCL_{MEWMA} is also determined based on Monte Carlo simulations. To construct the MEWMA charting statistic, the coefficients of the jth generated polynomial profile is first scaled as follows:

$$Z_j(\beta) = \frac{(\hat{\beta}_j - \beta_0)}{\sigma_0},$$
$$Z_j(\sigma) = \Phi^{-1}\{F((n-(k+1))\frac{\hat{\sigma}_j^2}{\sigma_0^2}; n-(k+1))\}, \quad (7)$$

where $\Phi^{-1}(.)$ is the inverse of the standard normal cumulative distribution function, and $F(.,n)$ is the chi-square cumulative distribution function with n degrees of freedom. The $(k+2)$-variate vector Z_j, which concatenates $Z_j(\beta)$ and $Z_j(\sigma)$ as $Z_j = [Z_j(\beta), Z_j(\sigma)]$, is a multivariate normally distributed with mean vector $\mathbf{0}$ and covariance matrix $\Sigma_m = \begin{pmatrix} (X'X)^{-1} & 0 \\ 0 & 1 \end{pmatrix}$, where Σ_m is a $(k+2) \times (k+2)$ square matrix. Eventually, W_j is defined as the MEWMA of the vector Z_j from the first to the jth profile:

$$\mathbf{W}_j = \theta \mathbf{Z}_j + (1-\theta)\mathbf{W}_{j-1}, \quad j=1,2,\ldots, \quad (8)$$

where θ is the EWMA smoothing parameter (here it is assumed to be fixed and equal to 0.2) and $\mathbf{W}_0$ is a $(k+2)$-dimensional vector (usually equal to the mean value of $\mathbf{Z}_j$ in the IC condition). The MEWMA chart signals if

$$U_j = \mathbf{W}_j^T \Sigma_m^{-1} \mathbf{W}_j > UCL_{MEWMA} = L\frac{\theta}{2-\theta}. \quad (9)$$

When there is autocorrelation effect in Phase II, the LMM approach, as well as the proposed statistics by Noorossana, Amiri and Soleimani [45], Soleimani, Noorossana and Amiri [46] and Ahmadi, Yeganeh and Shadman [47] can be utilized to remove the autocorrelation effect; note that this is similar to the Phase I analysis for the T^2 control chart. On the other hand, several researchers, such as Sheu and Lu [53] and Li, et al. [54], studied the robustness of EWMA control charts with the effect of autocorrelation in the Phase II analysis so that the MEWMA approach [42] can be directly employed with autocorrelated data. More discussions on the latter are presented in Appendix A.

3.5. Signal Interpretation

A Phase II signal in an industrial process is the indicator of an unnatural situation and it is highly recommended to stop the process after an OC signal to find and remove the assignable causes. As mentioned in Section 2, three aims are defined for the internet usage monitoring in this paper. In line with these aims, each OC signal is investigated and discussed to reach proper justifications and conclusions.

For better understanding, the above procedures and steps are summarized as a comprehensive framework in Figure 2. In this framework, the values of m (number of subgroups in Phase I) and ARL_0 are assigned by the user.

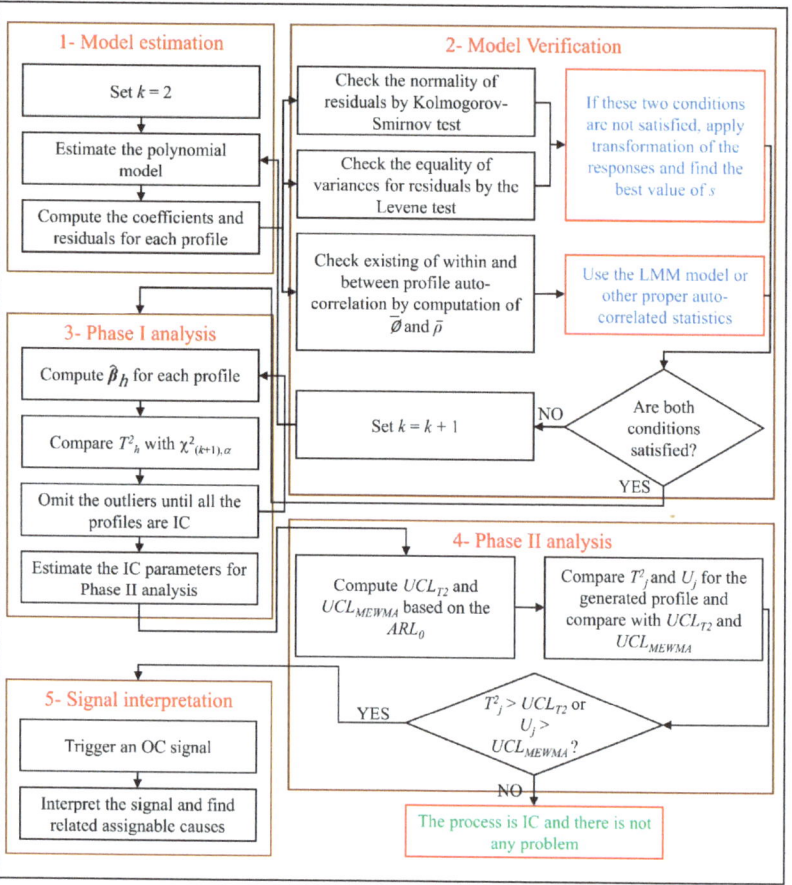

Figure 2. The proposed monitoring framework of this study including five steps (the values of m, α and ARL_0 are determined by the user).

4. Analysis of the Practical Implementation of the Proposed Framework

In this section, the proposed framework in Figure 2 is investigated through a telecom company's real internet usage data. After several discussions with the company's experts about the monitoring challenges and considering the predefined monitoring aims in Section 2, three different scenarios (timestamps) of internet usage are defined to better illustrate the practical application of the proposed methods. In these scenarios, an OC situation occurred in a specific time which is known to us. We want to demonstrate that had the proposed framework been used at that time, it would have identified this unnatural event by triggering an OC signal. To provide a realistic condition, we adjust the Phase I analysis before the change in a way that some of the days in the beginning of Phase II would also be IC. Using this approach, the ability of the proposed framework in the detection of false alarms (i.e., triggering an OC signal when in fact the process is IC) is also evaluated. The following subsections provide the details about each scenario. In all the scenarios, the values of m, α and ARL_0 are set at 20, 0.95 and 200, respectively.

4.1. Scenario I: Detection of Unnatural Patterns in Internet Usage

This subsection aims to investigate the unnatural pattern detection ability of the proposed method. Due to some repairs and maintenances actions in the network, there were some network problems in the internet usage from 24 September 2021 until 15 days later. Naturally, the internet usage profile was different during this interval. For brevity and the similarity of conclusions, the results of this subsection are only based on Instagram data and the results of WhatsApp are neglected.

Suppose the current time is 3 September 2021 and m is set at 20 so Phase I data involves 15 August–3 September 2021. Figure 3a depicts the profile of Instagram internet usage in each hour of the day in Phase I. It is obvious that each line depicts a specific day in this period. Considering steps 1 and 2 in the framework, i.e., model estimation and verification (Sections 3.1 and 3.2), the proper value of k is obtained as 8. Table 1 reports the procedure of obtaining the proper value of k. In this table, the procedure of increasing the value of k was performed on raw (without transformation columns) and transformed values. The p-value of the Kolmogorov–Smirnov and Levene tests are shown with p_{ks} and p_l. Additionally, the values of $\tilde{\varnothing}$ and $\tilde{\rho}$ were computed by adjusting the lag to 1. The nonacceptable values (lower than 0.05 for p-values and absolute values lower than 0.3 for autocorrelation coefficients) are shown in red. It is noteworthy to mention that the best value of s (the power value) was acquired as 0.5 in all cases.

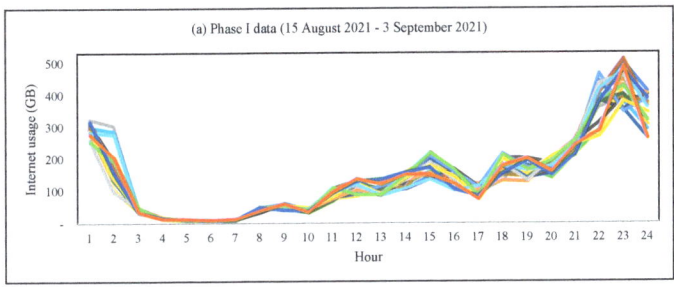

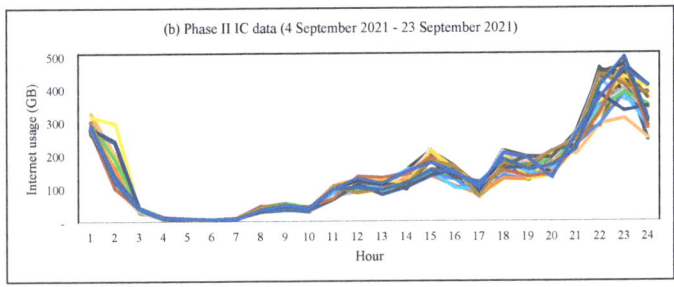

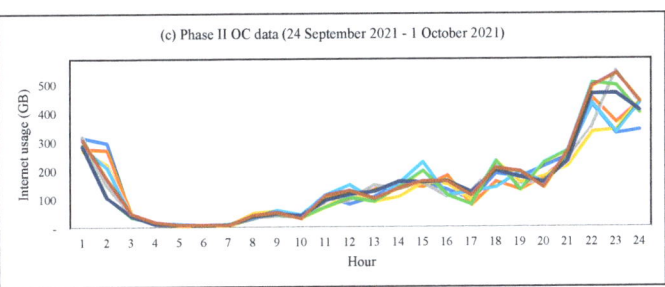

Figure 3. The Instagram internet usage of more frequently active subscribers for (**a**) Phase I, (**b**) IC samples (days) of Phase II and (**c**) the needed days to reach an OC signal with MEWMA approach (each line is indicator of a specific day).

Table 1. The procedure for reaching the best value of *k* as the model estimation and verification steps (red values are indicator of nonacceptable results)—the best selected value of *k* is 8.

k	Without Transformation				With Transformation			
	p_{ks}	p_l	$\bar{\varnothing}$	$\bar{\rho}$	p_{ks}	p_l	$\bar{\varnothing}$	$\bar{\rho}$
2	0	0.99	0.39	−0.11	0	1	0.59	−0.10
3	0	0.97	0.4	−0.14	0	1	0.57	−0.10
4	0	0.87	−0.04	−0.15	0	0.78	0.01	−0.09
5	0	0.81	−0.06	−0.16	0.0001	0.63	−0.14	−0.11
6	0	0.78	−0.4	−0.16	0.02	0.31	−0.18	−0.11
7	0	0.04	−0.22	−0.12	0.02	0.04	−0.25	−0.09
8	0	0.22	−0.32	−0.12	0.39	0.55	−0.30	−0.07
9	0	0.18	−0.31	−0.08	0.16	0.5	−0.30	−0.06

From Table 1, it could be concluded that the normal distribution is not followed in each situation when the raw values used as the *p*-values of the Kolmogorov–Smirnov test (p_{ks}) are near to 0. Having transformed the raw values by *s* = 0.5, we were able to reach normal error terms in *k* = 8 and 9. Due to the lower complexity, the value of *k* is assigned as 8. In this situation, the autocorrelation effects are neglected; hence, the absolute values of $\bar{\varnothing}$ and $\bar{\rho}$ are lower than 0.3 (it is equal to $\phi_0 = \rho_0 = 0$). It is also obvious that there is no problem regarding the equality of variance based on the Levene's test results (the values of p_l are greater than 0.05 in all cases except *k* = 7).

In the next step (Phase I analysis, i.e., Section 3.3), considering Equation (4), the charting statistics (T_h^2) are computed until there are no outliers in the data. For this scenario, it needed three iterations to have no outliers, which is depicted in Figure 4. In the first iteration (Figure 4a), the 14th sample exceeded the control limit ($\chi^2_{(9,0.95)}$ = 16.92) and was removed from the Phase I data. Then (Figure 4b) the Phase I computations were continued by *m* = 19, which led to another outlier and finally, in the third iteration, all the samples (*m* = 18) were in IC (Figure 4c). After this procedure, the IC parameters were estimated by the remaining eighteen profiles. The IC model was obtained as y_{ij} = 18.34 + 3.52x_i − 6.14x_i^2 − 1.89x_i^3 − 0.27x_i^4 + 0.02x_i^5 − 0.001x_i^6 + 0.000025x_i^7 − 0.00000026x_i^8 + ε_{ij}, where the standard deviation of errors is estimated as 1.38. Furthermore, the variance–covariance matrix is a 9 × 9 matrix which is not reported, for brevity.

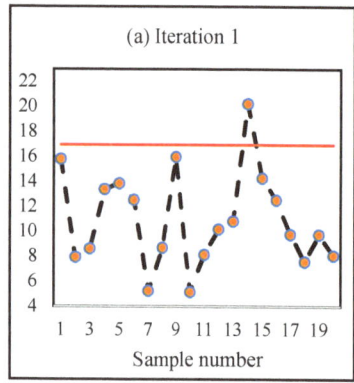
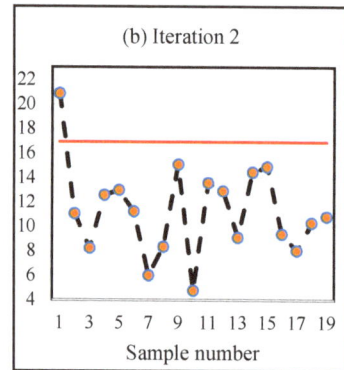
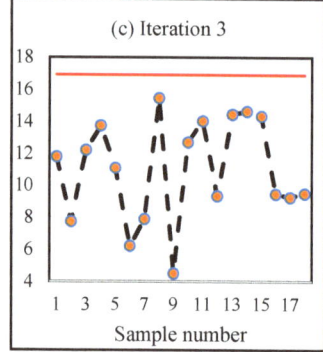

Figure 4. The Phase I chart statistics (T^2_h) in the three iterations. The red horizontal line is equal to $\chi^2_{(9,0.95)}$.

To start the Phase II analysis, UCL_{T^2} and UCL_{MEWMA} are calculated by simulations as 23.59 and 2.68, respectively. As the network problems, which are usually referred to as OC shift in the SPC literature, had happened on 24 September 2021, there are 20 natural (IC) days in Phase II, and it was not expected to have any signal during the period from 4 to 23 September 2021. The profile of these 20 IC samples (days) are shown in Figure 3b. It could be inferenced from Figure 3a,b that the Phase I and IC Phase II samples provided similar profiles.

To detect unnatural conditions, two control charts were used to monitor the internet usage in Phase II and these are depicted in Figure 5. As an interesting result in Figure 5a, the T^2 control chart had very weak performance in such a way that it triggered an OC signal with a 132-day delay. As mentioned in the initial part of this subsection, the problem in the network (or equivalent OC condition) lasted about 15 days, so it could be said that the T^2 control chart was not able to identify this situation. On the other hand, the MEWMA chart only needed 7 days to reach an OC signal (i.e., 1 October 2021) which is completely superior to the T^2 control chart. The increasing pattern in MEWMA statistics after the occurrence of the change (i.e., 24 September 2021), which is obvious in Figure 5b, indicates that the exact time of the shift was nearly seven days later. The reason behind this inference is related to another SPC field denoted as the change point estimation. As it is beyond the scope of this paper, interested readers are referred to Holland and Hawkins [55] and Montgomery [2].

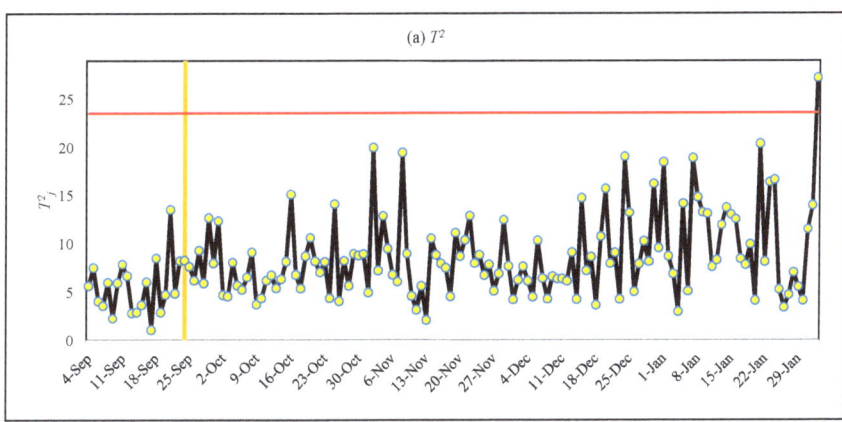

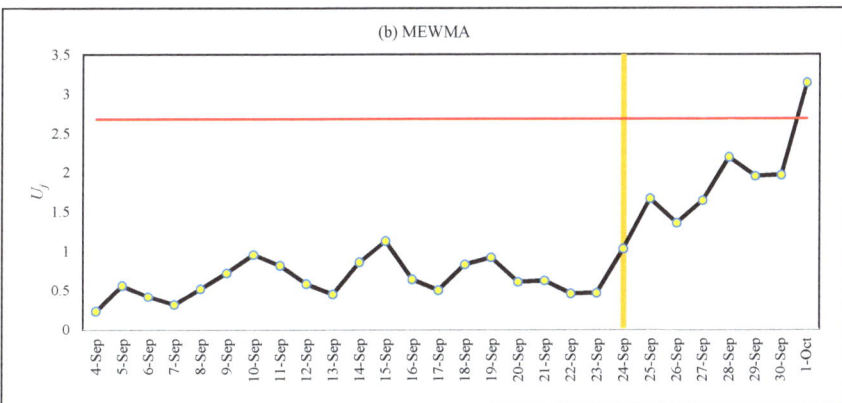

Figure 5. The statistics of T^2 (**a**) and MEWMA (**b**) control charts from the start of Phase II until an OC signal. The red horizontal lines are control limits. The yellow vertical lines indicate the network problem beginning.

As the T^2 control chart is weak in the detection of small shifts, it could be stated that the applied magnitude shift on 24 September 2021, due to a network problem, was small. On the other hand, the MEWMA approach, which has superior ability in small shifts, identified the OC situation after 7 days. To better show this assertion, the Instagram internet usage profiles of these 7 days (i.e., 24 September–1 October 2021) are illustrated in Figure 3c. Some differences in the pattern of this period can be seen comparing two other IC situations (Figure 3a,b) which could be an indicator of a small shift. However, decision making about the IC and OC situations with only the information from plotting the profiles is not an easy task, and a scientific and statistical approach is needed to provide acceptable results. As the last step, the provided signal should be interpreted by an expert to find the main reasons for its occurrence. That said, we know now it was caused by some repairs and maintenances actions in the network. Generally, the discussions and conclusions of this subsection show that the proposed framework is a proper choice for the identification of the unnatural patterns in the monitoring procedure of internet usage.

4.2. Scenario II: Detection the Effect of Discount Policy

The second scenario investigates the effect of a discount policy in the internet usage pattern using the proposed framework. The company had suggested an 80% discount on the internet prices due to holidays on 12–26 March 2022 between 0:00 and 7:00. So, it is expected to see a tangible increase in these hours' usage on the given days and also some changes in the patterns of other hours. The aim is to find how and when the proposed framework identifies this situation. Additionally, dealing with the autocorrelation effect is investigated. Similarly, to the previous subsection, the results of WhatsApp are here neglected for brevity.

Suppose the current time is 6 March 2022 and m is set at 20, so Phase I data involves 15 February–6 March 2022. The profiles of these twenty days are shown in Figure 6a. Obviously, there are five IC days in Phase II (7–11 March 2022) which are depicted in Figure 6b. Finally, the discount times (12–26 March 2022) in which it is supposed to reach an OC signal as soon as possible are illustrated in Figure 6c. The effect of the discount is obvious in Figure 6c, especially between 3:00 and 7:00, as the internet usage has almost quadrupled.

It is again necessary to transform the data due to the non-normality of the raw values; for example, Figure 7 depicts the histogram of the residual of the first ten profiles, which is an indicator of the existence of the skewness in all subgroups. After transforming the values, p_K and p_L were calculated as 0.22 and 0.51, respectively (the best value of k was again 8). However, the values of $\bar{\varnothing}$ and $\bar{\rho}$ were computed as 0.38 and 0.06, respectively, which indicates the between-profile autocorrelation, and thus the charting statistics in Section 3.3 are not appropriate. Figure 7 depicts a matrix of the residual of the first ten Phase I profiles as the scatter plots for each pair (the diagonal histograms are depicted based on the residual of raw values of each profile). The first lag autocorrelation and also non-normality of the raw values' residuals can be clearly observed in Figure 7.

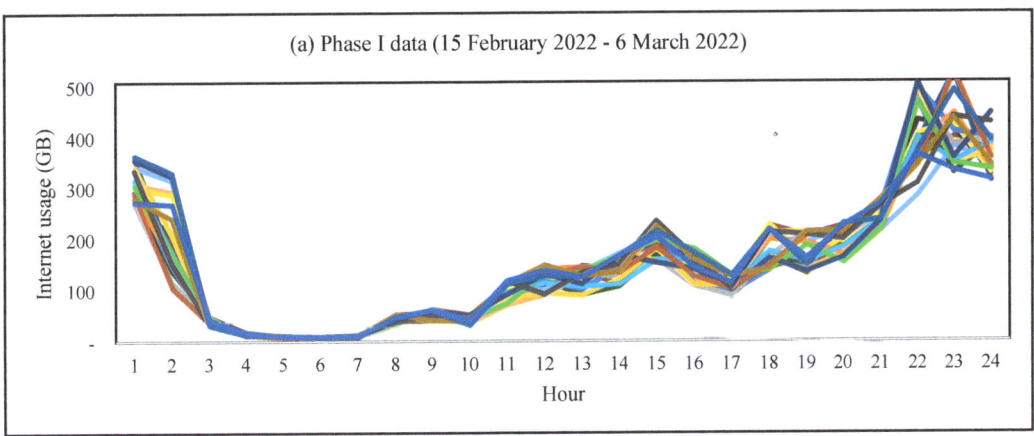
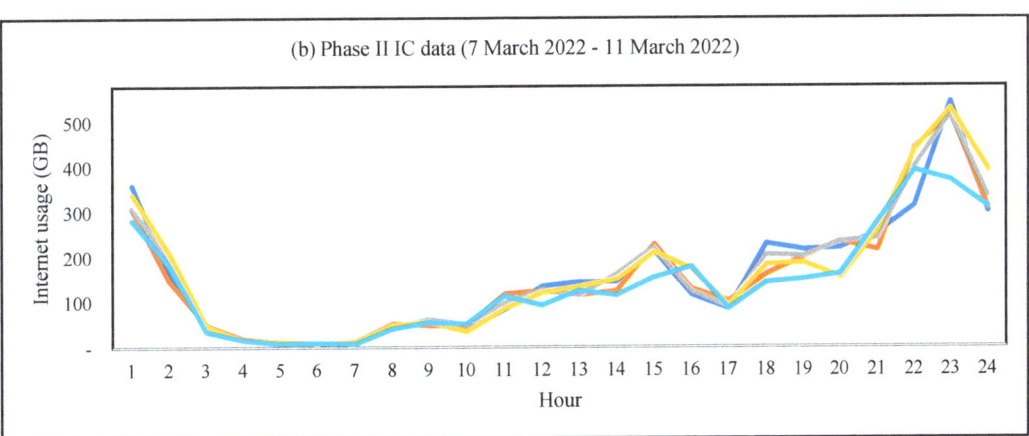
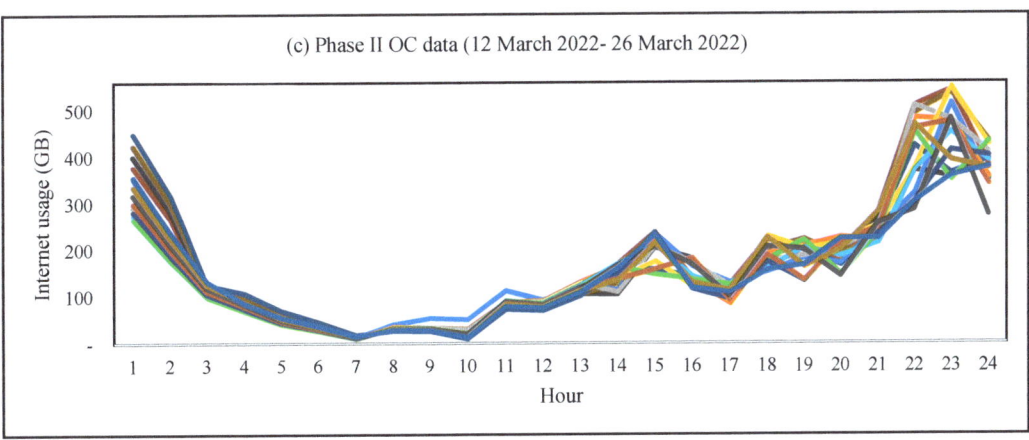

Figure 6. The Instagram internet usage of more frequently active subscribers for (**a**) Phase I (**b**), IC samples (days) of Phase II and (**c**) the discount time (each line is indicator of a specific day).

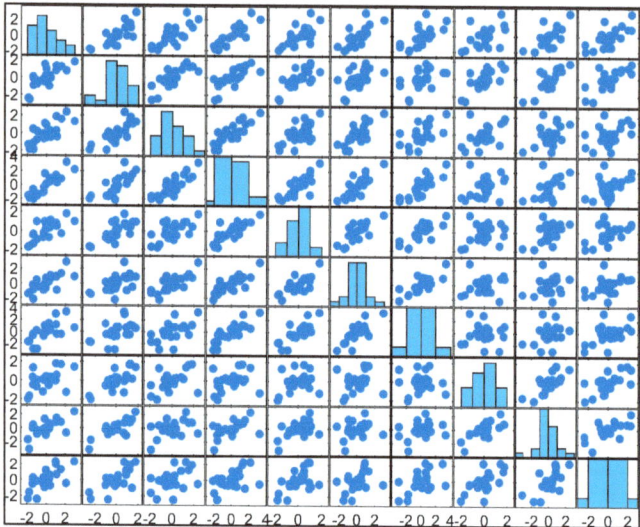

Figure 7. The scatter plot of the first ten Phase I profiles' residuals. The diagonal histograms were drawn by the raw values of residuals.

To remedy this challenge, we first tested the LMM approach implemented by Amiri, Jensen and Kazemzadeh [17] but it was not able to remove the autocorrelation effect. Then, we utilized the approach proposed by Noorossana, Amiri and Soleimani [45]. As it was developed for Phase II analysis, we applied some modification to be able to implement it in Phase I. For further illustration, the details are provided in Appendix A. Considering the formula provided in Appendix A, the UCL was obtained as $\chi^2_{(24,0.95)}$, which is equal to 36.42 and the charting statistics were computed as shown in Figure 8. From that, it could be concluded that all the profiles were IC and the parameters could be estimated. The IC model was obtained as $y_{ij} = 18.11 + 5x_i - 6.83x_i^2 + 2x_i^3 - 0.28x_i^4 + 0.02x_i^5 - 0.001x_i^6 + 0.000025x_i^7 - 0.00000025x_i^8 + \varepsilon_{ij}$, and the standard deviation of errors is estimated as 1.55. Furthermore, the variance–covariance matrix is a 9×9 matrix which is not reported, for brevity.

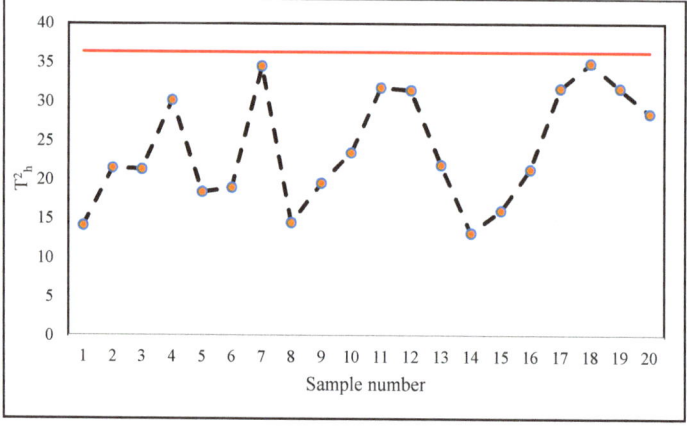

Figure 8. The Phase I chart statistics (T^2_h) based on the Noorossana, Amiri and Soleimani [45] approach. The red horizontal line is equal to $\chi^2_{(24,0.95)}$.

For Phase II analysis, the charting statistic of Noorossana, Amiri and Soleimani [45] was utilized as the T^2 control chart to consider the autocorrelation effect. Due to the robustness of EWMA control charts against the autocorrelation effect in Phase II analysis, we also used the MEWMA approach, proposed by Zou, Tsung and Wang [42], in this section with consideration of autocorrelation in the parameter estimation as discussed in Appendix A. For this aim, the UCL_{T^2} and UCL_{MEWMA} are calculated using simulations as 45.56 and 1.78, respectively. Figure 9 shows the charting statistics of each control chart from the beginning of Phase II until the end of the discount time (7–26 March 2022). The yellow vertical lines are the start of the discount. As it can be seen, both charts did not provide a false alarm in the first five days (7–11 March 2022). On the other hand, T^2 control charts had a better performance as they could detect the OC situation in the first day of the discount, but the MEWMA chart needed two days for a signal. It could be inferred that a shift with a large magnitude occurred in the profile parameters after the beginning of the discount, as the T^2 control chart could detect it earlier. Due to the recursive effect in the MEWMA statistic, we can see the rise in the U_j values during the discount period. From the managerial side, the operator should check and interpret the reason behind the signal. The investigation of the usage of internet in the signal days (T^2: 12 March 2022 and MEWMA: 13 March 2022) shows that there is an increase in the internet usage between 0:00 and 7:00 in this period.

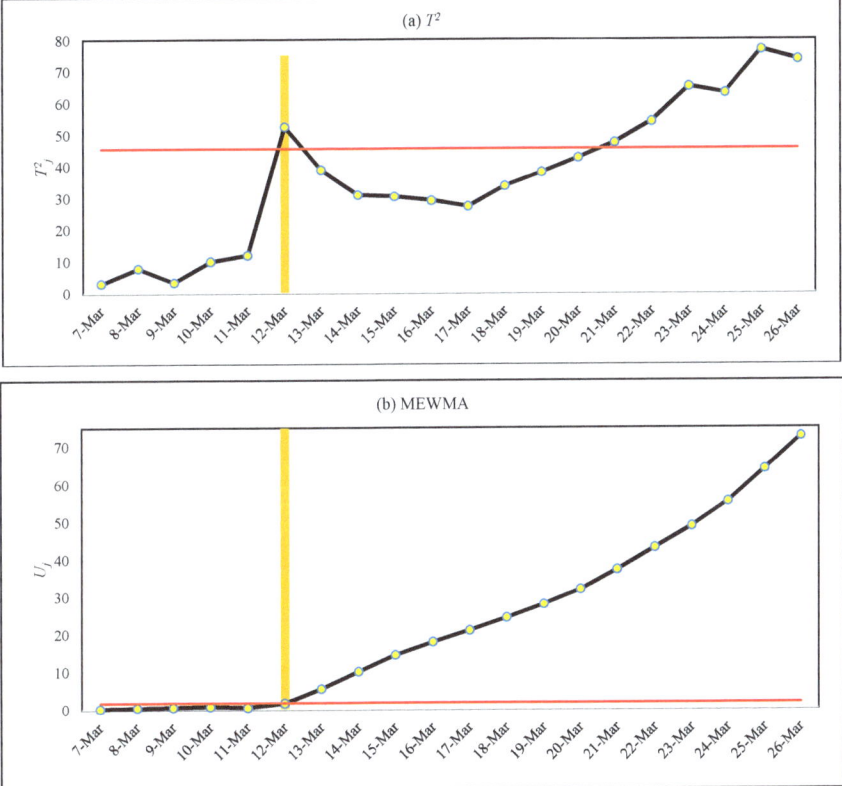

Figure 9. The statistics of T^2 (**a**) and MEWMA (**b**) control charts from start of Phase II until the end of the discount time. The red horizontal lines are control limits. The yellow vertical lines indicate the beginning of the discount (i.e., 12 March 2022).

4.3. Scenario III: Detection of Social Behavior Variations

Due to some protests in the company's country, the connectivity to some major social networks such as Telegram was lost between 24 February and 1 March 2021. On the other hand, the users were able to connect to WhatsApp without any filtering; this caused a tangible increase in the WhatsApp messages during these days. The difference between this situation and Scenario I (in Section 4.1) is that only one platform changed in Scenario III while all of them were influenced in Scenario I. The aim of this subsection is to illustrate the ability of the proposed framework to detect this social behaviour variation. Some of the similar results are not given in this subsection for the sake of brevity.

Suppose that the current time is 20 February 2021 and m is set at 20 so Phase I data involves 1–20 February 2021. The profiles of these twenty days are shown in Figure 10a. Three days in Phase II did not have any change (21–23 February 2021), which are depicted in Figure 10b. Finally, the period with a sudden increase in the WhatsApp internet usage (24 February–1 March 2021) is illustrated in Figure 10c; for example, the maximum usage between 22:00 and 23:00 was about 100 GB on 1 March, while the average usage in this hour was about 60 GB in Phase I, i.e., 1–20 February 2021.

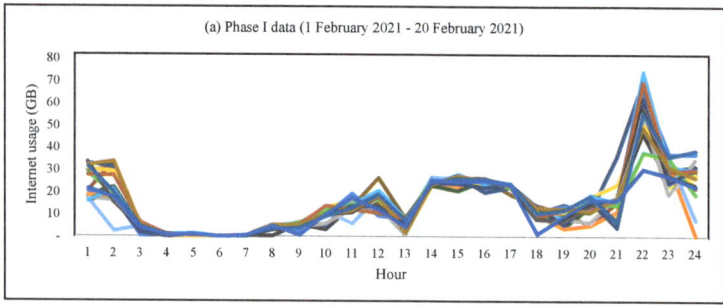

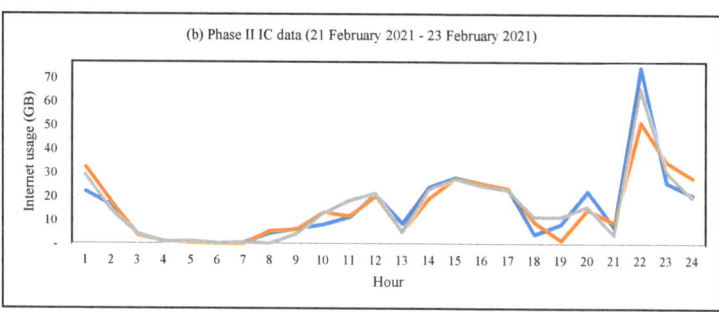

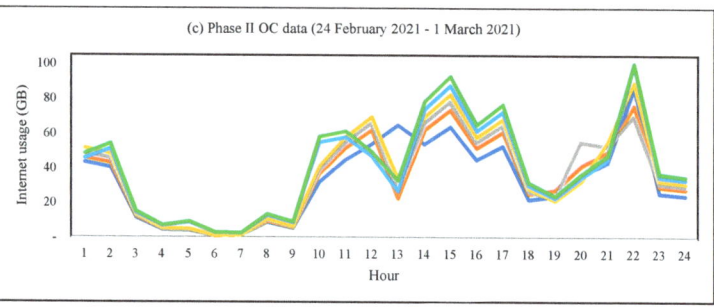

Figure 10. The WhatsApp internet usage of more frequently active subscribers for (**a**) Phase I (**b**), IC samples (days) of Phase II and (**c**) the sudden increase time (each line is indicator of a specific day).

By a similar procedure as Table 1, using $s = 0.55$, it was possible to reach a verified model with $k = 5$, $p_K = 0.06$, $p_L = 0.95$, $\bar{\varnothing} = -0.07$ and $\bar{\rho} = -0.15$. So, the autocorrelation effect was neglected in the subsequent computations. Then, the Phase I analysis was carried out with two iterations and the IC model was obtained as $y_{ij} = 9.12 - 4.19x_i - 0.73x_i^2 - 0.05x_i^3 + 0.002x_i^4 - 0.00002x_i^5 + \varepsilon_{ij}$, and the standard deviation of errors was estimated as 1.12. Furthermore, the variance–covariance matrix is a 9×9 matrix which is not reported, for brevity.

For Phase II analysis, the UCL_{T^2} and UCL_{MEWMA} were calculated using simulations as 18.54 and 2.18, respectively. Then, the charting statistics were computed for both approaches from the beginning of Phase II until twenty days later. The results are reported in Table 2. Both control charts could detect the unnatural situation as soon as it happened. The T^2 control chart retrieved from OC to IC on 5 March 2021 in which the WhatsApp internet usage was normal, while the MEWMA scheme still showed the OC situation until 13 March 2021. As the MEWMA statistic stores the previous points information, it needs more time to come back to the IC condition. To avoid false alarms, one can reset the initial parameters of MEWMA schemes (i.e., W_0 in Equation (8)) which is not in line with the aim of this study [56]. Another useful solution may be the use of smaller smoothing parameters for the MEWMA approach (θ in Equation (8)). Montgomery [2] evaluated the performance of the EWMA control chart under different smoothing parameter values. Knowing the magnitude of the OC shift, we can select the proper value for θ. The very large charting statistics in OC profiles are an indication of large shifts in the OC situation which can also be found by comparing different panels in Figure 10.

Table 2. The charting statistics of T^2 and MEWMA control charts for the phase II analysis of WhatsApp internet usage—the red values exceed the control limit.

j	Date	Process Condition	T_j^2	U_j
1	24 February 2021	IC	2.03	0.08
2	25 February 2021	IC	14.77	0.17
3	26 February 2021	IC	4.06	0.29
4	27 February 2021	OC	1163.49	3.61
5	28 February 2021	OC	1206.91	11.31
6	1 March 2021	OC	1422.83	22.11
7	2 March 2021	OC	1584.19	34.91
8	3 March 2021	OC	1763.52	48.04
9	4 March 2021	OC	2049.03	62.63
10	5 March 2021	IC	14.27	39.35
11	6 March 2021	IC	5.94	25.86
12	7 March 2021	IC	7.46	17.92
13	8 March 2021	IC	25.02	12.24
14	9 March 2021	IC	8.59	8.52
15	10 March 2021	IC	19.21	6.07
16	11 March 2021	IC	13.36	3.55
17	12 March 2021	IC	10.91	2.94
18	13 March 2021	IC	10.01	2.07
19	14 March 2021	IC	15.37	1.15
20	15 March 2021	IC	27.94	0.98

The results of the three scenarios show that the proposed framework can be considered as a practical application in the monitoring of unnatural situations and events in telecom companies. Using this procedure, it is not needed to monitor each hour of the day separately, as it incorporates all of them as a charting statistic. Due to the extension of the profile monitoring statistic, an acceptable detection ability is achieved such that the OC situations in all the scenarios are detected very quickly. Furthermore, consideration of non-normal and autocorrelated conditions provide a very powerful and accurate framework under different model assumptions.

5. Concluding Remarks

Although several novelties and ideas have been extended in the area of profile monitoring, papers about the practical application of profile monitoring are scarce. In line with this knowledge gap, in this paper, on the basis of a profile monitoring idea, a novel practical monitoring framework is proposed which includes five steps: (i) model estimation, (ii) model verification, (iii) Phase I analysis, (iv) Phase II analysis, and finally (v) signal interpretation. In this study, effort is made to consider different practical challenges and assumptions, such as the normality of data, the existence of autocorrelation, etc., and to provide useful solutions for them. The practical implementation of the proposed framework is demonstrated on the internet usage monitoring of telecom companies which has not been investigated in the literature. These companies encounter three major monitoring challenges entailing (i) detection of unnatural patterns, network failure and other abnormal situations in which the total trend of a day is affected and varied, (ii) identifying the impact of policies such as providing discounts or markups for specific hours and (iii) investigation of general social behaviour variations which could also be considered as the customer requirement. For this aim, a polynomial profile model between the hours of each day and the internet usage is defined and monitored over time. The proposed framework is applied to real historical data in three different timestamps and its monitoring ability is investigated. The results show that it can detect unnatural situations very quickly under the proper adjustments and setups.

As this paper proposes a general applicable framework, it is not limited to polynomial profiles and it can be implemented in other real applications with different IC models. Finally, future studies can apply other Phase I and II control charts under the proposed framework and compare them with this approach. The major limitation of this study relates to the violation of predefined assumption such as normality, independence and so forth. In a situation such as the latter, one useful solution is to extend the non-parametric methods in this area of study. Several non-parametric control charts have been proposed and the use of them in this area can be considered as a nice future research direction.

Author Contributions: Conceptualization, U.N., A.Y. and S.C.S.; Formal analysis, A.Y. and U.N.; Investigation, A.Y., U.N., S.C.S. and A.H., Validation, Resources, A.Y. and A.H.; Software, A.Y. and U.N.; Writing—review & editing, A.Y., U.N., S.C.S. and A.H.; Methodology, A.Y. and U.N.; Supervision, A.Y. and S.C.S.; Project Administration, S.C.S. All authors have read and agreed to the published version of the manuscript.

Funding: The Masters research work of U.N. is funded by National Research Foundation (NRF) in South Africa and the research work of A.Y. is funded by University of the Free State Postdoctoral Fellowship.

Institutional Review Board Statement: Not applicable.

Informed Consent Statement: Not applicable.

Data Availability Statement: The data of this study can be given to the interested researchers upon request.

Conflicts of Interest: The authors declare no conflict of interest.

Appendix A. The Proposed Charting Statistic by Noorossana, Amiri and Soleimani [45]

In this approach, it is assumed that profiles are not independent from each other over time. This dependence exists in successive profiles in such a way that the autocorrelation structure follows a first-order autoregressive model with a known coefficient ϕ between the profiles. Furthermore, the IC simple linear model parameters are shown by A_0, A_1 and σ_0^2 which are denoted by intercept, slope and error variance. For the jth generated profile in Phase II, the estimated response variables and errors are obtained based on the previous profile responses as

$$\hat{y}_{ij} = \phi y_{ij} + (1-\phi)(A_0 + A_1 x_i),$$
$$e_{ij} = y_{ij} - \hat{y}_{ij} = y_{ij} - \phi y_{ij} - (1-\phi)(A_0 + A_1 x_i). \quad (A1)$$

Then, the charting statistic is defined as

$$T_j^2 = \underline{e}_j \Sigma_{e_j}^{-1} \underline{e}_j^{-1},$$
$$\Sigma_{e_j} = \sigma_0^2 I, \quad (A2)$$

where I represents the $n \times n$ identity matrix and $\underline{e}_j$ indicates the $1 \times n$ vector of error terms. The UCL of the proposed statistic is computed by $\chi^2_{n,\alpha}$. Furthermore, the estimation of error variance is different from Equation (3), where for the jth generated profile, it is computed as follows:

$$\hat{\sigma}_j^2 = \frac{\sum_{i=1}^n e_{ij}^2}{n}. \quad (A3)$$

References

1. Viharos, Z.J.; Jakab, R. Reinforcement Learning for Statistical Process Control in Manufacturing. *Measurement* **2021**, *182*, 109616. [CrossRef]
2. Montgomery, D.C. *Introduction to Statistical Quality Control*; John Wiley & Sons: Hoboken, NJ, USA, 2020.
3. Yeganeh, A.; Abbasi, S.A.; Shongwe, S.C. A Novel Simulation-Based Adaptive MEWMA Approach for Monitoring Linear and Logistic Profiles. *IEEE Access* **2021**, *9*, 124268–124280. [CrossRef]
4. Liu, Y.; Zhu, J.; Lin, D.K.J. A generalized likelihood ratio test for monitoring profile data. *J. Appl. Stat.* **2021**, *48*, 1402–1415. [CrossRef] [PubMed]
5. Kang, L.; Albin, S.L. On-Line Monitoring When the Process Yields a Linear Profile. *J. Qual. Technol.* **2000**, *32*, 418–426. [CrossRef]
6. Kim, K.; Mahmoud, M.A.; Woodall, W.H. On the Monitoring of Linear Profiles. *J. Qual. Technol.* **2003**, *35*, 317–328. [CrossRef]
7. Woodall, W.H.; Spitzner, D.J.; Montgomery, D.C.; Gupta, S. Using Control Charts to Monitor Process and Product Quality Profiles. *J. Qual. Technol.* **2004**, *36*, 309–320. [CrossRef]
8. Yeganeh, A.; Shongwe, S.C. A novel application of statistical process control charts in financial market surveillance with the idea of profile monitoring. *PLoS ONE* **2023**, *18*, e0288627. [CrossRef]
9. Ding, D.; Tsung, F.; Li, J. Ordinal profile monitoring with random explanatory variables. *Int. J. Prod. Res.* **2017**, *55*, 736–749. [CrossRef]
10. Alevizakos, V.; Koukouvinos, C.; Lappa, A. Comparative study of the Cp and Spmk indices for logistic regression profile using different link functions. *Qual. Eng.* **2019**, *31*, 453–462. [CrossRef]
11. Mohammadzadeh, M.; Yeganeh, A.; Shadman, A. Monitoring logistic profiles using variable sample interval approach. *Comput. Ind. Eng.* **2021**, *158*, 107438. [CrossRef]
12. He, S.; Song, L.; Shang, Y.; Wang, Z. Change-point detection in Phase I for autocorrelated Poisson profiles with random or unbalanced designs. *Int. J. Prod. Res.* **2020**, 1–18. [CrossRef]
13. Steiner, S.; Jensen, W.A.; Grimshaw, S.D.; Espen, B. Nonlinear Profile Monitoring for Oven-Temperature Data. *J. Qual. Technol.* **2016**, *48*, 84–97. [CrossRef]
14. Pacella, M.; Semeraro, Q. Monitoring roundness profiles based on an unsupervised neural network algorithm. *Comput. Ind. Eng.* **2011**, *60*, 677–689. [CrossRef]
15. Ghosh, M.; Li, Y.; Zeng, L.; Zhang, Z.; Zhou, Q. Modeling multivariate profiles using Gaussian process-controlled B-splines. *IISE Trans.* **2021**, *53*, 787–798. [CrossRef]
16. Kazemzadeh, R.B.; Noorossana, R.; Amiri, A. Phase I monitoring of polynomial profiles. *Commun. Stat.—Theory Methods* **2008**, *37*, 1671–1686. [CrossRef]
17. Amiri, A.; Jensen, W.A.; Kazemzadeh, R.B. A case study on monitoring polynomial profiles in the automotive industry. *Qual. Reliab. Eng. Int.* **2010**, *26*, 509–520. [CrossRef]
18. Zhang, Y.; He, Z.; Shan, L.; Zhang, M. Directed control charts for detecting the shape changes from linear profiles to quadratic profiles. *Int. J. Prod. Res.* **2014**, *52*, 3417–3430. [CrossRef]
19. Zhang, Y.; He, Z.; Zhang, M.; Wang, Q. A Score-test-based EWMA Control Chart for Detecting Prespecified Quadratic Changes in Linear Profiles. *Qual. Reliab. Eng. Int.* **2016**, *32*, 921–931. [CrossRef]
20. Zhang, Y.; Shang, Y.; He, Z.; Wang, Q. CUSUM Schemes for Monitoring Prespecified Changes in Linear Profiles. *Qual. Reliab. Eng. Int.* **2017**, *33*, 579–594. [CrossRef]

21. Yao, C.; Li, Z.; He, C.; Zhang, J. A Phase II control chart based on the weighted likelihood ratio test for monitoring polynomial profiles. *J. Stat. Comput. Simul.* **2020**, *90*, 676–698. [CrossRef]
22. Jamal, A.; Mahmood, T.; Riaz, M.; Al-Ahmadi, H.M. GLM-based flexible monitoring methods: An application to real-time highway safety surveillance. *Symmetry* **2021**, *13*, 362. [CrossRef]
23. Gupta, S.; Montgomery, D.C.; Woodall, W.H. Performance evaluation of two methods for online monitoring of linear calibration profiles. *Int. J. Prod. Res.* **2006**, *44*, 1927–1942. [CrossRef]
24. Abbas, T.; Mahmood, T.; Riaz, M.; Abid, M. Improved linear profiling methods under classical and Bayesian setups: An application to chemical gas sensors. *Chemom. Intell. Lab. Syst.* **2020**, *196*, 103908. [CrossRef]
25. Jeong, Y.-S.; Kim, B.; Ko, Y.-D. Exponentially weighted moving average-based procedure with adaptive thresholding for monitoring nonlinear profiles: Monitoring of plasma etch process in semiconductor manufacturing. *Expert Syst. Appl.* **2013**, *40*, 5688–5693. [CrossRef]
26. Azarnoush, B.; Paynabar, K.; Bekki, J.; Runger, G. Monitoring Temporal Homogeneity in Attributed Network Streams. *J. Qual. Technol.* **2016**, *48*, 28–43. [CrossRef]
27. Fotuhi, H.; Amiri, A.; Taheriyoun, A.R. A novel approach based on multiple correspondence analysis for monitoring social networks with categorical attributed data. *J. Stat. Comput. Simul.* **2019**, *89*, 3137–3164. [CrossRef]
28. Yeganeh, A.; Shadman, A.; Shongwe, S.C.; Abbasi, S.A. Employing evolutionary artificial neural network in risk-adjusted monitoring of surgical performance. *Neural Comput. Appl.* **2023**, *35*, 10677–10693. [CrossRef]
29. Wang, Y.; Li, J.; Ma, Y.; Song, L.; Wang, Z. Nonparametric monitoring schemes in Phase II for ordinal profiles with application to customer satisfaction monitoring. *Comput. Ind. Eng.* **2022**, *165*, 107931. [CrossRef]
30. Barbeito, I.; Zaragoza, S.; Tarrío-Saavedra, J.; Naya, S. Assessing thermal comfort and energy efficiency in buildings by statistical quality control for autocorrelated data. *Appl. Energy* **2017**, *190*, 1–17. [CrossRef]
31. Kim, J.; Lim, C. Customer complaints monitoring with customer review data analytics: An integrated method of sentiment and statistical process control analyses. *Adv. Eng. Inform.* **2021**, *49*, 101304. [CrossRef]
32. Chen, K.-Y.; Shaw, Y.-C. Applying back propagation network to cold chain temperature monitoring. *Adv. Eng. Inform.* **2011**, *25*, 11–22. [CrossRef]
33. Freitas, L.L.G.; Kalbusch, A.; Henning, E.; Walter, O.M.F.C. Using Statistical Control Charts to Monitor Building Water Consumption: A Case Study on the Replacement of Toilets. *Water* **2021**, *13*, 2474. [CrossRef]
34. Gebert, S.; Pries, R.; Schlosser, D.; Heck, K. Internet Access Traffic Measurement and Analysis. In *Proceedings of Traffic Monitoring and Analysis*; Springer: Berlin/Heidelberg, Germany, 2012; pp. 29–42.
35. Jovic, J.; Pantovic-Stefanovic, M.; Mitkovic-Voncina, M.; Dunjic-Kostic, B.; Mihajlovic, G.; Milovanovic, S.; Ivkovic, M.; Fiorillo, A.; Latas, M. Internet use during coronavirus disease of 2019 pandemic: Psychiatric history and sociodemographics as predictors. *Indian J. Psychiatry* **2020**, *62*, S383. [CrossRef] [PubMed]
36. Subudhi, R.; Palai, D. Impact of internet use during COVID lockdown. *Horiz. J. Hum. Soc. Sci* **2020**, *2*, 59–66.
37. Zwetsloot, I.M.; Jones-Farmer, L.A.; Woodall, W.H. Monitoring univariate processes using control charts: Some practical issues and advice. *Qual. Eng.* **2023**, 1–13. [CrossRef]
38. Chuang, S.-C.; Hung, Y.-C.; Tsai, W.-C.; Yang, S.-F. A framework for nonparametric profile monitoring. *Comput. Ind. Eng.* **2013**, *64*, 482–491. [CrossRef]
39. Nawaz, M.; Maulud, A.S.; Zabiri, H.; Taqvi, S.A.A.; Idris, A. Improved process monitoring using the CUSUM and EWMA-based multiscale PCA fault detection framework. *Chin. J. Chem. Eng.* **2021**, *29*, 253–265. [CrossRef]
40. Han, D.; Tsung, F. A reference-free cuscore chart for dynamic mean change detection and a unified framework for charting performance comparison. *J. Am. Stat. Assoc.* **2006**, *101*, 368–386. [CrossRef]
41. Motasemi, A.; Alaeddini, A.; Zou, C. An Area-based Methodology for the Monitoring of General Linear Profiles. *Qual. Reliab. Eng. Int.* **2017**, *33*, 159–181. [CrossRef]
42. Zou, C.; Tsung, F.; Wang, Z. Monitoring General Linear Profiles Using Multivariate Exponentially Weighted Moving Average Schemes. *Technometrics* **2007**, *49*, 395–408. [CrossRef]
43. Yeganeh, A.; Abbasi, S.A.; Pourpanah, F.; Shadman, A.; Johannssen, A.; Chukhrova, N. An ensemble neural network framework for improving the detection ability of a base control chart in non-parametric profile monitoring. *Expert Syst. Appl.* **2022**, *204*, 117572. [CrossRef]
44. Malela-Majika, J.-C.; Shongwe, S.C.; Chatterjee, K.; Koukouvinos, C. Monitoring univariate and multivariate profiles using the triple exponentially weighted moving average scheme with fixed and random explanatory variables. *Comput. Ind. Eng.* **2022**, *163*, 107846. [CrossRef]
45. Noorossana, R.; Amiri, A.; Soleimani, P. On the monitoring of autocorrelated linear profiles. *Commun. Stat.—Theory Methods* **2008**, *37*, 425–442. [CrossRef]
46. Soleimani, P.; Noorossana, R.; Amiri, A. Simple linear profiles monitoring in the presence of within profile autocorrelation. *Comput. Ind. Eng.* **2009**, *57*, 1015–1021. [CrossRef]
47. Ahmadi, A.; Yeganeh, A.; Shadman, A. Monitoring simple linear profiles in the presence of within- and between-profile autocorrelation. *Qual. Reliab. Eng. Int.* **2023**, *39*, 752–775.
48. Yeh, A.B.; Huwang, L.; Li, Y.-M. Profile monitoring for a binary response. *IIE Trans.* **2009**, *41*, 931–941. [CrossRef]

49. Williams, J.D.; Woodall, W.H.; Birch, J.B. Statistical monitoring of nonlinear product and process quality profiles. *Qual. Reliab. Eng. Int.* **2007**, *23*, 925–941. [CrossRef]
50. Williams, J.D.; Woodall, W.H.; Birch, J.B.; Sullivan, J.H. Distribution of Hotelling's T 2 statistic based on the successive differences estimator. *J. Qual. Technol.* **2006**, *38*, 217–229. [CrossRef]
51. Yeganeh, A.; Chukhrova, N.; Johannssen, A.; Fotuhi, H. A network surveillance approach using machine learning based control charts. *Expert Syst. Appl.* **2023**, *219*, 119660. [CrossRef]
52. Yeganeh, A.; Shadman, A.; Amiri, A. A novel run rules based MEWMA scheme for monitoring general linear profiles. *Comput. Ind. Eng.* **2021**, *152*, 107031. [CrossRef]
53. Sheu, S.-H.; Lu, S.-L. Monitoring autocorrelated process mean and variance using a GWMA chart based on residuals. *Asia-Pac. J. Oper. Res.* **2008**, *25*, 781–792. [CrossRef]
54. Li, Y.; Pan, E.; Xiao, Y. On autoregressive model selection for the exponentially weighted moving average control chart of residuals in monitoring the mean of autocorrelated processes. *Qual. Reliab. Eng. Int.* **2020**, *36*, 2351–2369. [CrossRef]
55. Holland, M.D.; Hawkins, D.M. A control chart based on a nonparametric multivariate change-point model. *J. Qual. Technol.* **2014**, *46*, 63–77. [CrossRef]
56. Mitra, A.; Lee, K.B.; Chakraborti, S. An adaptive exponentially weighted moving average-type control chart to monitor the process mean. *Eur. J. Oper. Res.* **2019**, *279*, 902–911. [CrossRef]

Disclaimer/Publisher's Note: The statements, opinions and data contained in all publications are solely those of the individual author(s) and contributor(s) and not of MDPI and/or the editor(s). MDPI and/or the editor(s) disclaim responsibility for any injury to people or property resulting from any ideas, methods, instructions or products referred to in the content.

Article

A Combined Runs Rules Scheme for Monitoring General Inflated Poisson Processes

Eftychia Mamzeridou [1] and Athanasios C. Rakitzis [2,*]

[1] Department of Statistics & Actuarial-Financial Mathematics, University of the Aegean, 83200 Samos, Greece; emam@aegean.gr
[2] Department of Statistics and Insurance Science, University of Piraeus, 18534 Piraeus, Greece
* Correspondence: arakitz@unipi.gr; Tel.: +30-2104142452

Abstract: In this work, a control chart with multiple runs rules is proposed and studied in the case of monitoring inflated processes. Usually, Shewhart-type control charts for attributes do not have a lower control limit, especially when the in-control process mean level is very low, such as in the case of processes with a low number of defects per inspected unit. Therefore, it is not possible to detect a decrease in the process mean level. A common solution to this problem is to apply a runs rule on the lower side of the chart. Motivated by this approach, we suggest a Shewhart-type chart, supplemented with two runs rules; one is used for detecting decreases in process mean level, and the other is used for improving the chart's sensitivity in the detection of small and moderate increasing shifts in the process mean level. Using the Markov chain method, we examine the performance of various schemes in terms of the average run length and the expected average run length. Two illustrative examples for the use of the proposed schemes in practice are also discussed. The numerical results show that the considered schemes can detect efficiently various shifts in process parameters in either direction.

Keywords: attributes control chart; average run length count data; expected average run length; inflated Poisson distribution; statistical process monitoring

MSC: 62P30

1. Introduction

Statistical process monitoring (SPM) is a collection of methods and techniques which focus on the monitoring of a process and the timely detection of changes in it. The most frequently used SPM method is the control chart. The most common chart is the Shewhart chart, suggested by Walter A. Shewhart in the 1920's, while there are two other main types of control charts, namely the cumulative sum (CUSUM) chart, proposed by Page [1], and the exponentially weighted moving average (EWMA) chart, proposed by Roberts [2]. A Shewhart chart is useful for the detection of sudden and of large magnitude shifts, while the CUSUM and EWMA charts are better than Shewhart charts in the detection of shifts of small magnitude. See [3] for more details on the properties and applications of the main types of control charts.

The superiority of CUSUM and EWMA charts in the detection of shifts of small magnitude is attributed to their inherent memory, i.e., the respective charting statistics consist of information not only from the most recent sample (or recent observation) but also from the past ones. However, even though their practical implementation is nowadays a routine application, there are still some difficulties in their statistical design, mainly in how to choose the most appropriate values for their chart parameters. From this point of view, intermediate solutions, such as supplementing a Shewhart chart with additional stopping rules based on runs (i.e., runs rules), are still popular in practical problems. We refer to [4,5] for thorough reviews of control charts with runs rules.

Control charts for attributes are used when count data are available from the process. This situation occurs when it is not possible to find a critical-to-quality (CTQ) characteristic X (random variable, r.v.) that follows a continuous distribution. Usually, in the case of attributes charts, the distribution of X can either take values on $\{0, 1, \ldots\}$ (e.g., in the case of monitoring the number of defects that follow a Poisson distribution) or on $\{0, 1, \ldots, n\}$ (e.g., in the case of monitoring the number of defective in a sample of size n, which follows a binomial distribution). Even though Poisson distribution and the associated Shewhart-type chart, the c-chart, are frequently used in the monitoring of count data, there are many cases where both are not appropriate. For example, this is the case of count data that exhibit over-dispersion, i.e., the variance of the distribution of X is much larger than its mean. Recall that in Poisson distribution, mean and variance are equal. Therefore, a solution to this problem is to adopt a distributional model for the available count data from the process that can capture this deviation from the ordinary Poisson model.

Inflated probability distributions have been studied by several authors due to their flexibility in modeling over-dispersion in the data. A common sub-class of the family of inflated distributions is that of zero-inflated distributions (see, for example, Chapter 8 in [6]), where the probability for the occurrence of a zero value is much larger than the respective probability under the non-inflated distribution. Even though inflated distributions (not necessarily only at zero) have been studied in the past (Yoneda [7]), in recent years, there has also been interest in extending them to model the inflation of two or more values. See, for example, [8–11] and references therein for inflated distributions in exactly two values: at 0 and at another one non-zero value. Also, Sun et al. [12] suggested a zero-one-two inflated distribution, a distribution inflated at exactly three values. Begum et al. [13] and Rakitzis et al. [14] proposed a general inflated Poisson model that takes into account the inflation on the first $r+1$ values (i.e., the $\{0, 1, \ldots, r\}$) of the distribution. Also, Rakitzis et al. [14] proposed and studied a two-parameter mixture model, namely the r-geometrically inflated Poisson (GIP_r) distribution, which can model the inflation not only on the zero-values but also on other values of the Poisson distribution, while it has only two parameters.

Control charts for inflated distributions have been studied quite extensively in the recent literature, especially those related to the monitoring of zero-inflated processes. See the overview provided in [15]. Even though in the literature exists almost every main type of chart for inflated distributions (such as Shewhart, CUSUM, and EWMA), there are very few control charts with supplementary runs rules for monitoring this type of process. Actually, the case of attributes control charts with supplementary runs rules has not been paid much attention compared to their variable counterparts. Supplementing a Shewhart chart with runs rules is an easy-to-apply solution for the problem of their insensitivity in detecting small to moderate shifts in process parameter(s). We should also mention that it is very common for attribute charts to have no lower control limit, especially in the case of high-quality processes, where the fraction of non-conforming items is very low. For example, the lower control limit of a c-chart with 3 sigma limits and an IC mean lower than 9 is negative. In this case, it is suggested to use only an upper control limit on the chart (see [16]). Nelson [17], Acosta-Mejia [18], Lucas et al. [19], and Chang and Gan [20] studied attributes control charts with no lower control limit, and instead, they applied a runs rule on the lower side of the chart to detect a decrease in the process mean level.

Usually, high-quality processes can be modeled according to a discrete inflated probability distribution. Motivated by the works in [19,20], we propose and study a control chart with multiple runs rules to monitor a GIP_r process and detect increases as well as decreases in its mean level. The aim is to suggest a control chart for attributes that retains the simplicity of a Shewhart-type chart with runs rules while it has an improved performance compared to its competitors. To the best of our knowledge, in the literature, there are no two-sided control charts with multiple runs rules for monitoring a GIP_r processes. Also, it is worth noting that even though the properties of the proposed scheme are investigated in the case of monitoring a GIP_r process, after some straightforward but necessary modifications, it can be used for the monitoring of any other process which is modeled according to a discrete probability distribution. These are the main scopes and motivations of this work.

The structure of this work is as follows: In Section 2, we present the main properties of the GIP_r distribution. In Section 3, we introduce the proposed scheme and present the measures for evaluating its performance. In Section 4, we present the findings of an extensive numerical study regarding the statistical design and the performance of the proposed chart. Practical guidelines for applying the chart in practice are given as well. In Section 5, we present two illustrative examples (using real data), which show how to implement the proposed charts in practice. Finally, in Section 6, the conclusions and main findings of this work are summarized, while topics for future research are also given.

2. The r-Geometrically Inflated Poisson Distribution

We start this section by presenting the basic properties of the r-geometrically inflated Poisson distribution. Further details can be found in Rakitzis et al. [14]. Let X be a discrete random variable with support $S = \{0, 1, 2, \ldots\}$. If the probability mass function (pmf) of X is

$$f_{GIP_r}(x|\phi, \lambda) = \frac{\phi^{x+1}}{r+1} I_{\{0,1,\ldots,r\}}(x) + \frac{1}{r+1}(r+1 - g_0(r, \phi))\frac{e^{-\lambda}\lambda^x}{x!}, \quad x \in \{0, 1, \ldots\}, \quad (1)$$

then we say that X follows the r-geometrically inflated Poisson distribution with parameters $\phi \in (0,1)$ (the inflation parameter), $\lambda > 0$ and $r \in \{0, 1, 2, \ldots\}$ (i.e., $X \sim GIP_r(\phi, \lambda)$). Also, $g_0(x, \phi) = \phi(1 - \phi^{x+1})/(1 - \phi)$ satisfies the inequality $r + 1 - g_0(r, \phi) > 0$ for every $\phi \in (0,1)$ (so as $f_{GIP_r}(x|\phi, \lambda)$ is a true pmf), while $I_A(x)$ is the usual indicator function. For $r = 0$, the $GIP_r(\phi, \lambda)$ distribution reduces to the ZIP distribution with parameters ϕ and λ.

The cumulative distribution function (cdf) $F_{GIP_r}(x|\phi, \lambda)$ of the $GIP_r(\phi, \lambda)$ distribution is given by

$$F_{GIP_r}(x|\phi, \lambda) = \begin{cases} 0, & x < 0 \\ \frac{1}{r+1}(g_0(\lfloor x \rfloor, \phi) + (r+1 - g_0(r, \phi))F_P(x|\lambda)), & 0 \leq x \leq r \\ \frac{1}{r+1}(g_0(r, \phi) + (r+1 - g_0(r, \phi))F_P(x|\lambda)), & x \geq r+1 \end{cases} \quad (2)$$

The $\lfloor \ldots \rfloor$ represents the smallest integer contained in $\lfloor \ldots \rfloor$ and $F_P(x|\lambda)$ is the cdf of the Poisson distribution with parameter λ. In addition, the mean of the $GIP_r(\phi, \lambda)$ distribution equals

$$\mu_{GIP_r} = E(X) = \frac{g_1(r, \phi) + (r+1 - g_0(r, \phi))\lambda}{r+1}, \quad (3)$$

where, for $\phi \in (0,1)$, it is $g_1(x, \phi) = \left(\frac{x\phi^{x+1} - (x+1)\phi^x + 1}{(1-\phi)^2}\right)\phi^2$.

For a given r value, the parameters ϕ and λ can be estimated by using either the maximum likelihood estimation method or the method of moments; see [14] for further details on estimation methods for the parameters of the $GIP_r(\phi, \lambda)$ distribution. In the present work, we assume that process parameters are known, or that they have been estimated from a (sufficiently) large Phase I sample.

3. The Proposed Monitoring Scheme

3.1. Operation of the Proposed Monitoring Scheme

In this section, we present the operation of the proposed monitoring scheme that can be used for the monitoring of a GIP_r process. Following the setup in [21], we assume that the value of r is predetermined and remains unaffected by the presence of assignable causes. Moreover, in this work the aim is to detect either upward or downward shifts in the in-control (IC) process mean level, which is denoted as μ_{0,GIP_r}. The control charts studied by [21,22] can detect only upward shifts in μ_{0,GIP_r}. Let us also assume that at each sampling stage, an individual observation is obtained from the GIP_r process. We denote as λ_0, ϕ_0 the IC values of the process parameters. When the process is OOC, we assume that the OOC values of the process parameters are $\lambda_1 = \delta \cdot \lambda_0$ and $\phi_1 = \tau \cdot \phi_0$ with $\delta > 0$ and $\tau \in (0, 1/\phi_0]$. When $(\tau, \delta) = (1, 1)$ the process is IC.

Both μ_{0,GIP_r} and μ_{1,GIP_r} are obtained via Equation (3) for the respective IC and OOC values of ϕ and λ. It is worth mentioning that in practice, it is of great importance to detect an increase in the mean of the process, that is, a change from μ_{0,GIP_r} to $\mu_{1,GIP_r} > \mu_{0,GIP_r}$, because it is related to process deterioration. For example, control chart operators are interested in detecting an increase in the expected number of nonconformities of the inspected units or an increase in the expected weekly number of confirmed new infections from a specific disease. In addition, in modern statistical process monitoring (see, for example, [23–26]), the case of decreases is considered very important because it is related to process improvement. In such cases, the interest is in detecting a decrease in the expected number of non-conformities in the inspected units, which can be considered an indication that the attempts for process improvement have been successful. Note also that this improvement can be attributed to the occurrence of assignable causes, such as the recruitment of well-trained personnel or the use of improved raw materials. Consequently, for continuous improvement, it becomes essential to consider the improvement case. In this work, we consider both OOC cases: $\mu_{1,GIP_r} > \mu_{0,GIP_r}$ or $\mu_{1,GIP_r} < \mu_{0,GIP_r}$.

Let l, m, k be positive integers with $2 \leq l \leq m$ and $k \geq 2$. Then, the proposed monitoring scheme, to be denoted as $CRR_{l,m}$, gives an OOC signal when at least one of the following events occurs:

i. A single value X_t is beyond an upper control limit UCL (Region 1);
ii. l-out-of-m successive values are in the interval $(UWL, UCL]$ (Region 2) with the intermediate $m - l$ values being in the interval $(LWL, UWL]$ (Region 3);
iii. k successive values are in the interval $[0, LWL]$ (Region 4);

whichever of the (i)–(iii) occurs first.

The suggested scheme can be viewed as an extension of the schemes studied by [19,20]. Specifically, in the above-mentioned works, the rule for detecting a decreasing shift in μ_{0,GIP_r} is of the type k-out-of-k successive zero values. For the detection of increasing shifts in μ_{1,GIP_r}, Lucas et al. [19] applied only the rule of a single value X_t above an upper control limit UCL while Chang and Gan [20] applied a rule of type l-out-of-m consecutive points above an UCL.

The regions of the $CRR_{l,m}$ chart are given in Figure 1. To apply the proposed control chart in practice, the values of LWL, UWL, UCL, and k must be determined for the given integer values l, m ($2 \leq l \leq m$). This is carried out via an appropriate design procedure; the values (LWL, UWL, UCL, k) are determined so the $CRR_{l,m}$ scheme has the desired IC performance, and it is sensitive enough in the detection of specific shifts in process parameters.

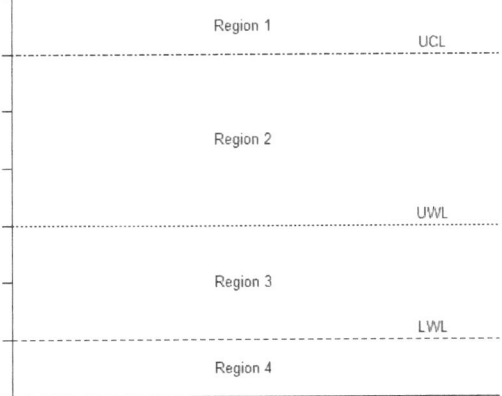

Figure 1. Regions of the $CRR_{l,m}$ control chart.

3.2. Performance of the Proposed Monitoring Scheme

The performance of a process monitoring scheme is evaluated by the run length (RL) distribution, which is defined as the number of points plotted on the chart until it gives for the first time an OOC signal.

The computation of the RL properties of the $CRR_{l,m}$ scheme is feasible using the finite Markov chain embedding technique of Fu and Koutras [27] (see also [28,29] and references therein). Further details on the method can be found in the Appendix A. Using this method, it is possible to calculate the entire RL distribution, as well as its expected value $E(RL)$, which is known as the average run length (ARL). The ARL is the most common performance measure of a control chart.

Apart from the ARL, there are other performance measures that can be viewed as measures of the overall performance of the chart. Specifically, the ARL evaluates the performance of a chart at a specific shift (τ, δ) in process parameters ϕ_0 and λ_0, respectively. In practice, these shifts are rarely known, or practitioners may want to know how the chart performs for a range of shifts. For this reason, in this work, we use also the expected average run length ($EARL$), which is defined as (see, for example, [30,31])

$$EARL = \frac{1}{(\tau_{max} - \tau_{min})(\delta_{max} - \delta_{min})} \int_{\tau_{min}}^{\tau_{max}} \int_{\delta_{min}}^{\delta_{max}} ARL(\tau, \delta) d\delta d\tau,$$

where $[\delta_{min}, \delta_{max}]$ and $[\tau_{min}, \tau_{max}]$ are the intervals for the shifts δ and τ in λ_0 and ϕ_0, respectively. The chart with the minimum $EARL$ is the one with the best overall performance at the specific range of shifts.

4. Numerical Results

In this section, we present the results of a numerical study regarding the performance of the $CRR_{l,m}$ chart. Specifically, we consider the following specific values for l, m (see Table 1):

Table 1. The $CRR_{l,m}$ schemes under study.

Scheme	1	2	3	4	5	6	7
l	2	2	2	2	3	4	5
m	2	3	4	5	4	5	5
Notation	$CRR_{2,2}$	$CRR_{2,3}$	$CRR_{2,4}$	$CRR_{2,5}$	$CRR_{3,4}$	$CRR_{4,5}$	$CRR_{5,5}$

These values correspond to very well-known runs rules such as the 2-of-2, the 2-of-3, and the 4-of-5, to mention a few.

Next, in Table 2 we provide six indicative scenarios for different IC GIP_r processes. Cases 5 and 6 correspond to a zero-inflated Poisson (ZIP) process. The IC process mean level is given in the third row of Table 2 and varies from 0.4 to 2.6250 for the different 6 scenarios. Thus, they can be considered cases of high-quality processes.

Table 2. Scenarios for the IC GIP_r Processes.

Case	1	2	3	4	5	6
(r, ϕ_0, λ_0)	$(3, 0.7, 3)$	$(3, 0.7, 1.5)$	$(2, 0.9, 3)$	$(1, 0.5, 4)$	$(0, 0.8, 2)$	$(0, 0.9, 6)$
μ_{0, GIP_r}	2.1442	1.3091	1.3170	2.6250	0.4000	0.6000

For each IC scenario in Table 2, we determined the values of the design parameters LWL, UWL, UCL, and k for the seven schemes in Table 1 to have comparable IC performance. For illustrative purposes, we considered an IC ARL approximately equal to 100. Any other value can be used depending on practitioners' needs.

The design procedure is based on an extensive grid search on the possible positive integer values for LWL, UWL, UCL, and k which satisfy the following two constraints: (I) $0 \leq LWL < UWL < UCL \leq 15$ and (II) $k \in \{7, 8, \ldots, 50\}$. Then, any combination of LWL, UWL, UCL, and k that gives an IC ARL approximately equal to the pre-specified $ARL_0 = 100$ value, i.e., in the interval $(98, 102)$, is considered accepted.

Once all the possible combinations for (LWL, UWL, UCL, k) that give the desired IC performance have been determined, we evaluated the OOC ARL of the proposed charts for several pairs of shifts (τ, δ). The considered pairs are $(\tau, \delta) \in \{0.6, 0.8, 1, 1.1\} \times \{0.5, 0.8, 1, 1.2, 1.5\}$ while for $(\tau, \delta) = (1, 1)$, the process is IC.

In Tables 3–8 we provide the best charts, i.e., the ones with the minimum OOC ARL values, for various shifts in ϕ_0 and λ_0 when the IC ARL is (approximately) equal to 100. The shifts are given in columns 'τ' and 'δ', while the OOC values λ_1 and ϕ_1 of the process parameters are given in the respective columns. The column 'μ_{1,GIP_r}' gives the OOC process mean level, while the column '$\mu_{1,GIP_r}/\mu_{0,GIP_r}$' gives the change in IC process mean level. Note that when $\frac{\mu_{1,GIP_r}}{\mu_{0,GIP_r}} < 1$, there is a decrease in the process mean level, otherwise, there is an increase. Also, the results in Tables 3–8 have been sorted according to the values of this column to discriminate the decreases from the increases.

Table 3. Best charts in terms of ARL, Case 1, $ARL_0 \approx 100$.

τ	δ	λ_1	ϕ_1	μ_{1,GIP_r}	$\frac{\mu_{1,GIP_r}}{\mu_{0,GIP_r}}$	ARL_1	LWL	UWL	UCL	k	Scheme	SH $UCL_{SH}=7$	η/η $\eta=3$	Combined $UCL_{SH}=7, \eta=4$
1	0.5	1.5	0.70	1.3092	0.61	18.72	3	6	10	14	$CRR_{2,2}$		51.84	173.69
0.8	0.5	1.5	0.56	1.3096	0.61	19.07	2	3	15	8	$CRR_{4,5}$		51.89	173.11
1.1	0.5	1.5	0.77	1.3260	0.62	17.73	3	6	10	14	$CRR_{2,2}$		54.06	184.28
0.6	0.5	1.5	0.42	1.3414	0.63	19.63	2	3	15	8	$CRR_{4,5}$		56.87	194.78
1.1	0.8	2.4	0.77	1.7376	0.81	34.07	3	6	10	14	$CRR_{2,2}$		100.60	261.13
1	0.8	2.4	0.70	1.8102	0.84	42.90	3	6	10	14	$CRR_{2,2}$		111.30	258.62
0.8	0.8	2.4	0.56	1.9514	0.91	63.42	3	6	10	14	$CRR_{2,2}$		145.35	264.59
1.1	1	3	0.77	2.0119	0.94	65.00	3	6	10	14	$CRR_{2,2}$		126.47	140.21
0.6	0.8	2.4	0.42	2.0835	0.97	85.58	3	6	10	14	$CRR_{2,2}$		208.52	276.92
1	1	3	0.70	2.1442	1.00							150.89	149.31	125.37
1.1	1.2	3.6	0.77	2.2863	1.07	48.53	0	5	7	7	$CRR_{2,4}$	71.03		64.58
0.8	1	3	0.56	2.3793	1.11	59.11	2	3	9	12	$CRR_{3,4}$	117.80		107.85
1	1.2	3.6	0.70	2.4783	1.16	37.64	0	5	7	7	$CRR_{2,4}$	58.34		54.94
0.6	1	3	0.42	2.5783	1.20	41.15	2	3	9	12	$CRR_{3,4}$	101.87		97.99
1.1	1.5	4.5	0.77	2.6978	1.26	17.90	0	5	7	7	$CRR_{2,4}$	25.26		24.50
0.8	1.2	3.6	0.56	2.8072	1.31	26.17	1	4	9	10	$CRR_{2,2}$	45.55		44.44
1	1.5	4.5	0.70	2.9793	1.39	14.05	0	5	7	7	$CRR_{2,4}$	20.74		20.37
0.6	1.2	3.6	0.42	3.0730	1.43	18.46	2	3	9	12	$CRR_{3,4}$	39.39		39.03
0.8	1.5	4.5	0.56	3.4490	1.61	10.30	0	5	7	7	$CRR_{2,4}$	16.20		16.09
0.6	1.5	4.5	0.42	3.8152	1.78	8.55	0	5	7	7	$CRR_{2,4}$	14.01		13.98

Table 4. Best charts in terms of ARL, Case 2, $ARL_0 \approx 100$.

τ	δ	λ_1	ϕ_1	μ_{1,GIP_r}	$\frac{\mu_{1,GIP_r}}{\mu_{0,GIP_r}}$	ARL_1	LWL	UWL	UCL	k	Scheme	SH $UCL_{SH}=4$	η/η $\eta=4$	Combined $UCL_{SH}=5, \eta=4$
0.6	0.5	0.75	0.42	0.7229	0.55	20.09	1	4	5	8	$CRR_{2,5}$		31.10	31.00
0.8	0.5	0.75	0.56	0.7748	0.59	23.43	1	4	5	8	$CRR_{2,5}$		35.06	34.94
1	0.5	0.75	0.70	0.8916	0.68	33.77	1	2	5	8	$CRR_{5,5}$		46.58	46.42
1.1	0.5	0.75	0.77	0.9831	0.75	46.23	1	2	5	8	$CRR_{5,5}$		59.02	58.82
0.6	0.8	1.2	0.42	1.0940	0.84	52.88	1	4	5	8	$CRR_{2,5}$		97.64	87.12
0.8	0.8	1.2	0.56	1.0957	0.84	55.47	1	4	5	8	$CRR_{2,5}$		96.27	87.28
1	0.8	1.2	0.70	1.1421	0.87	69.62	1	2	5	8	$CRR_{5,5}$		108.80	99.74
1.1	0.8	1.2	0.77	1.1889	0.91	86.51	1	2	5	8	$CRR_{5,5}$		124.06	114.33
1	1	1.5	0.7	1.3091	1.00							96.70	176.58	122.79
0.8	1	1.5	0.56	1.3096	1.00	84.94	1	3	6	8	$CRR_{2,5}$	75.49		113.21
1.1	1	1.5	0.77	1.3260	1.01	98.12	1	2	7	9	$CRR_{3,4}$	117.73		135.38

Table 4. Cont.

τ	δ	λ_1	ϕ_1	μ_{1,GIP_r}	$\frac{\mu_{1,GIP_r}}{\mu_{0,GIP_r}}$	ARL_1	LWL	UWL	UCL	k	Scheme	SH $UCL_{SH}=4$	η/η $\eta=4$	Combined $UCL_{SH}=5, \eta=4$
0.6	1	1.5	0.42	1.3414	1.02	81.71	1	3	6	8	$CRR_{2,5}$	65.28		115.36
1.1	1.2	1.8	0.77	1.4632	1.12	77.65	1	2	7	9	$CRR_{3,4}$	60.07		117.02
1	1.2	1.8	0.70	1.4762	1.13	77.82	1	2	7	9	$CRR_{3,4}$	49.34		105.01
0.8	1.2	1.8	0.56	1.5236	1.16	73.34	1	2	7	9	$CRR_{3,4}$	38.52		93.48
0.6	1.2	1.8	0.42	1.5888	1.21	63.63	1	3	6	8	$CRR_{2,5}$	33.31		89.78
1.1	1.5	2.25	0.77	1.6690	1.27	46.85	1	2	7	9	$CRR_{3,4}$	28.03		66.36
1	1.5	2.25	0.70	1.7267	1.32	42.50	1	2	7	9	$CRR_{3,4}$	23.02		57.06
0.8	1.5	2.25	0.56	1.8445	1.41	33.58	1	3	6	8	$CRR_{2,5}$	17.96		47.17
0.6	1.5	2.25	0.42	1.9598	1.50	26.71	1	3	6	8	$CRR_{2,5}$	15.54		42.26

Table 5. Best charts in terms of ARL, Case 3, $ARL_0 \approx 100$.

τ	δ	λ_1	ϕ_1	μ_{1,GIP_r}	$\frac{\mu_{1,GIP_r}}{\mu_{0,GIP_r}}$	ARL_1	LWL	UWL	UCL	k	Scheme	SH $UCL_{SH}=6$	η/η $\eta=4$	Combined $UCL_{SH}=6, \eta=5$
1.1	0.5	1.5	0.99	1.0034	0.76	73.09	1	5	7	8	$CRR_{2,5}$		118.61	307.21
1.1	0.8	2.4	0.99	1.0212	0.78	76.25	1	5	7	8	$CRR_{2,5}$		121.98	341.42
1.1	1	3	0.99	1.0332	0.78	76.81	1	5	7	8	$CRR_{2,5}$		123.05	299.94
1	0.5	1.5	0.90	1.0365	0.79	68.60	1	3	7	8	$CRR_{3,4}$		109.88	307.21
1.1	1.2	3.6	0.99	1.0451	0.79	75.63	1	5	7	8	$CRR_{2,5}$		123.64	234.44
1.1	1.5	4.5	0.99	1.0630	0.81	70.47	1	5	7	8	$CRR_{2,5}$		124.07	166.75
0.8	0.5	1.5	0.72	1.1158	0.85	70.10	1	2	7	8	$CRR_{4,5}$		108.15	279.85
1	0.8	2.4	0.90	1.2048	0.91	100.78	1	3	7	8	$CRR_{3,4}$		143.59	229.34
0.6	0.5	1.5	0.54	1.2076	0.92	82.37	1	2	7	8	$CRR_{4,5}$		124.94	308.88
1	1	3	0.90	1.3170	1.00							159.59	156.73	121.55
1	1.2	3.6	0.90	1.4292	1.09	51.35	1	3	6	14	$CRR_{2,2}$	72.98		64.33
0.8	0.8	2.4	0.72	1.5323	1.16	64.24	1	3	11	9	$CRR_{2,4}$	186.37		150.42
1	1.5	4.5	0.90	1.5975	1.21	25.34	1	3	6	14	$CRR_{2,2}$	31.65		29.98
0.8	1	3	0.72	1.8100	1.37	28.91	0	3	6	7	$CRR_{2,2}$	64.49		60.84
0.6	0.8	2.4	0.54	1.8109	1.38	35.98	1	3	11	9	$CRR_{2,4}$	128.67		119.23
0.8	1.2	3.6	0.72	2.0877	1.59	15.52	0	3	6	7	$CRR_{2,2}$	29.49		28.84
0.6	1	3	0.54	2.2131	1.68	16.13	1	3	7	10	$CRR_{2,3}$	44.52		43.84
0.8	1.5	4.5	0.72	2.5042	1.90	8.34	0	3	6	7	$CRR_{2,2}$	12.79		12.68
0.6	1.2	3.6	0.54	2.6153	1.99	9.19	0	3	6	7	$CRR_{2,2}$	20.36		20.26
0.6	1.5	45	0.54	3.2186	2.44	5.15	0	3	6	7	$CRR_{2,2}$	8.83		8.82

Table 6. Best charts in terms of ARL, Case 4, $ARL_0 \approx 100$.

τ	δ	λ_1	ϕ_1	μ_{1,GIP_r}	$\frac{\mu_{1,GIP_r}}{\mu_{0,GIP_r}}$	ARL_1	LWL	UWL	UCL	k	Scheme	SH $UCL_{SH}=8$	η/η $\eta=3$	Combined $UCL_{SH}=9, \eta=4$
1.1	0.5	2	0.55	1.2988	0.49	14.14	3	7	10	9	$CRR_{2,5}$		33.68	98.08
1	0.5	2	0.5	1.3750	0.52	14.79	3	7	10	9	$CRR_{2,5}$		38.62	118.01
0.8	0.5	2	0.4	1.5200	0.58	15.56	4	11	15	12	$CRR_{2,5}$		52.67	179.34
0.6	0.5	2	0.3	1.6550	0.63	16.08	4	11	15	12	$CRR_{2,5}$		199.06	295.48
1.1	0.8	3.2	0.55	1.9873	0.76	31.97	4	11	15	12	$CRR_{2,5}$		52.22	151.12
1	0.8	3.2	0.5	2.1250	0.81	35.51	4	11	15	12	$CRR_{2,5}$		64.64	188.76
0.8	0.8	3.2	0.4	2.3840	0.91	43.49	4	11	15	12	$CRR_{2,5}$		106.26	293.53
1.1	1	4	0.55	2.4463	0.93	78.41	4	11	15	12	$CRR_{2,5}$		58.74	105.87
0.6	0.8	3.2	0.3	2.6210	1.00	52.64	4	11	15	12	$CRR_{2,5}$		199.06	428.80
1	1	4	0.5	2.6250	1.00							74.89	74.41	116.96
1.1	1.2	4.8	0.55	2.9053	1.11	37.28	0	6	10	9	$CRR_{2,4}$	31.23		53.02
0.8	1	4	0.4	2.9600	1.13	72.66	0	4	9	7	$CRR_{2,3}$	65.01		133.62
1	1.2	4.8	0.5	3.1250	1.19	32.10	0	6	10	9	$CRR_{2,4}$	28.67		52.96
0.6	1	4	0.3	3.2650	1.24	54.41	0	4	9	7	$CRR_{2,3}$	58.15		139.62
0.8	1.2	4.8	0.4	3.5360	1.35	24.96	0	6	10	9	$CRR_{2,4}$	24.88		51.20
1.1	1.5	6	0.55	3.5938	1.37	13.01	0	6	10	9	$CRR_{2,4}$	11.41		19.08
1	1.5	6	0.5	3.8750	1.48	11.41	0	6	10	9	$CRR_{2,4}$	10.47		18.03
0.6	1.2	4.8	0.3	3.9090	1.49	19.71	0	4	9	7	$CRR_{2,3}$	22.26		48.20
0.8	1.5	6	0.4	4.4000	1.68	9.16	0	6	10	9	$CRR_{2,4}$	9.09		16.19
0.6	1.5	6	0.3	4.8750	1.86	7.69	0	6	10	9	$CRR_{2,4}$	8.13		14.70

Table 7. Best charts in terms of ARL, Case 5, $ARL_0 \approx 100$.

τ	δ	λ_1	ϕ_1	μ_{1,GIP_r}	$\frac{\mu_{1,GIP_r}}{\mu_{0,GIP_r}}$	ARL_1	LWL	UWL	UCL	k	Scheme	SH $UCL_{SH}=4$	η/η $\eta=15$	Combined $UCL_{SH}=5, \eta=17$
1.1	0.5	1	0.88	0.1200	0.30	30.50	1	4	6	21	$CRR_{2,5}$		29.86	98.08
1.1	0.8	1.6	0.88	0.1920	0.48	42.33	1	4	6	21	$CRR_{2,5}$		36.83	151.12
1	0.5	1	0.8	0.2000	0.50	40.23	1	4	6	21	$CRR_{2,5}$		52.16	118.01
1.1	1	2	0.88	0.2400	0.60	50.72	1	4	6	21	$CRR_{2,5}$		40.21	105.87
1.1	1.2	2.4	0.88	0.2880	0.72	54.11	0	2	7	18	$CRR_{2,2}$		42.69	53.02
1	0.8	1.6	0.8	0.3200	0.80	70.27	3	4	8	47	$CRR_{2,5}$		78.81	188.76
0.8	0.5	1	0.64	0.3600	0.90	55.64	3	4	8	47	$CRR_{2,5}$		206.93	179.34
1.1	1.5	3	0.88	0.3600	0.90	45.33	0	2	5	22	$CRR_{2,5}$		45.14	19.08
1	1	2	0.8	0.4000	1.00							94.96	93.99	116.96
1	1.2	2.4	0.8	0.4800	1.20	63.72	0	2	5	22	$CRR_{2,5}$	52.15		52.96
0.6	0.5	1	0.48	0.5200	1.30	60.16	3	4	8	47	$CRR_{2,5}$	525.45		295.48
0.8	0.8	1.6	0.64	0.5760	1.44	100.18	3	4	8	47	$CRR_{2,5}$	117.29		293.53
1	1.5	3	0.8	0.6000	1.50	35.55	0	2	5	22	$CRR_{2,5}$	27.07		18.03
0.8	1	2	0.64	0.7200	1.80	51.68	0	2	5	22	$CRR_{2,5}$	52.76		133.62
0.6	0.8	1.6	0.48	0.8320	2.08	55.88	0	2	5	22	$CRR_{2,5}$	81.20		428.80
0.8	1.2	2.4	0.64	0.8640	2.16	29.68	0	2	5	22	$CRR_{2,5}$	28.97		51.20
0.6	1	2	0.48	1.0400	2.60	26.71	0	2	5	22	$CRR_{2,5}$	36.52		139.62
0.8	1.5	3	0.64	1.0800	2.70	16.55	0	2	5	22	$CRR_{2,5}$	15.04		16.19
0.6	1.2	2.4	0.48	1.2480	3.12	16.01	0	2	5	22	$CRR_{2,5}$	20.06		48.20
0.6	1.5	3	0.48	1.5600	3.90	9.49	0	2	5	22	$CRR_{2,5}$	10.41		14.70

Table 8. Best charts in terms of ARL, Case 6, $ARL_0 \approx 100$.

τ	δ	λ_1	ϕ_1	μ_{1,GIP_r}	$\frac{\mu_{1,GIP_r}}{\mu_{0,GIP_r}}$	ARL_1	LWL	UWL	UCL	k	Scheme	SH $UCL_{SH}=9$	η/η $\eta=23$	Combined $UCL_{SH}=10, \eta=27$
1.1	0.5	3	0.99	0.030	0.05	25.84	0	1	14	23	$CRR_{4,5}$		25.84	30.94
1.1	0.8	4.8	0.99	0.048	0.08	25.98	0	1	14	23	$CRR_{4,5}$		25.98	31.04
1.1	1	6	0.99	0.060	0.10	25.99	0	1	14	23	$CRR_{4,5}$		26.00	30.75
1.1	1.2	7.2	0.99	0.072	0.12	25.95	0	1	14	23	$CRR_{4,5}$		26.00	30.11
1.1	1.5	9	0.99	0.090	0.15	25.73	0	1	14	23	$CRR_{4,5}$		26.01	28.56
1	0.5	3	0.9	0.300	0.50	38.21	5	12	15	33	$CRR_{2,4}$		94.07	144.80
1	0.8	4.8	0.9	0.480	0.80	63.51	6	9	15	40	$CRR_{2,5}$		101.32	136.47
1	1	6	0.9	0.600	1.00							119.16	102.37	95.51
1	1.2	7.2	0.9	0.720	1.20	47.29	1	6	9	49	$CRR_{2,5}$	52.53		57.09
0.8	0.5	3	0.72	0.840	1.40	23.51	0	1	13	45	$CRR_{2,5}$	3239.43		6913.29
1	1.5	9	0.9	0.900	1.50	23.11	1	6	9	49	$CRR_{2,5}$	24.24		28.11
0.8	0.8	4.8	0.72	1.344	2.24	17.64	0	1	13	45	$CRR_{2,5}$	142.06		337.92
0.6	0.5	3	0.54	1.380	2.30	9.58	0	1	13	45	$CRR_{2,5}$	1971.82		7431.91
0.8	1	6	0.72	1.680	2.80	16.61	0	1	13	45	$CRR_{2,5}$	42.56		83.51
0.8	1.2	7.2	0.72	2.016	3.36	15.32	1	6	9	49	$CRR_{2,5}$	18.76		31.48
0.6	0.8	4.8	0.54	2.208	3.68	7.39	0	1	13	45	$CRR_{2,5}$	86.47		208.69
0.8	1.5	9	0.72	2.520	4.20	7.91	1	6	9	49	$CRR_{2,5}$	8.66		12.14
0.6	1	6	0.54	2.760	4.60	7.03	0	1	13	45	$CRR_{2,5}$	25.90		51.01
0.6	1.2	7.2	0.54	3.312	5.52	6.80	0	1	13	45	$CRR_{2,5}$	11.42		19.18
0.6	1.5	9	0.54	4.140	6.90	4.58	1	6	9	49	$CRR_{2,5}$	5.27		7.39

For comparison purposes, we included the ARL profiles of the following schemes: The upper one-sided GIP_r Shewhart chart in Mamzeridou and Rakitzis [21] (in column 'SH'), a lower one-sided scheme, which is the add-on procedure suggested by Lucas et al. [19] and gives an OOC signal when η-out-of-η ($\eta \geq 2$) consecutive zero values occur (in column 'η/η') and a two-sided scheme (in column 'Combined') that combines the previously mentioned one-sided schemes. This combined scheme is the procedure suggested by Lucas et al. [19]. Note also that the GIP_r Shewhart chart has one upper control limit UCL_{SH}; thus, its ARL performance is evaluated only for increasing shifts in μ_{0,GIP_r}. The value of UCL_{SH} is determined so as its IC performance is as close as possible to 100 in order to have a fair comparison with the other schemes. Given the values of UCL_{SH}, ϕ_1, and λ_1, the ARL is calculated by

$$ARL_U = \frac{1}{1 - F_{GIP_r}(UCL_{SH}|\phi_1, \lambda_1)}$$

Furthermore, the ARL for the η/η scheme is given by (see [19])

$$ARL_L = \frac{1 - p_0^\eta}{p_0^\eta(1 - p_0)}$$

where $p_0 = F_{GIP_r}(0|\phi_1, \lambda_1)$. Also, its ARL performance is evaluated only for decreasing shifts in μ_{0,GIP_r}. The value of η is determined so the IC ARL_L is also as close as possible to 100. Finally, the ARL of the two-sided that combines the two one-sided procedures equals

$$ARL_C = \frac{1 - p_0^\eta}{1 - p_0 - p_1\left(1 - p_0^\eta\right)}$$

where $p_1 = F_{GIP_r}(UCL_{SH}|\phi_1, \lambda_1) - F_{GIP_r}(0|\phi_1, \lambda_1)$. Again, the values of (UCL_{SH}, η) are determined so the IC ARL_C is as close as possible to 100.

The numerical results for the $CRR_{l,m}$ charts show that there is not a single chart that has the best performance, neither for all shifts nor for all different cases. For the upper side of the chart, we suggest using the rule 2-out-of-m, $m \in \{2, 3, 4, 5\}$, depending on the IC values of the process parameters as well as on the values of the shifts (τ, δ). For the lower side of the chart, larger values for k are needed when $r = 0$ than when $r \geq 1$. For the latter, a value around 11–15 is sufficient in most of the cases, but the exact value depends on the shift we want to detect.

The comparison with two one-sided charts and the combined two-sided chart shows that there is at least one $CRR_{l,m}$ chart that has better ARL performance, in almost all the considered cases. It should be mentioned that due to the discrete nature of the process, the IC ARL of the upper GIP_r Shewhart chart, the lower-sided η/η scheme, and the combined two-sided chart is not always comparable with that of the $CRR_{l,m}$. Therefore, conclusions should be made with caution. Another interesting point is that the rule of η-out-of-η consecutive zero values works well in the case of zero-inflated Poisson processes (Cases 5 and 6) but its performance is weak in the remaining ones. Thus, the add-on procedure suggested by Lucas et al. [19] is still effective when there is an excessive number of zero values in the data but when there is inflation in zero and non-zero values, a modification is needed. This modification is offered by the proposed $CRR_{l,m}$ chart.

Another interesting finding is that there are cases where a change in one of the process parameters results in a $\mu_{1,GIP_r} \approx \mu_{0,GIP_r}$. See, for example, Case 2 for $(\tau, \delta) = (0.8, 1)$ and $(1.1, 1)$ or Case 3 for $(\tau, \delta) = (1, 0.8)$. The charts can only marginally detect this change while they cannot distinguish the type of shift. This is one of the limitations of this study. A solution is to use, if possible, rational samples of size $n \geq 2$ (instead of individual observations), as well as an appropriate statistic (e.g., the MLEs from each sample) to improve chart's ability to discriminate between the different types of shifts. This topic needs a separate careful investigation, and it is left for a future study.

Next, in Tables 9 and 10, we provide the $EARL$ values for the seven charts in Table 1 and the combined two-sided chart for each of the six cases in Table 2. For the shifts (τ, δ) in process parameters, we considered the following two scenarios:

- Scenario 1: $[\tau_{min}, \tau_{max}] = [0.6, 1.1]$ and $[\delta_{min}, \delta_{max}] = [0.5, 1.5]$
- Scenario 2: $[\tau_{min}, \tau_{max}] = [0.3, 1.1]$ and $[\delta_{min}, \delta_{max}] = [0.3, 2.0]$

Table 9. Best charts based on $EARL$, $ARL_0 \approx 100$, Scenario 1.

Case	$CRR_{2,2}$	$CRR_{2,3}$	$CRR_{2,4}$	$CRR_{2,5}$	$CRR_{3,4}$	$CRR_{4,5}$	$CRR_{5,5}$	Combined
1	65.31 2, 6, 7, 10	64.89 2, 6, 7, 10	59.30 2, 5, 8, 9	58.76 2, 5, 8, 9	61.34 2, 3, 10, 11	64.19 2, 3, 12, 8	63.94 2, 3, 7, 10	142.59 7, 4
2	68.67 1, 4, 5, 8	68.16 1, 4, 5, 8	64.57 1, 3, 6, 8	63.92 1, 3, 6, 8	67.95 1, 3, 5, 8	69.72 1, 3, 5, 8	68.53 1, 2, 5, 8	84.88 5, 4

Table 9. Cont.

Case	$CRR_{2,2}$	$CRR_{2,3}$	$CRR_{2,4}$	$CRR_{2,5}$	$CRR_{3,4}$	$CRR_{4,5}$	$CRR_{5,5}$	Combined
3	134.41 1, 3, 6, 14	64.67 1, 3, 9, 9	63.47 1, 3, 11, 9	87.35 1, 4, 6, 10	76.26 1, 3, 7, 8	74.22 1, 2, 7, 8	90.79 1, 2, 7, 8	130.75 6, 5
4	64.39 3, 7, 10, 9	50.53 4, 6, 11, 15	51.31 4, 6, 12, 15	51.17 4, 6, 12, 15	51.56 4, 5, 10, 14	54.06 4, 5, 9, 15	56.68 3, 4, 9, 10	144.35 9, 4
5	78.45 1, 2, 6, 25	127.77 1, 3, 7, 21	94.11 0, 2, 6, 19	93.71 0, 2, 6, 19	116.05 3, 4, 6, 49	116.44 3, 4, 6, 49	116.45 3, 4, 6, 49	95.68 5, 17
6	42.13 0, 1, 14, 43	42.17 0, 1, 13, 45	42.16 0, 1, 13, 45	42.17 0, 1, 13, 45	137.79 7, 8, 13, 50	143.94 3, 4, 9, 45	144.82 7, 8, 13, 50	413.46 10, 27

Table 10. Best charts based on $EARL$, $ARL_0 \approx 100$, Scenario 2.

Case	$CRR_{2,2}$	$CRR_{2,3}$	$CRR_{2,4}$	$CRR_{2,5}$	$CRR_{3,4}$	$CRR_{4,5}$	$CRR_{5,5}$	Combined
1	46.17 2, 5, 7, 11	44.67 2, 6, 7, 10	40.29 2, 5, 8, 9	39.84 2, 5, 8, 9	38.92 2, 3, 10, 11	43.07 2, 3, 7, 11	43.46 2, 3, 7, 10	104.55 7, 4
2	48.90 1, 4, 5, 8	48.32 1, 4, 5, 8	43.39 1, 3, 6, 8	42.86 1, 3, 6, 8	47.71 1, 2, 7, 9	49.89 1, 3, 5, 8	48.76 1, 2, 5, 8	60.22 5, 4
3	101.20 1, 3, 6, 14	43.21 1, 3, 9, 9	42.83 1, 3, 11, 9	61, 64 1, 4, 6, 10	52.69 1, 3, 7, 8	48.28 1, 2, 7, 8	131.31 1, 2, 7, 8	100.17 6, 5
4	46.58 3, 7, 10, 9	34.15 4, 6, 11, 15	34.58 4, 6, 12, 15	34.47 4, 6, 12, 15	34.72 4, 5, 10, 14	37.15 4, 5, 9, 15	39.54 3, 4, 9, 10	141.35 9, 4
5	58.78 1, 2, 6, 25	85.73 1, 3, 5, 24	85.23 1, 3, 5, 24	85.11 1, 3, 5, 24	95.02 3, 4, 6, 49	94.99 3, 4, 6, 49	95.05 3, 4, 6, 49	269.10 5, 17
6	27.17 0, 1, 14, 43	26.50 0, 1, 13, 45	26.49 0, 1, 13, 45	26.49 0, 1, 13, 45	116.18 7, 8, 13, 50	90.86 0, 1, 14, 23	118.89 7, 8, 13, 50	6091.24 10, 27

The first line of each cell consists of the $EARL$ value (in two decimals accuracy), while in the second line we provide the values of the chart's design parameters in the form (LWL, UWL, UCL, k). For the combined chart, the parameters are given as (UCL_{SH}, η).

The results in Tables 9 and 10 reveal that there is not one scheme among the examined ones that outperforms the others in terms of $EARL$. However, in none of the six cases do the $CRR_{4,5}$ and $CRR_{5,5}$ charts achieve the minimum $EARL$ value, while the $CRR_{3,4}$ is the best chart in Case 1 and only under Scenario 2. On the other hand, one of the schemes $CRR_{2,m}$, $m \in \{2, 3, 4, 5\}$ is the best scheme in almost all Cases, under both Scenarios. We also notice that the combined scheme does not have a competitive $EARL$ performance except for Case 5 under Scenario 1.

5. Examples

In this section, we present two real data examples regarding the application of the proposed charts in practice. First, we discuss an application of the proposed chart in the monitoring of changes in the number of unintentional needle-stick occurrences per day in a hospital. The second example is from the area of public health and focuses on the monitoring of the monthly number of poliomyelitis cases in the USA. Both examples highlight the usefulness of control charts as powerful tools in biosurveillance and the detection of abnormalities. Consequently, authorities (such as hospital's administration or the center for disease prevention and control) can be warned based on the signals given by the charts and take actions to improve the quality of the services offered by the professionals in health-care units.

5.1. Monitoring the Daily Number of Unintentional Needle-Stick Occurrences

In the first example, we consider a case where we are interested in the quality of the provided healthcare services. An increase on the average daily number of unintentional needle-stick occurrences is an indication of process deterioration, while a decrease shows improved performance by the hospital staff.

We use the dataset provided by Fatahi et al. [32] and analyzed further by Aly et al. [33]. Following [33], we assume that the data are from a ZIP process, and we use the first 40 observations as a Phase I sample to estimate the process parameters. Using the maximum likelihood method, we obtain $\hat{\lambda} \approx 2.38$, $\hat{\phi} \approx 0.56$ (in two decimals accuracy). Then, we apply the Shewhart chart as a Phase I analysis method. For illustrative purposes we assume that the desired IC ARL value is (approximately) equal to 200, yielding a UCL for the Shewhart chart equal to 6. The theoretical IC performance of the chart implies that on average, a false alarm signal would be triggered once in 6.5 months. Note also that the (theoretical) IC ARL equals 204.39.

The respective chart is given in Figure 2 from which we notice that none of the 40 points exceeds the UCL. Therefore, the process was IC, when these values have been collected, and we can proceed to the Phase II analysis by assuming that the IC values of the process parameters equal the respective estimates, that is $\lambda_0 = 2.38$, $\phi_0 = 0.56$.

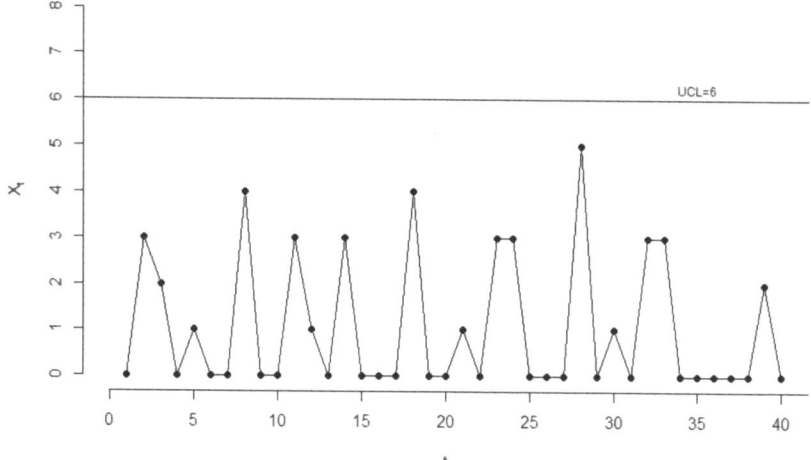

Figure 2. The ZIP Shewhart chart for the Phase I data of the number of unintentional needle-stick occurrences per day in a hospital.

Since there are no further details regarding the future behavior of the process and the shifts that are considered important and must be detected quickly, we will use for Phase II monitoring the best chart in terms of the $EARL$, with a desired $ARL_0 \approx 200$. Again, we consider the two scenarios in Section 4. In Table 11, we provide for each $CRR_{l,m}$ chart the minimum $EARL$ value (first line), the IC ARL (second line), and the set of values (LWL, UWL, UCL, k), in the third line, which gives the minimum $EARL$. Even though the $CRR_{4,5}$ chart attains the minimum $EARL$, the IC ARL is not very close to the desired nominal value of 200. Thus, we choose the $CRR_{2,3}$ as the best chart in terms of $EARL$, for both scenarios, since its IC ARL performance is much closer to the desired one.

Table 11. Best charts based on $EARL$, for a ZIP process with $(\lambda_0, \phi_0) = (2.38, 0.56)$ and $ARL_0 \approx 200$.

Scenario	$CRR_{2,2}$	$CRR_{2,3}$	$CRR_{2,4}$	$CRR_{2,5}$	$CRR_{3,4}$	$CRR_{4,5}$	$CRR_{5,5}$
1	164.18	154.79	260.01	255.96	257.17	152.35	325.06
	204.85	202.87	204.20	203.76	198.37	215.46	214.97
	1, 4, 7, 14	1, 4, 9, 13	0, 4, 9, 10	0, 4, 10, 10	0, 3, 7, 10	1, 2, 7, 14	0, 2, 8, 9
2	132.30	121.59	310.54	301.20	309.93	107.60	398.30
	204.85	202.87	204.20	203.76	198.37	215.46	214.97
	1, 4, 7, 14	1, 4, 9, 13	0, 4, 9, 10	0, 4, 10, 10	0, 3, 7, 10	1, 2, 7, 14	0, 2, 8, 9

In Figure 3, we provide the $CRR_{2,3}$ chart with $(LWL, UWL, UCL, k) = (1, 4, 9, 13)$ for the 50 Phase II observations. This chart gives an OOC signal at point 35 since points 33 and 35 are plotted in Region 2, while point 34 is plotted in Region 3. These points are marked with ∗. It is worth noting that the $CRR_{2,3}$ gives this OOC signal at (almost) the same period with the AEWMA chart in [33].

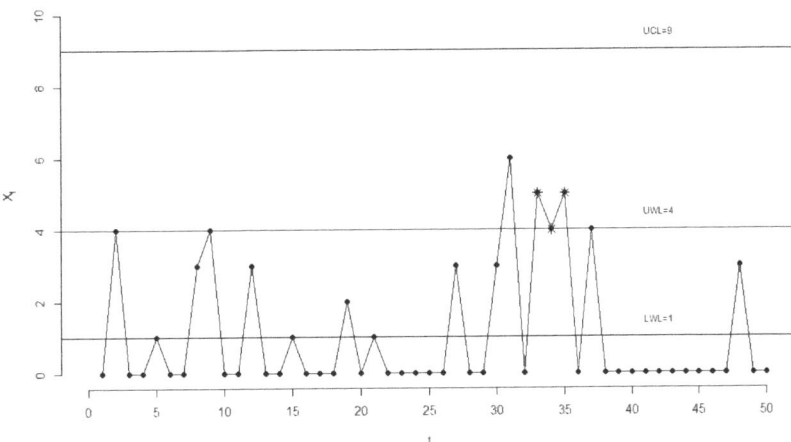

Figure 3. The $CRR_{2,3}$ chart with $(LWL, UWL, UCL, k) = (1, 4, 9, 13)$.

5.2. Monitoring the Monthly Number of Poliomyelitis Cases

In the second example, we consider the monthly number of polio cases in the USA (see also [21]). Rakitzis et al. [22] used 131 observations (from January 1973 to November 1983) where observations 1–100 are the Phase I dataset, and the remaining ones are the Phase II dataset. According to the results of the Phase I analysis conducted by Rakitzis et al. [22], the $GIP_1(0.604, 1.54)$ is the model with the best fit in the data when the process is IC. We assume that the IC values of process parameters are known and proceed directly to Phase II monitoring using the observations during the period May 1981–November 1983.

For illustrative purposes, we assume that the desired IC ARL value is (approximately) equal to 20. This means that, on average, every 20 months, the chart gives a false alarm signal. Larger IC ARL, such as the well-known textbook value 370.4, might not be very useful because, in this case, the chart would give, on average, a false alarm (i.e., a false indication of a change in the average monthly number of polio cases) every 370 months (>30 years). This is a (very) long period for an infectious disease to retain its initial characteristics.

For Phase II monitoring, we will use the best chart in terms of the $EARL$ and IC ARL as close as possible to 20. Again, we consider the two scenarios in Section 4. In Table 12, we provide for each $CRR_{l,m}$ chart the minimum $EARL$ value (first line), the IC ARL (second line), and the set of values (LWL, UWL, UCL, k), in the third line, which gives the minimum

$EARL$. We choose the $CRR_{2,2}$ as the best chart since it has the minimum $EARL$ and its ARL performance is very close to the desired one.

Table 12. Best charts based on $EARL$, for a GIP_1 process with $(\lambda_0, \phi_0) = (1.54, 0.604)$ and $ARL_0 \approx 20$.

	$CRR_{2,2}$	$CRR_{2,3}$	$CRR_{2,4}$	$CRR_{2,5}$	$CRR_{3,4}$	$CRR_{4,5}$	$CRR_{5,5}$
Scenario 1	17.782	23.110	23.108	23.108	18.200	18.337	18.352
	20.084	20.184	20.184	20.184	20.044	20.178	20.188
	1, 2, 4, 8	3, 4, 6, 15	3, 4, 6, 15	3, 4, 6, 15	1, 2, 3, 11	1, 2, 3, 11	1, 2, 3, 11
Scenario 2	14.286	25.995	25.973	25.971	14.483	14.613	14.631
	20.084	20.184	20.184	20.184	20.044	20.178	20.188
	1, 2, 4, 8	3, 4, 6, 15	3, 4, 6, 15	3, 4, 6, 15	1, 2, 3, 11	1, 2, 3, 11	1, 2, 3, 11

In Figure 4, we provide the $CRR_{2,2}$ chart with $(LWL, UWL, UCL, k) = (1, 2, 4, 8)$ for the 31 Phase II observations. The occurrence of 8 successive points in Region 4 (points 6–13, marked with *) indicates an OOC situation at point 13, which is related to a possible decrease in the average monthly number of polio cases. It is worth noting that for the same dataset, the charts proposed in [21,22] are not able to detect this OOC situation since they are upper one-sided charts and can only detect increases in the process mean level. Also, another OOC signal is given at point 31 (marked with *) since this point is above the UCL (Region 1).

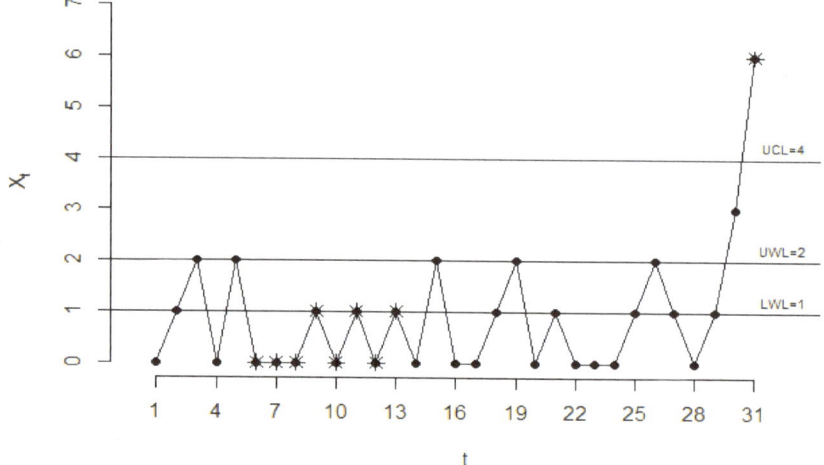

Figure 4. The $CRR_{2,2}$ chart with $(LWL, UWL, UCL, k) = (1, 2, 4, 8)$.

6. Conclusions

In this work, we proposed and studied a control chart for attributes with multiple runs rules that can be used for detecting either increasing or decreasing shifts in the process mean level. The suggested scheme is preferable when the lower control limit of the usual Shewhart-type chart is negative and thus is not capable of detecting a decrease in the process mean level. The proposed chart was studied in the case of monitoring zero- or general inflated Poisson processes. However, it can be used (after some necessary but straightforward modifications) for the monitoring of any other process with count data. The Markov chain method was used for studying its in-control and out-of-control performance, and practical guidelines were provided. The out-of-control performance of the chart was evaluated in terms of ARL and $EARL$. The numerical results showed that there is not a unique scheme among the examined ones, with the best performance for all shifts, under either performance measures. However, for improving the performance in the detection of increasing shifts we suggest the use of a runs rule of type 2-out-of-m. For the case of decreasing shifts, larger k values are needed in the case of zero-inflated Poisson processes than in the case of general inflated Poisson processes.

It should be also mentioned that in this work, we assumed that the value of r is known. Clearly, r, as well as ϕ_0 and λ_0, must be determined before the application of the charts. Therefore, the suggested approach is to first conduct a detailed model selection procedure among the possible candidate models (e.g., various GIP_r distributions with different r values) and then, by applying model selection criteria and/or goodness-of-fit tests, to choose the model with the best fit in the data. With this procedure, practitioners can determine the appropriate value for r.

Topics for future research consist of studying the proposed charts in the case of estimated parameters, as well as the case of a $CRR_{l,m}$ chart with variable sampling intervals and variable sample sizes. For both topics, the research is in progress.

Finally, the programs for reproducing the results provided in this work have been written in R [34], version 4.3.0 and are available from the authors upon request.

Author Contributions: Methodology, E.M. and A.C.R.; Software, E.M. and A.C.R.; writing—original draft preparation, E.M.; writing—review and editing, E.M. and A.C.R. All authors have read and agreed to the published version of the manuscript.

Funding: This research received no external funding.

Data Availability Statement: The data in Section 5.1 can be found in Reference [33], while the data in Section 5.2 can be found in the R package gamlss.data, version 6.0-2.

Acknowledgments: The authors would like to thank the three anonymous reviewers for their valuable comments and suggestions, which helped us to improve the paper.

Conflicts of Interest: The authors declare no conflict of interest.

Appendix A

Here, we briefly describe the Markov chain method that we used to study the proposed two-sided control chart $CRR_{l,m}$. Let us consider a two-sided chart where four regions are defined, as in Figure 1. The probability that a single point falls on Region j is denoted as p_j, $j = 1, 2, 3, 4$, and for $\lambda = \lambda_1 = \delta\lambda_0$, $\phi = \phi_1 = \tau\phi_0$, equals

$$p_4 = P(X \leq LWL) = F_{GIP_r}(LWL|\phi_1, \lambda_1),$$

$$p_3 = P(LWL < X \leq UWL) = F_{GIP_r}(UWL|\phi_1, \lambda_1) - F_{GIP_r}(LWL|\phi_1, \lambda_1),$$

$$p_2 = P(UWL < X \leq UCL) = F_{GIP_r}(UCL|\phi_1, \lambda_1) - F_{GIP_r}(UWL|\phi_1, \lambda_1),$$

$$p_1 = P(X > UCL) = 1 - F_{GIP_r}(UCL|\phi_1, \lambda_1)$$

where $X \sim GIP_r(\phi_1, \lambda_1)$. The c.d.f. $F_{GIP_r}(\cdot|\phi, \lambda)$ is given in Equation (2).

For illustrative purposes, we consider the case of the two-sided $CRR_{2,3}$ chart. Using the finite Markov chain imbedding (FMCI) method, it is possible to study any other $CRR_{l,m}$ chart and derive its entire run length distribution. Let $\{Y_i, i \geq 1\}$ be a sequence of i.i.d. multistate random variables taking values in the set $\mathcal{L} = \{1, 2, 3, 4\}$ such that $P(Y_i = j) = p_j$, $j = 1, 2, 3, 4$, where p_j are the previously defined probabilities. The $CRR_{2,3}$ chart gives an OOC signal when either a single point is above the UCL or if two-out-of-three successive points are in the interval $(UWL, UCL]$ and at most one intermediate point is in the interval $(LWL, UWL]$ or if k consecutive points are in the interval $[0, LWL]$.

Let W be the waiting time (random variable) until the first occurrence of the compound pattern:

$$\mathcal{E} = \{1, 22, 232, \underbrace{44\ldots 4}_{k}\}.$$

This means that W is the distribution of the number of trials until the occurrence of one of the following simple patterns for the first time:

$$\text{"1", "22", "232" and "}\underbrace{44\ldots 4}_{k}\text{"}.$$

Then, the distribution of W coincides with the run length distribution of the $CRR_{2,3}$ chart and can be obtained as follows. First, the simple patterns are decomposed into the following sub-patterns, defining the $k+3$ states of the embedding Markov chain:

$$1 \equiv \text{"4"}, 2 \equiv \text{"44"}, 3 \equiv \text{"444"}, \ldots, k-1 \equiv \underbrace{\text{"44}\ldots 4\text{"}}_{k-1}, k \equiv \text{"3"}, k+1 \equiv \text{"2"},$$

$$k+2 \equiv \text{"23"}, k+3 \equiv \{1, 22, 232, \underbrace{44\ldots 4}_{k}\}$$

The first $k+2$ states are the transient states of the Markov chain while state $k+3$ is the unique absorbing state. Let also $\{Z_i, i \geq 1\}$ be the imbedded Markov chain with state space $\mathcal{T} = \{1, 2, 3, \ldots, k+2, k+3\}$. Then, Z_i is in transient state j ($j \in \{1, 2, \ldots, k+2\}$) if the maximum ending block of the first v trials $Y_1, Y_2, \ldots, Y_v$ (counting backward) is identified to be the block corresponding to value j.

Therefore, the transition probability matrix $\mathbf{P}$ of the imbedded Markov chain is

$$\mathbf{P} = \begin{pmatrix} \mathbf{Q} & (\mathbf{I}-\mathbf{Q})\mathbf{1}^\top \\ \mathbf{0} & 1 \end{pmatrix} = \begin{array}{c} \\ 1 \\ 2 \\ \vdots \\ k-1 \\ k \\ k+1 \\ k+2 \\ k+3 \end{array} \begin{pmatrix} \begin{array}{cccccccc} 1 & 2 & 3 & \cdots & k & k+1 & k+2 & k+3 \end{array} \\ \begin{pmatrix} 0 & p_4 & 0 & \cdots & p_3 & p_2 & 0 & p_1 \\ 0 & 0 & p_4 & \cdots & p_3 & p_2 & 0 & p_1 \\ \vdots & \vdots & \vdots & \ddots & \vdots & \vdots & \vdots & \vdots \\ 0 & 0 & 0 & \cdots & p_3 & p_2 & 0 & p_4+p_1 \\ p_4 & 0 & 0 & \cdots & p_3 & p_2 & 0 & p_1 \\ p_4 & 0 & 0 & \cdots & 0 & 0 & p_3 & p_2+p_1 \\ p_4 & 0 & 0 & \cdots & p_3 & 0 & 0 & p_2+p_1 \\ 0 & 0 & 0 & \cdots & 0 & 0 & 0 & 1 \end{pmatrix} \end{array}$$

where $\mathbf{Q}$ is the $(k+2) \times (k+2)$ matrix of the transition probabilities between the transient states of the Markov chain, $\mathbf{I}$ is the $(k+2) \times (k+2)$ identity matrix and $\mathbf{1}^\top$ is a $(k+2) \times 1$ vector of ones.

From the theory of Markov chains, the average run length of the chart equals

$$E(W) = ARL = 1 + \mathbf{v}(\mathbf{I} - \mathbf{Q})^{-1} \mathbf{1}^\top$$

where $\mathbf{v}$ is the $1 \times (k+2)$ initial probabilities vector, with

$$\mathbf{v} = (P(Y_1 \in 1), P(Y_1 \in 2), \ldots, P(Y_1 \in k), P(Y_1 \in k+1), P(Y_1 \in k+2))$$
$$= (p_4, 0, \ldots, p_3, p_2, 0)$$

References

1. Page, E.S. Continuous inspection schemes. *Biometrika* **1954**, *41*, 100–115. [CrossRef]
2. Roberts, S.W. Control chart tests based on geometric moving averages. *Technometrics* **1959**, *1*, 239–250. [CrossRef]
3. Montgomery, D.C. *Introduction to Statistical Quality Control*, 8th ed.; John Wiley & Sons: New York, NY, USA, 2020.
4. Bersimis, S.; Koutras, M.V.; Rakitzis, A.C. Run and scan rules in statistical process monitoring. In *Handbook of Scan Statistics*; Glaz, J., Koutras, M., Eds.; Springer: New York, NY, USA, 2020. [CrossRef]
5. Jalilibal, Z.; Karavigh, M.H.A.; Amiri, A.; Khoo, M.B.C. Run rules schemes for statistical process monitoring: A literature review. *Qual. Technol. Quant. Manag.* **2023**, *20*, 21–52. [CrossRef]
6. Johnson, N.L.; Kemp, A.W.; Kotz, S. *Univariate Discrete Distributions*, 3rd ed.; John Wiley & Sons: Hoboken, NJ, USA, 2005.
7. Yoneda, K. Estimations in some modified Poisson distributions. *Yokohama Math. J.* **1962**, *10*, 73–96.
8. Murat, M.; Snyzal, D. Non-zero inflated modified power series distributions. *Commun. Stat.-Theory Methods* **1998**, *27*, 3047–3064. [CrossRef]

9. Melkersson, M.; Rooth, D.O. Modeling female fertility using inflated count data models. *J. Popul. Econ.* **2000**, *13*, 189–203. [CrossRef]
10. Lin, T.H.; Tsai, M.H. Modeling health survey data with excessive zero and K responses. *Stat. Med.* **2013**, *32*, 1572–1583. [CrossRef]
11. Tang, Y.; Liu, W.; Xu, A. Statistical inference for zero-and-one inflated Poisson models. *Stat. Theory Relat. Fields* **2017**, *1*, 216–226. [CrossRef]
12. Sun, Y.; Zhao, S.; Tian, G.L.; Tang, M.L.; Li, T. Likelihood-based methods for the zero-one-two inflated Poisson model with applications to biomedicine. *J. Stat. Comput. Simul.* **2023**, *93*, 956–982. [CrossRef]
13. Begum, A.; Mallick, A.; Pal, N. A generalized inflated Poisson distribution with application to modeling fertility data. *Thail. Stat.* **2014**, *12*, 135–139.
14. Rakitzis, A.C.; Castagliola, P.; Maravelakis, P.E. A two-parameter general inflated Poisson distribution: Properties and applications. *Stat. Methodol.* **2016**, *29*, 32–50. [CrossRef]
15. Mahmood, T.; Xie, M. Models and monitoring of zero-inflated processes: The past and current trends. *Qual. Reliab. Eng. Int.* **2019**, *35*, 2540–2557. [CrossRef]
16. Paulino, S.; Morais, M.C.; Knoth, S. An ARL-unbiased c-chart. *Qual. Reliub. Eng. Int.* **2016**, *32*, 2847–2858. [CrossRef]
17. Nelson, L.S. Supplementary runs tests for *np* control charts. *J. Qual. Technol.* **1997**, *29*, 225–227. [CrossRef]
18. Acosta-Mejia, C.A. Improved *p* charts to monitor process quality. *IIE Trans.* **1999**, *31*, 509–516. [CrossRef]
19. Lucas, J.M.; Davis, D.J.; Saniga, E.M. Detecting improvement using Shewhart attribute control charts when the lower control limit is zero. *IIE Trans.* **2006**, *38*, 699–709. [CrossRef]
20. Chang, T.C.; Gan, F.F. Modified Shewhart charts for high yield processes. *J. Appl. Stat.* **2007**, *34*, 857–877. [CrossRef]
21. Mamzeridou, E.; Rakitzis, A.C. Synthetic-type control charts for monitoring general inflated Poisson processes. *Commun. Stat.-Simul. Comput.* **2022**, *in press*. [CrossRef]
22. Rakitzis, A.C.; Castagliola, P.; Maravelakis, P.E. Cumulative sum control charts for monitoring geometrically inflated Poisson processes: An application to infectious disease counts data. *Stat. Methods Med. Res.* **2018**, *27*, 622–641. [CrossRef] [PubMed]
23. Woodall, W.H.; Adams, B.M.; Benneyan, J.C. The use of control charts in healthcare. In *Statistical Methods in Healthcare*; Faltin, F.W., Kenett, R.S., Ruggeri, F., Eds.; John Wiley & Sons: Oxford, UK, 2012; pp. 251–267.
24. Reynolds, M.R., Jr. The Bernoulli CUSUM chart for detecting decreases in a proportion. *Qual. Reliab. Eng. Int.* **2013**, *29*, 529–534. [CrossRef]
25. Bourke, P.D. Detecting a downward shift in a proportion using a geometric CUSUM chart. *Qual. Eng.* **2020**, *32*, 75–90. [CrossRef]
26. Chakraborti, S.; Kumar, N.; Rakitzis, A.C.; Sparks, R.S. Time between events monitoring with control charts. In *Wiley StatsRef: Statistics Reference Online*; Balakrishnan, N., Colton, T., Everitt, B., Piegorsch, W., Ruggeri, F., Teugels, J.L., Eds.; John Wiley & Sons: Hoboken, NJ, USA, 2023; pp. 1–13.
27. Fu, J.C.; Koutras, M.V. Distribution theory of runs: A Markov chain approach. *J. Am. Stat. Assoc.* **1994**, *89*, 1050–1058. [CrossRef]
28. Fu, J.C.; Lou, W.W. *Distribution Theory of Runs and Patterns and Its Applications: A Finite Markov Chain Imbedding Approach*, 1st ed.; World Scientific: Singapore, 2003.
29. Fu, J.C.; Shmueli, G.; Chang, Y.M. A unified Markov chain approach for computing the run length distribution in control charts with simple or compound rules. *Stat. Probab. Lett.* **2003**, *65*, 457–466. [CrossRef]
30. Machado, M.A.G.; Costa, A.F.B. A side-sensitive synthetic chart combined with an X chart. *Int. J. Prod. Res.* **2014**, *52*, 3404–3416. [CrossRef]
31. Mukherjee, A.; Sen, R. Optimal design of Shewhart-Lepage type schemes and its application in monitoring service quality. *Eur. J. Oper. Res.* **2018**, *266*, 147–167. [CrossRef]
32. Fatahi, A.A.; Noorossana, R.; Dokouhaki, P. Zero inflated Poisson EWMA control chart for monitoring rare health-related events. *J. Mech. Med. Biol.* **2012**, *12*, 1250065. [CrossRef]
33. Aly, A.A.; Saleh, N.A.; Mahmoud, M.A. An adaptive EWMA control chart for monitoring zero-inflated Poisson processes. *Commun. Stat.-Simul. Comput.* **2022**, *51*, 1564–1577. [CrossRef]
34. R Core Team. *R: A Language and Environment for Statistical Computing*; R Foundation for Statistical Computing: Vienna, Austria, 2023. Available online: https://www.r-project.org/ (accessed on 29 September 2023).

Disclaimer/Publisher's Note: The statements, opinions and data contained in all publications are solely those of the individual author(s) and contributor(s) and not of MDPI and/or the editor(s). MDPI and/or the editor(s) disclaim responsibility for any injury to people or property resulting from any ideas, methods, instructions or products referred to in the content.

Article

Archimedean Copulas-Based Estimation under One-Parameter Distributions in Coherent Systems

Ioannis S. Triantafyllou

Department of Statistics and Insurance Science, University of Piraeus, 18534 Piraeus, Greece; itriantafyllou@unipi.gr

Abstract: In the present work we provide a signature-based framework for delivering the estimated mean lifetime along with the variance of the continuous distribution of a coherent system consisting of exchangeable components. The dependency of the components is modelled by the aid of well-known Archimedean multivariate copulas. The estimated results are calculated under two different copulas, namely the so-called Frank copula and the Joe copula. A numerical experimentation is carried out for illustrating the proposed procedure under all possible coherent systems with three components.

Keywords: moment estimator; Frank copula; Joe copula; maximal signatures; exchangeable components

MSC: 62F10; 62H12; 62N05

Citation: Triantafyllou, I.S. Archimedean Copulas-Based Estimation under One-Parameter Distributions in Coherent Systems. *Mathematics* **2024**, *12*, 334. https://doi.org/10.3390/math12020334

Academic Editors: Arne Johannssen and Nataliya Chukhrova

Received: 31 December 2023
Revised: 15 January 2024
Accepted: 18 January 2024
Published: 19 January 2024

Copyright: © 2024 by the author. Licensee MDPI, Basel, Switzerland. This article is an open access article distributed under the terms and conditions of the Creative Commons Attribution (CC BY) license (https://creativecommons.org/licenses/by/4.0/).

1. Introduction

In the field of Statistical Reliability Modeling, several studies have been carried out under the assumption that the components of the underlying structures are independent. However, this condition is not always fulfilled in real-life problems. Therefore, it is of some research interest to investigate reliability systems consisting of exchangeable components, namely components which are identically distributed but are (possibly) dependent to each other. For instance, [1] delivered a signature-based analysis of m-consecutive k-out-of-n: F systems with exchangeable components. Moreover, [2] proved that the lifetime of any coherent system with dependent components can be expressed as generalized mixture of series (or parallel) subsystem lifetime distributions.

For a reliability coherent system with n exchangeable components, the dependency between them can be well modelled by the aid of appropriately chosen copulas. It is noteworthy that copulas have been proved to be a useful tool for studying the joint distribution of the random lifetimes of the components of a reliability model (see, e.g., [3–5]). For example, a copula-based approach can be applied in order to evaluate the reliability characteristics such as availability, reliability, and mean time to failure of a coherent system [4,5].

In addition, the parameter's estimation for the common continuous distribution of the components of the underlying reliability system is of high importance. Having at hand a point or interval estimation of the distributional parameter, one may readily deduce several results and conclusions concerning the behavior not only of the components, but also of the whole structure (see, e.g., [4]).

In the present paper, we provide a theoretical framework for providing the estimated mean lifetime (along with its variance) for a reliability structure and also for establishing the moment estimator of the parameter of the common continuous distribution of its components. However, the present work focuses on the reliability study of coherent systems with exchangeable components. More precisely, it aims to draw conclusions about their expected lifetimes and the respective variances. Thereof, we provide just a short discussion about the parameter's estimation of the common continuous distribution of its components at the end of Section 3. All necessary notions and formulae about the copulas which shall be used later on are presented in Section 2. A short introduction referring to the maximal signature of a coherent reliability system is also provided.

In Section 3, the proposed procedure is described in detail. The main results of the paper refer to the lifetime of a reliability system having exchangeable components and the parameter's estimation of the underlying components' distribution. This goal is fulfilled by the aid of explicit expressions, which are introduced and proved under different Archimedean copulas-based models. More precisely, the Frank copula model and the Joe copula model are considered and studied in some detail. It is evident that the shapes of these two copulas are quite similar, but the Frank Archimedean copula lacks an asymptote at $-\infty$, whereas the Joe Archimedean copula does not. In addition, it is known that Joe Archimedean copulas have an exponential functional form, while Frank Archimedean copulas have a logarithmic functional form. Among other reasons, we chose to consider the specific models due to their wide applicability in several fields. For instance, one may refer to the utilization of Frank Archimedean copulas for studying the linked risk factors, while Joe Archimedean copulas are frequently employed to model negatively related risk factors. For more details about the Archimedean copulas, the interested reader is referred to [6–11] and the references therein.

In Section 4, an extensive numerical experimentation is carried out and the implementation of the proposed estimation procedure is illustrated. In order to provide adequate numerical evidence about the ability of the proposed technique to estimate the desired quantities, all possible coherent systems with three components are taken into account. In addition, Monte Carlo simulations are also realized for studying the distribution of the resulting moment estimator. Finally, the Discussion section summarizes the contribution of the present paper, while some practical concluding remarks are also highlighted therein.

2. General Notions and Notations

In this section, we present the necessary notions and notations for establishing the proposed estimation procedure. In what follows, some basic results referring to the copula models and maximal signatures shall be discussed in order to pave the way for delivering the main results of the paper in the next section.

Let us first consider a coherent system consisting of n exchangeable components with common continuous distribution function F. The dependence between the components could be readily modelled by the aid of appropriate Archimedean copulas. A sufficient incentive to choose Archimedean copulas over other types of copulas is their simple form and ease with which they can be constructed. Moreover, the great variety of families of copulas which belong to this class gives the Archimedean copulas a central role and great applicability.

Generally speaking, the copulas are useful tools for determining the joint distribution of random variables, since they are functions that join (or couple) multivariate distribution functions to their one-dimensional marginal distribution functions. For recent advances and applications of the Archimedean copulas, one may refer to [12–17].

The main findings of the present work refer to the expected lifetime of a coherent system consisting of n exchangeable components under the assumption that the dependence of the components is modelled by the aid of specific copula models. In most of real-life engineering systems, such as transportation systems, communication networks, aerospace systems, healthcare delivery systems, and manufacturing processes, the dependence among the components is inevitable due to the common random production and operating environments.

Let us first denote by $T_1, T_2, \ldots, T_n$, the lifetimes of the components of underlying reliability structure. If $G_j(t_j) = P(T_j \leq t_j)$ corresponds to the cumulative distribution function of the variable T_j, $j = 1, 2, \ldots, n$, then $H(t_1, \ldots, t_n) = P(T_1 \leq t_1, T_2 \leq t_2, \ldots, T_n \leq t_n)$ is simply the joint distribution function of the lifetimes of the components of the system.

Each vector $(t_1, t_2, \ldots, t_n)$ of real numbers leads to a point $(G_1(t_1), G_2(t_2), \ldots, G_n(t_n))$ in the unit region $[0,1]^n$, while these ordered coordinates correspond to a number $H(t_1, \ldots, t_n)$ in $[0,1]$. The aforementioned correspondence, which assigns the value of the joint distribution function to each ordered vector of values of the individual distribution functions, is actually the copula function C. Generally speaking, note that the probability density function of the copulas can be derived from the corresponding cumulative density function by the aid of appropriate derivatives of the copula function.

Due to the exchangeability of the components of the underlying reliability structure, the following holds true

$$G_j(t) = G(t), \; j = 1, 2, \ldots, n. \tag{1}$$

Therefore, if $C(u_1, \ldots, u_n)$ is the copula function related to $H(t_1, \ldots, t_n)$, we deduce that

$$H(t_1, \ldots, t_n) = C(G(t_1), G(t_2), \ldots, G(t_n)). \tag{2}$$

An Archimedean copula behaves like a binary operation on the interval $[0,1]$. In other words, the copula function C assigns to each pair (u,v) in $[0,1]$ a number $C(u,v)$ in $[0,1]$. In addition, the function C is commutative, associative, and preserves order, e.g., $u_1 \leq u_2$ and $v_1 \leq v_2$ implies $C(u_1, v_1) \leq C(u_2, v_2)$.

Throughout the course of the present work, we shall consider two different copula functions for modeling the dependence of the components in the underlying reliability system. Both models implemented in the next lines are members of the well-known class of Archimedean copulas (see, e.g., [3,18,19] and references therein). Kindly note that under the assumption of exchangeability, these models have never been studied before for modeling the dependency of the components of a system. More precisely, we shall consider the following multivariate Archimedean copulas:

- The Frank family of n-copulas.

The generator function of the bivariate Frank copulas is given by

$$\varphi_\theta(t) = -\ln\left[\frac{e^{-\theta t} - 1}{e^{-\theta} - 1}\right]. \tag{3}$$

For $\theta > 0$ and $n \geq 2$, the copula function of the multivariate Frank class of n-copulas is expressed as

$$C_\theta^n(u_1, u_2, \ldots, u_n) = -\frac{1}{\theta} \ln\left(1 + \frac{(e^{-\theta u_1} - 1) \cdot (e^{-\theta u_2} - 1) \cdots (e^{-\theta u_n} - 1)}{(e^{-\theta} - 1)^{n-1}}\right). \tag{4}$$

- The Joe family of n-copulas.

The generator function of the bivariate Joe copulas is given by

$$\varphi_\theta(t) = -\ln\left[1 - (1-t)^\theta\right], \; \theta \geq 1. \tag{5}$$

For $\theta \geq 1$ and $n \geq 2$, the copula function of the multivariate Joe class of n-copulas is written as

$$C_{*,\theta}^n(u_1, u_2, \ldots, u_n) = n - 1 - \left[\sum_{i=1}^n (1-u_i)^\theta - \prod_{i=1}^n (1-u_i)^\theta\right]^{1/\theta}. \tag{6}$$

If T corresponds to the lifetime of a coherent system with n exchangeable components and $T_{i:n}, i = 1, 2, \ldots, n$ is the i-th ordered component's lifetime, then the reliability function of the system is expressed as

$$P(T > t) = \sum_{i=1}^n \beta_i P(T_{i:i} > t) = \sum_{i=1}^n \beta_i (1 - P(T_1 \leq t, T_2 \leq t, \ldots, T_i \leq t)), \tag{7}$$

where β_i, $i = 1, 2, \ldots, n$ satisfy the condition $\sum_{i=1}^{n} \beta_i = 1$. Note that the vector $(\beta_1, \beta_2, \ldots, \beta_n)$ is the maximal signature of the coherent system (see, e.g., [2,20]). It is worth mentioning that, under specified reliability models, several general results have been proved in the literature for determining the β_i's. For instance, one may refer to the exact closed formulae for the maximal signatures of an m-consecutive-k-out-of-n: F system, which have been delivered in [1]. Throughout the lines of the next sections, we shall focus on specific models of coherent systems, such as parallel structures, series structures, or consecutive-type systems. For the latter ones, it is known that a consecutive k-out-of-n: F system is a structure made up of n components ordered sequentially and fails if and only if at least k consecutive components fail (see, e.g., [1]).

3. Main Results

In this section, we study the lifetime of a reliability structure consisting of n exchangeable components. The dependency between the components is modelled by the aid of two specific copula models. More precisely, the Frank and the Joe copulas are considered. Generally speaking, Frank Archimedean copulas are more sensitive to positive association than Joe Archimedean copulas in terms of association sensitivity.

The main result refers to the expected lifetime of such a structure, while the corresponding variance is determined. Based on these outcomes, the estimation of the parameter of the underlying components' distribution can also be achieved.

Let us next consider a reliability system consisting of n exchangeable components with a common continuous distribution G. We denote by $T_1, T_2, \ldots, T_n$ the lifetimes of the components, while $T = \varphi(T_1, T_2, \ldots, T_n)$ corresponds to the lifetime of the resulting structure.

The next proposition offers expressions for determining the expected lifetime and its corresponding variance for a system under the Frank copula-based dependency.

Proposition 1. *Let us consider a reliability system with n exchangeable and exponentially distributed components with parameter λ. If the dependency of the components is described by the Frank copula model, the following ensue,*

(i) *The expected lifetime of the system is given by*

$$E(T) = \sum_{i=1}^{n} \beta_i \cdot \frac{i}{\lambda} \times \int_0^\infty \frac{t \cdot e^{-t/\lambda - \theta \cdot (1 - e^{-t/\lambda})} \cdot \left(e^{-\theta} - 1\right)^{1-i} \cdot \left(e^{-\theta \cdot (1 - e^{-t/\lambda})} - 1\right)^{i-1}}{1 + \left(e^{-\theta} - 1\right)^{1-i} \cdot \left(e^{-\theta \cdot (1 - e^{-t/\lambda})} - 1\right)^i} dt, \quad (8)$$

(ii) *The variance of the lifetime of the system is given by*

$$Var(T) = \sum_{i=1}^{n} \beta_i \cdot \frac{i}{\lambda} \cdot \int_0^\infty \frac{t^2 \cdot e^{-\frac{t}{\lambda} - \theta \cdot (1 - e^{-\frac{t}{\lambda}})} \cdot \left(e^{-\theta} - 1\right)^{1-i} \cdot A_{\theta,\lambda}(t, i-1)}{1 + \left(e^{-\theta} - 1\right)^{1-i} \cdot A_{\theta,\lambda}(t, i)} dt$$

$$- \left(\sum_{i=1}^{n} \beta_i \cdot \frac{i}{\lambda} \cdot \int_0^\infty \frac{t \cdot e^{-t/\lambda - \theta \cdot (1 - e^{-t/\lambda})} \cdot \left(e^{-\theta} - 1\right)^{1-i} \cdot \left(e^{-\theta \cdot (1 - e^{-t/\lambda})} - 1\right)^{i-1}}{1 + \left(e^{-\theta} - 1\right)^{1-i} \cdot \left(e^{-\theta \cdot (1 - e^{-t/\lambda})} - 1\right)^i} dt \right)^2. \quad (9)$$

Proof. (i) Given that the components of the system are exponentially distributed with parameter λ, e.g., $G(t) = 1 - e^{-t/\lambda}$, $t > 0$, the joint distribution function of their lifetimes under the Frank copula model can be written as (see (2) and (4))

$$H(t_1, \ldots, t_n) = -\frac{1}{\theta} \ln \left(1 + \frac{\left(e^{-\theta \cdot (1 - e^{-t_1/\lambda})} - 1\right) \cdot \left(e^{-\theta \cdot (1 - e^{-t_2/\lambda})} - 1\right) \cdots \left(e^{-\theta \cdot (1 - e^{-t_n/\lambda})} - 1\right)}{\left(e^{-\theta} - 1\right)^{n-1}} \right). \quad (10)$$

It is known that the expected value of the lifetime T of a reliability system can be determined by the aid of the following formula:

$$E(T) = \int_0^\infty t\,dP(T \leq t). \tag{11}$$

Recalling (7), the last expression can be rewritten as

$$E(T) = \sum_{i=1}^n \beta_i \int_0^\infty t\,dP(T_{i:i} \leq t) = \sum_{i=1}^n \beta_i \cdot E(T_{i:i}) \tag{12}$$

where the vector $(\beta_1, \beta_2, \ldots, \beta_n)$ is the maximal signature of the system. Moreover, the event $\{T_{i:i} \leq t\}$ practically means that the maximum of the components' lifetimes $T_1, T_2, \ldots, T_i$, $i = 1, 2, \ldots, n$, does not exceed the value t, while no restriction is stated for the $n - i$ remaining lifetimes. Therefore, we may readily deduce that

$$P(T_{i:i} \leq t) = P(T_1 \leq t, T_2 \leq t, \ldots, T_i \leq t) = C_\theta^n(\underbrace{1 - e^{-t/\lambda}, \ldots, 1 - e^{-t/\lambda}}_{i}, \underbrace{1, \ldots, 1}_{n-i}) \tag{13}$$

where the copula function C_θ^n is defined in (4). Combining Formulae (4) and (13), we conclude that

$$P(T_{i:i} \leq t) = -\frac{1}{\theta} \ln\left(1 + \frac{(e^{-\theta \cdot (1 - e^{-t/\lambda})} - 1)^i}{(e^{-\theta} - 1)^{i-1}}\right) \tag{14}$$

and the expected value of the random variable $T_{i:i}$ is now determined as

$$E(T_{i:i}) = \int_0^\infty t\,dP(T_{i:i} \leq t) = \frac{i}{\lambda} \cdot \int_0^\infty \frac{t \cdot e^{-t/\lambda - \theta \cdot (1 - e^{-t/\lambda})} \cdot (e^{-\theta} - 1)^{1-i} \cdot (e^{-\theta \cdot (1 - e^{-t/\lambda})} - 1)^{i-1}}{1 + (e^{-\theta} - 1)^{1-i} \cdot (e^{-\theta \cdot (1 - e^{-t/\lambda})} - 1)^i} dt, \tag{15}$$

The result we are chasing for is effortlessly derived by replacing the last expression in (12).

(ii) The variance of the system's lifetime T shall be determined by applying the well-known identity

$$Var(T) = E(T^2) - [E(T)]^2 \tag{16}$$

where the expected value of T is given by (8). In addition, the 2nd moment of T can be expressed as

$$E(T^2) = \int_0^\infty t^2\,dP(T \leq t) = \sum_{i=1}^n \beta_i \cdot E(T_{i:i}^2) = \sum_{i=1}^n \beta_i \cdot \int_0^\infty t^2\,dP(T_{i:i} \leq t). \tag{17}$$

Following a parallel argumentation with the one implemented at part (i), the integral expression in (18) leads, by the aid of (7) and (14), to the desired result. □

The following proposition offers expressions for determining the expected lifetime and its corresponding variance for a system under the Joe copula-based dependency.

Proposition 2. *Let us consider a reliability system with n exchangeable and exponentially distributed components with parameter λ. If the dependency of the components is described by the Joe copula model, the following ensue,*

(i) *The expected lifetime of the system is given by*

$$E(T) = \lambda \cdot \sum_{i=1}^n \beta_i \cdot i^{1/\theta}, \tag{18}$$

(ii) *The variance of the lifetime of the system is given by*

$$Var(T) = 2 \cdot \lambda^2 \cdot \sum_{i=1}^{n} \beta_i \cdot i^{1/\theta} - \left(\lambda \cdot \sum_{i=1}^{n} \beta_i \cdot i^{1/\theta} \right)^2. \quad (19)$$

Proof. (i) Given that the components of the system are exponentially distributed with parameter λ, e.g., $G(t) = 1 - e^{-t/\lambda}$, $t > 0$, the joint distribution function of their lifetimes under the Joe copula model can be written as (see (2) and (6))

$$\begin{aligned} H(t_1, \ldots, t_n) &= n - 1 - \left[e^{-\theta \cdot t_1/\lambda} + \cdots + e^{-\theta \cdot t_n/\lambda} - e^{-\theta \cdot t_1/\lambda} \cdots e^{-\theta \cdot t_n/\lambda} \right]^{1/\theta} \\ &= n - 1 - \left(\sum_{j=1}^{n} e^{-\theta \cdot t_j/\lambda} - e^{-\theta \cdot \sum_{j=1}^{n} t_j/\lambda} \right)^{1/\theta} \end{aligned} \quad (20)$$

Since the copula function for the Joe model is given by (6), the following ensues (see also (13)),

$$P(T_{i:i} \leq t) = n - 1 - \left(\sum_{j=1}^{i} e^{-\theta \cdot t/\lambda} \right)^{1/\theta}. \quad (21)$$

The expected value of the random variable $T_{i:i}$ is now determined as

$$E(T_{i:i}) = \int_0^\infty t \, dP(T_{i:i} \leq t) = \frac{i^{1/\theta}}{\lambda} \cdot \int_0^\infty t \cdot e^{-t/\lambda} dt = \lambda \cdot i^{1/\theta}. \quad (22)$$

We next combine (12) and (22) and the desired result is straightforward.

(ii) The variance of the system's lifetime T shall be determined by applying the well-known identity (16). The 2nd moment of T can be expressed as

$$E(T^2) = \int_0^\infty t^2 dP(T \leq t) = \sum_{i=1}^{n} \beta_i \cdot E(T_{i:i}^2) = \sum_{i=1}^{n} \beta_i \cdot \int_0^\infty t^2 dP(T_{i:i} \leq t) = 2 \cdot \lambda^2 \cdot \sum_{i=1}^{n} \beta_i \cdot i^{1/\theta}. \quad (23)$$

The result we are chasing for is now immediately derived. □

It is evident that Propositions 1 and 2 of the present paper provide general results for any coherent system consisting of n exchangeable and exponentially distributed components. These results can be easily modified under different distributional assumptions for the components' lifetimes of the underlying structure.

It is noteworthy that the aforementioned results, which have been proved in Propositions 1 and 2, may contribute to deliver the estimation of the parameter of the common distribution of the components' lifetimes. According to the well-known moment estimation procedure, the theoretical moments provided by the previous propositions should be equated to the corresponding sample moments.

For instance, let us consider the same case with the one studied in Propositions 1 and 2, namely, we assume that the components of the system share a common exponential distribution with parameter λ. In order to estimate the distribution's mean λ, we need to determine the corresponding sample mean lifetime $\overline{T}$ of the resulting system. More precisely, if $E(T)$ denotes the 1st moment of system's lifetime T, then the desired estimation of parameter λ shall be delivered by solving the equation $E(T) = \overline{T}$ with respect to λ.

Since we study reliability systems having exchangeable components, whose dependency is modelled by an appropriately chosen Archimedean copula, the computation of the sample mean lifetime of the resulting system calls for a sampling procedure from the underlying copula. The challenge of efficiently sampling exchangeable Archimedean copulas has been already addressed in the literature (see, e.g., [9–11]).

One of the most powerful tools for sampling exchangeable Archimedean copulas is provided by the algorithm of Marshall and Olkin (algorithm MO, hereafter). According to the algorithmic procedure MO, we may simulate a random sample of size n from a specific Archimedean copula with generator ψ and continuous joint cumulative distribution function $H(t_1, \ldots, t_n)$ if we follow the next steps (see, e.g., [9]):

Step 1. Sample $V \sim F = \mathcal{LS}^{-1}(\psi)$, where $\mathcal{LS}^{-1}(\psi)$ denotes the Laplace-Stieltjes transform of ψ.

Step 2. Sample independent and identically distributed random variables $U_i, i = 1, 2, \ldots, n$ from the Uniform distribution in $[0, 1]$, namely $U_i \sim U[0, 1]$, $i = 1, 2 \ldots, n$.

Step 3. Determine $X_i = \psi(-\ln(U_i)/V)$, $i = 1, 2, \ldots, n$.

Step 4. The random variables $G^{-1}(X_i)$, $= 1, 2, \ldots, n$ constitute a sample from the exchangeable joint distribution function H, where G corresponds to the marginal cumulative distribution function of H.

4. Numerical Results

In the present section, we compute the expected mean lifetime and the corresponding variance for all possible coherent reliability structures with three exchangeable components, which are exponentially distributed components with parameter λ. The dependency between the components is modelled by either the Frank or the Joe copula. The numerical results and graphical illustrations that appeared in this section are all produced by the aid of the theoretical outcomes proved in the previous section of the present manuscript.

Let us denote by X_1, X_2, X_3 the lifetimes of the components of a reliability structure with three components. All possible reliability systems consisting of three exchangeable components are listed below (see, also [6]):

- **RS1**: Series system. The particular system fails if and only if at least one component fails. Thereof, the lifetime of a series system with three exchangeable components can be expressed as $S_1 = \min(X_1, X_2, X_3)$, while the corresponding maximal signature vector is given as $(\beta_1, \beta_2, \beta_3) = (3, -3, 1)$.
- **RS2**: Series-parallel system. The particular system fails if and only if either the 1st component fails or both the other two (e.g., the 2nd and the 3rd component) components fail. Thereof, the lifetime of a series-parallel system with three exchangeable components can be expressed as $S_2 = \min(X_1, \max(X_2, X_3))$, while the corresponding maximal signature vector is given as $(\beta_1, \beta_2, \beta_3) = (1, 1, -1)$.
- **RS3**: 2-out-of-3 system. The particular system fails if and only if at least two components fail. Thereof, the lifetime of a 2-out-of-3 system with three exchangeable components can be expressed as $S_3 = \max_{1 \leq i < j \leq 3} \min(X_i, X_j)$, while the corresponding maximal signature vector is given as $(\beta_1, \beta_2, \beta_3) = (0, 3, -2)$.
- **RS4**: Parallel-Series system. The particular system fails if and only if the 1st component fails and one of the other two (e.g., either the 2nd or the 3rd component) components fail. Thereof, the lifetime of a parallel-series system with three exchangeable components can be expressed as $S_4 = \max(X_1, \min(X_2, X_3))$, while the corresponding maximal signature vector is given as $(\beta_1, \beta_2, \beta_3) = (0, 2, -1)$.
- **RS5**: Parallel system. The particular system fails if and only if all components fail. Thereof, the lifetime of a parallel system with three exchangeable components can be expressed as $S_5 = \min(X_1, X_2, X_3)$, while the corresponding maximal signature vector is given as $(\beta_1, \beta_2, \beta_3) = (0, 0, 1)$.

We first compute the expected lifetimes and the corresponding variances for the above-mentioned structures under the Frank and Joe copula-based dependence. The common distribution of all components is assumed to be the exponential with parameter λ. The numerical results provided at Table 1 have been produced by the aid of Propositions 1 and 2 (see Section 3 of the present manuscript).

Table 1. Expected lifetime and its variance of all possible coherent systems with three exchangeable components under Frank and Joe copula-based dependency.

System	(λ,θ)	Frank Copula Expected Lifetime	Frank Copula Variance	Joe Copula Expected Lifetime	Joe Copula Variance
RS1		0.448826	0.219162	0.489410	0.739298
RS2		0.720882	0.438656	0.682163	0.898979
RS3	(1, 2)	0.856910	0.492892	0.778539	0.950955
RS4		1.136030	0.978296	1.096380	0.990712
RS5		1.694260	1.481690	1.732050	0.464102
RS1		0.897652	0.876646	0.978820	2.957190
RS2		1.441760	1.754630	1.364330	3.595920
RS3	(2, 2)	1.713820	1.971570	1.557080	3.803820
RS4		2.272050	3.913230	2.192750	3.962850
RS5		3.388530	5.926690	3.464100	1.856410
RS1		1.346480	1.972450	1.468230	6.653680
RS2		2.162650	3.947880	2.046490	8.090820
RS3	(3, 2)	2.570730	4.436030	2.335620	8.558600
RS4		3.408080	8.804730	3.289130	8.916400
RS5		5.082790	13.33510	5.196150	4.176910
RS1		1.795300	3.506600	1.957640	11.82880
RS2		2.883530	7.018480	2.728650	14.38370
RS3	(4, 2)	3.427640	7.886270	3.114160	15.21530
RS4		4.544110	15.65280	4.385510	15.85140
RS5		6.777060	23.70680	6.928200	7.535630
RS1		0.501188	0.267310	0.662486	0.886085
RS2		0.747671	0.485253	0.817671	0.966756
RS3	(1, 3)	0.870913	0.548659	0.895264	0.989020
RS4		1.123240	1.001210	1.077590	0.993979
RS5		1.627900	1.524290	1.442250	0.804415
RS1		1.002380	1.069230	1.324970	3.544340
RS2		1.495340	1.941020	1.635340	3.867030
RS3	(2, 3)	1.741830	2.194620	1.790530	3.956120
RS4		2.246480	4.004860	2.155190	3.975920
RS5		3.255800	6.097180	2.884500	3.217660
RS1		1.503560	2.405800	1.987460	7.974760
RS2		2.243010	7.162340	2.453010	8.700810
RS3	(3, 3)	2.612740	4.937920	2.685790	8.901270
RS4		3.369730	9.010870	3.232780	8.945810
RS5		4.883700	13.71860	4.326750	7.239740
RS1		2.004750	4.276790	2.649950	14.17740
RS2		2.990680	7.764080	3.270690	15.46810
RS3	(4, 3)	3.483650	8.778550	3.581060	15.82450
RS4		4.492970	16.01930	4.310370	15.90370
RS5		6.511600	24.38870	5.769000	12.87060

The upper (lower) entry of each cell corresponds to the mean lifetime (its variance) of the underlying structure.

Based on the numerical results provided in Table 1, we may readily deduce that the expected lifetime of reliability structure with exchangeable components under either Frank or Joe copula-based dependency increases,

- for fixed θ as the parameter λ increases
- for fixed λ as the parameter θ decreases

In addition, the numerical results displayed in Table 1, confirm, as it was expected,

- the superiority of RS5 against the structures RS1–RS4, as it has the largest expected lifetime among all systems taken into consideration under the same designs,

- the inferiority of RS1 against the structures RS2–RS5, as it has the smallest expected lifetime under the same designs.

It is also evident that the parallel-series system provides a more reliable structure compared to the 2-out-of-3 system in terms of expectation. At the same time, the 2-out-of-3 system seems to be better than the series-parallel system, which in turn overperforms the series system consisting of three exchangeable components.

Figures 1–4 provide some illustration for the behavior of the expected lifetime and the corresponding variance under the Frank copula-based dependency. More precisely, the mean lifetimes (along with their estimated variances) of systems RS2 and RS3 are depicted in Figures 1–4 for different values of the design parameter θ.

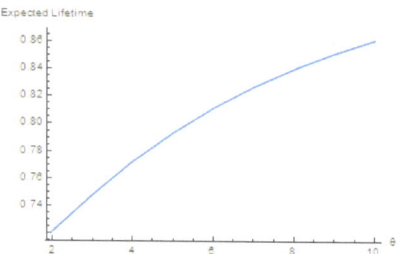

Figure 1. Expected lifetime of RS2 system versus parameter θ under Frank copula ($\lambda = 1$).

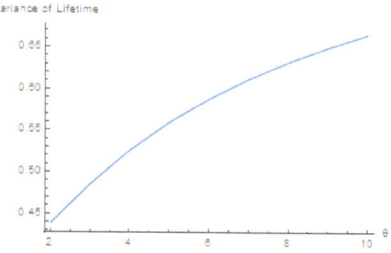

Figure 2. Variance of lifetime of RS2 system versus parameter θ under Frank copula ($\lambda = 1$).

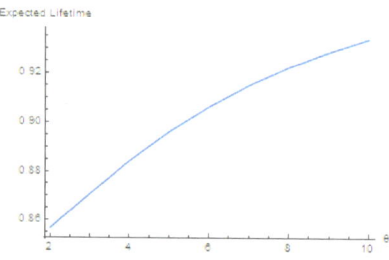

Figure 3. Expected lifetime of RS3 system versus parameter θ under Frank copula ($\lambda = 1$).

It is also of some interest to investigate the impact of parameter λ on the lifetime of the resulting reliability schemes. For this reason, we next construct some relative illustrations (see Figures 5–8), where the systems RS2 and RS3 have been taken into consideration once again.

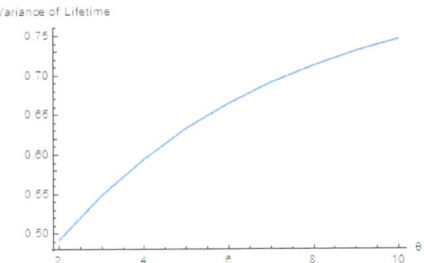

Figure 4. Variance of lifetime of RS3 system versus parameter θ under Frank copula ($\lambda = 1$).

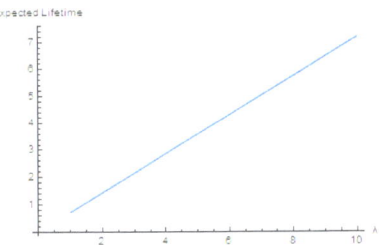

Figure 5. Expected lifetime of RS2 system versus parameter λ under Frank copula ($\theta = 2$).

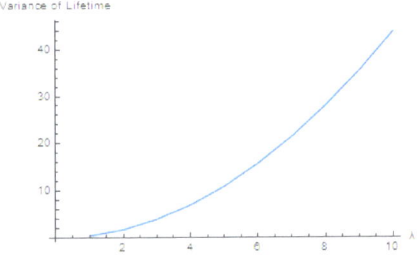

Figure 6. Variance of lifetime of RS2 system versus parameter λ under Frank copula ($\theta = 2$).

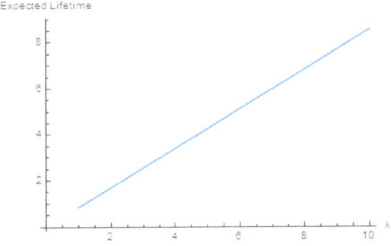

Figure 7. Expected lifetime of RS3 system versus parameter λ under Frank copula ($\theta = 2$).

Based on Figures 5–8, it is easily observed that the expected lifetime of the resulting reliability system increases in a linear way in terms of the parameter λ under the assumption that θ remains unchanged.

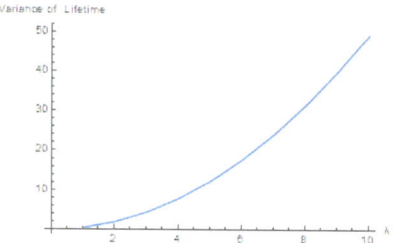

Figure 8. Variance of lifetime of RS3 system versus parameter λ under Frank copula ($\theta = 2$).

In addition, Figures 9–16 provide some illustration for the behavior of the expected lifetime and the corresponding variance under the Joe copula-based dependency. More precisely, we focus now on the RS1 and RS5 cases, and the respective mean lifetimes (along with their estimated variances) are displayed at Figures 9–12 for different values of the design parameter θ.

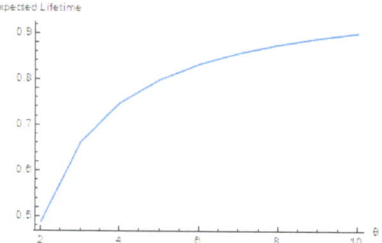

Figure 9. Expected lifetime of RS1 system versus parameter θ under Joe copula ($\lambda = 1$).

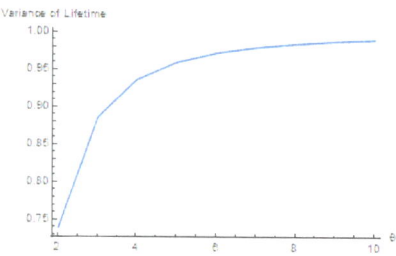

Figure 10. Variance of lifetime of RS1 system versus parameter θ under Joe copula ($\lambda = 1$).

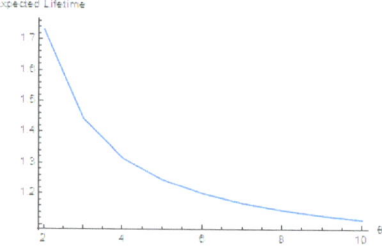

Figure 11. Expected lifetime of RS5 system versus parameter θ under Joe copula ($\lambda = 1$).

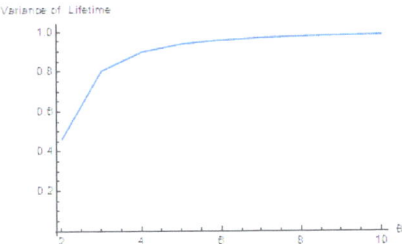

Figure 12. Variance of lifetime of RS5 system versus parameter θ under Joe copula ($\lambda = 1$).

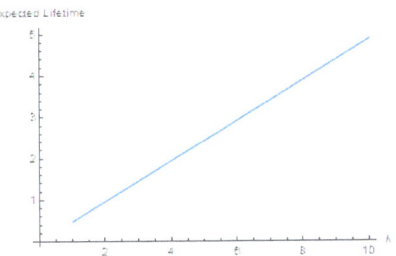

Figure 13. Expected lifetime of RS1 system versus parameter λ under Joe copula ($\theta = 2$).

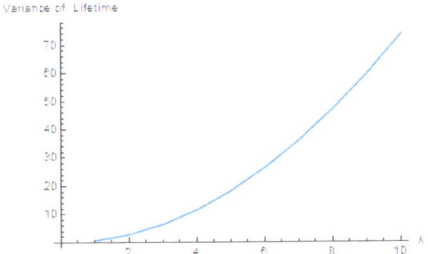

Figure 14. Variance of lifetime of RS1 system versus parameter λ under Joe copula ($\theta = 2$).

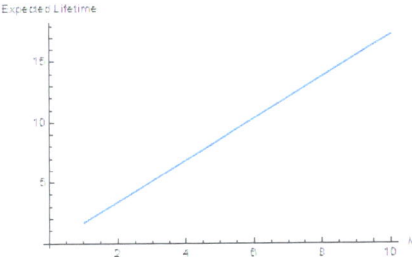

Figure 15. Expected lifetime of RS5 system versus parameter λ under Joe copula ($\theta = 2$).

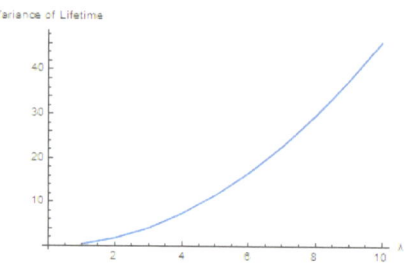

Figure 16. Variance of lifetime of RS5 system versus parameter λ under Joe copula ($\theta = 2$).

It goes without saying that systems RS1 and RS5 do not share a common behavior in terms of parameter θ under Joe copula-based dependency of their components. More specifically, we observe that,

- the expected lifetime of RS1 system becomes larger as θ increases, while
- the expected lifetime of RS5 system becomes larger as θ decreases.

The impact of parameter λ on the lifetime of the resulting reliability schemes is taken into consideration in Figures 13–16.

Based on Figures 13–16, it is deduced that the expected lifetime of the resulting reliability system increases in a linear way in terms of λ under the assumption that θ remains unchanged.

According to the numerical results provided previously, it is evident in both copulas that the expected lifetime of the underlying structure becomes larger as the corresponding parameter θ increases. More precisely, it seems that under the Frank copula the increase is more pronounced in comparison with the one observed under Joe copula. On the other hand, the numerical results do not point out that the Frank copula results in larger lifetimes than the Joe copula under the same reliability structure. In fact, there are cases where the Frank copula seems to formulate a structure with a larger expected lifetime than the corresponding one under Joe copula, while in other cases it holds the opposite conclusion. In other words, the numerical results seem to be quite robust under these two models.

5. Discussion

In the present article, a signature-based framework is provided for delivering the estimated mean lifetime (along with its variance) for a reliability structure with exchangeable components under the assumption that their dependency is modelled by the aid of well-known Archimedean copulas. The theoretical results contribute to the reliability study of such structures, while their usefulness can be also extended to the estimation of the parameter of the corresponding (common) distribution of their components. It is noteworthy that, since the present work deals with two specific copulas, the applicability of the results requires that the connection between the components of the underlying system can be described well by the copulas, which have been taken into consideration. It is noteworthy that the proposed framework does not strongly depend on the specific structure of the underlying reliability model. Therefore, similar steps could be followed in order to deliver respective results for any coherent systems with exchangeable components.

Moreover, the proposed framework seems to result in simple explicit integral expressions for the expected lifetime of the system and its corresponding variance. This conclusion seems to be evident under the assumption that the common distribution of the components has a quite simple form (as the exponential does). Therefore, in such cases, which are the most common in practice, the computational effort of the proposed methodology is manageable.

However, several bivariate copulas cannot (or at least are difficult to) be extended to multivariate models and therefore their applicability remains low. It is clear that this limitation deprives the implementation of the proposed approach to reliability structures consisting of more than two exchangeable components if the dependency among them is modeled by the aid of such copulas. For future research, it is of some interest to investigate different Archimedean copulas for modeling the dependency of the components of the underlying structure.

Funding: This research received no external funding.

Data Availability Statement: No new data were created or analyzed in this study. Data sharing is not applicable to this article.

Acknowledgments: The authors wish to thank four anonymous referees for their useful comments and suggestions, which improved the article.

Conflicts of Interest: The authors declare no conflicts of interest.

References

1. Eryilmaz, S.; Koutras, M.V.; Triantafyllou, I.S. Signature based analysis of m-consecutive k-out-of-n: F systems with exchangeable components. *Nav. Res. Logist.* **2011**, *58*, 344–354. [CrossRef]
2. Navarro, J.; Ruiz, J.M.; Sandoval, C.J. Properties of coherent systems with dependent components. *Commun. Stat. Theory Methods* **2007**, *36*, 175–191. [CrossRef]
3. Nelsen, R.B. *An Introduction to Copulas*, 2nd ed.; Springer Series in Statistics; Springer: New York, NY, USA, 2006.
4. Eryilmaz, S. Estimation in coherent reliability systems through copulas. *Reliab. Eng. Syst. Saf.* **2011**, *96*, 564–568. [CrossRef]
5. Tyagi, V.; Arora, V.; Ram, M.; Triantafyllou, I.S. Copula based Measures of Repairable Parallel System with Fault Coverage. *Int. J. Math. Eng. Manag. Sci.* **2021**, *6*, 322–344.
6. Joe, H. *Multivariate Models and Dependence Concepts*; Chapman & Hall: London, UK, 1997.
7. Genest, C. Frank's family of bivariate distributions. *Biometrika* **1987**, *74*, 549–555. [CrossRef]
8. Genest, C.; Rivest, L.-P. Statistical inference procedures for bivariate Archimedean copulas. *J. Americ. Stat. Assoc.* **1993**, *88*, 1034–1043. [CrossRef]
9. Nelsen, R.B. Concordance and copulas: A survey. In *Distributions with Given Marginals and Statistical Modelling*; Cuadras, C.M., Fortiana, J., Rodríguez Lallena, J.A., Eds.; Kluwer: Dordrecht, The Netherlands, 2002; pp. 169–177.
10. Yan, J. Multivariate Modeling with Copulas and Engineering Applications. In *Springer Handbook of Engineering Statistics*; Pham, H., Ed.; Springer Handbooks; Springer: London, UK, 2023; pp. 931–945.
11. Verdier, G. Application of copulas to multivariate control charts. *J. Stat. Plan. Inference* **2013**, *143*, 2151–2159. [CrossRef]
12. Sohrabian, B. Geostatistical prediction through convex combination of Archimedean copulas. *Spat. Stat.* **2021**, *41*, 100488. [CrossRef]
13. Kularatne, T.D.; Li, J.; Pitt, D. On the use of Archimedean copulas for insurance modelling. *Ann. Actuar. Sci.* **2020**, *15*, 57–81. [CrossRef]
14. Kasper, T.M. On convergence and singularity of conditional copulas of multivariate Archimedean copulas, and conditional dependence. *J. Multiv. Anal.* **2023**, 105275, in press. [CrossRef]
15. Alzaid, A.A.; Alhadlaq, W.M. A New Family of Archimedean Copulas: The Half-Logistic Family of Copulas. *Mathematics* **2024**, *12*, 101. [CrossRef]
16. Yang, Y.; Li, S. On a Family of Log-Gamma-Generated Archimedean Copulas. *N. Am. Actuar. J.* **2022**, *26*, 123–142. [CrossRef]
17. Alzaid, A.A.; Alhadlaq, W.M. A New Family of Archimedean Copulas: The Truncated-Poisson Family of Copulas. *Bull. Malays. Math. Sci. Soc.* **2022**, *45*, 477–504. [CrossRef]
18. Hofert, M. Sampling Archimedean copulas. *Comp. Stat. Data Anal.* **2008**, *52*, 5163–5174. [CrossRef]
19. Marshall, A.W.; Olkin, I. Families of multivariate distributions. *J. Am. Stat. Assoc.* **1988**, *83*, 834–841. [CrossRef]
20. Eryilmaz, S. Mixture representations for the reliability of consecutive-k systems. *Math. Comp. Model.* **2010**, *51*, 405–412. [CrossRef]

Disclaimer/Publisher's Note: The statements, opinions and data contained in all publications are solely those of the individual author(s) and contributor(s) and not of MDPI and/or the editor(s). MDPI and/or the editor(s) disclaim responsibility for any injury to people or property resulting from any ideas, methods, instructions or products referred to in the content.

Article

Composite and Mixture Distributions for Heavy-Tailed Data—An Application to Insurance Claims

Walena Anesu Marambakuyana and Sandile Charles Shongwe *

Department of Mathematical Statistics and Actuarial Science, Faculty of Natural and Agricultural Sciences, University of the Free State, Bloemfontein 9301, South Africa
* Correspondence: shongwesc@ufs.ac.za

Abstract: This research provides a comprehensive analysis of two-component non-Gaussian composite models and mixture models for insurance claims data. These models have gained attraction in actuarial literature because they provide flexible methods for curve-fitting. We consider 256 composite models and 256 mixture models derived from 16 popular parametric distributions. The composite models are developed by piecing together two distributions at a threshold value, while the mixture models are developed as convex combinations of two distributions on the same domain. Two real insurance datasets from different industries are considered. Model selection criteria and risk metrics of the top 20 models in each category (composite/mixture) are provided by using the 'single-best model' approach. Finally, for each of the datasets, composite models seem to provide better risk estimates.

Keywords: claims; composite models; Danish fire loss; heavy-tailed; loss distribution; mixture models; risk measures; single best model approach; skewed

MSC: 62E15

Citation: Marambakuyana, W.A.; Shongwe, S.C. Composite and Mixture Distributions for Heavy-Tailed Data—An Application to Insurance Claims. *Mathematics* 2024, 12, 335. https://doi.org/10.3390/math12020335

Academic Editors: Arne Johannssen and Nataliya Chukhrova

Received: 21 December 2023
Revised: 18 January 2024
Accepted: 18 January 2024
Published: 19 January 2024

Copyright: © 2024 by the authors. Licensee MDPI, Basel, Switzerland. This article is an open access article distributed under the terms and conditions of the Creative Commons Attribution (CC BY) license (https://creativecommons.org/licenses/by/4.0/).

1. Introduction

In the area of loss modelling, basic classical distributions such as the lognormal, Weibull, gamma, Pareto, and Burr distributions are increasingly becoming less popular as composite and mixture models are gaining more attention because of their flexibility. Composite models are developed by piecing together two distributions (which are termed head and tail distributions) at a threshold value so that small and moderate losses are modelled by the head distribution, whereas large losses are modelled by the tail distribution. On the other hand, mixture models are developed as convex combinations of distributions defined on the same overlapping domain, i.e., the positive real line. The different combinations of models that can be constructed provide a large degree of flexibility for modelling heavy-tailed loss data.

The first composite model used to model actuarial data was proposed by Cooray and Ananda [1]. This model has paved the way for more composite model research in the actuarial and risk management curriculum. The idea behind the model was to use the lognormal distribution to model the behaviour of small and moderate losses (high frequency/low severity) and the Pareto distribution to model the behaviour of the large losses (low frequency/high severity). However, this model was criticised by Scollnik [2] as it can be interpreted as a two-component mixture model with fixed and a priori known mixing weights. Scollnik [2] then proposed two models with unrestricted mixing weights. Unlike the model proposed by [1], these models provided more flexibility due to the accommodation of different proportions of the two distributions of the composite model. The models discussed in Scollnik [2] were extended by Pigeon and Denuit [3] for when the threshold is assumed to vary among observations. Pigeon and Denuit [3] proposed two examples of distributions which can be used for the threshold—this resulted in the gamma-distributed threshold and lognormaldistributed threshold. Next, Nadarajah and Bakar [4] introduced the composite lognormal-Burr model, where it was observed that in the case of the Danish fire loss data, this model performed better than the composite lognormal-Pareto family. Parallel to the studies based on the composite lognormal models at that time, Ciumara [5] introduced a model with the Weibull distribution to model the behaviour of small and moderate losses and the Pareto distribution to model the behaviour of large losses. Scollnik and Sun [6] also criticised the restrictive nature of the fixed and a priori known mixing weights of the model discussed in Ciumara [5]. Scollnik and Sun [6] proposed two additional models with unrestricted weights. Abu Bakar et al. [7] extended the class of Weibull composite distributions by proposing seven models with the tail belonging to the family of transformed beta distributions. The new composite models proposed were the composite Weibull-Burr, the composite Weibull-Loglogistic, the composite Weibull-Paralogistic, the composite Weibull-Generalised Pareto, the composite Weibull-Inverse Burr, the composite Weibull-Inverse Pareto and the composite Weibull-Inverse paralogistic models. At the time of their study, [7] found that the composite models with the Weibull as the head distribution performed better for the Danish fire loss data compared to other composite models in the literature. The extension of the framework for composite models was provided by Grün and Miljkovic [8], where they conducted a thorough analysis of 256 distinct composite curve-fitting models which emerged from piecing together two distributions (i.e., head and tail distributions) from the list of 16 widely used parametric distributions—these are provided in Table A1 in Appendix A.

For the Danish fire loss data, Grün and Miljkovic [8] identified the top 20 composite models that fit the data the best and examined the goodness-of-fit characteristics and risk assessments for those 20 models. The composite Weibull-Inverse Weibull, composite Paralogistic-Inverse Weibull, and composite Inverse Burr-Inverse Weibull, respectively, were the top three models based on the Bayesian Information Criterion (BIC). Among the 256 composite models assessed, none of the top 20 best-fitting had the lognormal distribution in the head. Contrarily, using the Weibull, paralogistic, and inverse Burr distributions in the head was proven to be the most practical approach for simulating the small- and moderate-sized claims of Danish fire loss data. The best choices for modelling the long tail of the loss data were the inverse Weibull, inverse paralogistic, loglogistic, Burr, inverse gamma, and paralogistic. Neither the Pareto nor the generalised Pareto distributions were among the top 20 based on the BIC. Calderin-Ojeda and Kwok [9] suggested the use of composite models where the mode is the splice point (or the truncation point). This method is known as the mode matching procedure, and it was used to construct the composite lognormal-Stoppa and the composite Weibull-Stoppa, where the composite Weibull-Stoppa model had the best performance up to date for the Danish data.

Keatinge [10] introduced the use of the mixture of exponentials as a semiparametric approach. Klugman and Rioux [11] stated that a drawback of the mixture of exponentials is its zero mode, and they proposed the augmented mixture of exponentials distribution which consisted of the mixture of exponentials, the gamma or lognormal distribution, and the Pareto distribution, respectively. Lee and Lin [12] stated that a drawback of the augmented mixture of exponentials is that it has a maximum of three modes, and they proposed a mixture of Erlang distributions with the same scale parameter. It is said that the mixture of Erlang distributions is dense in the space of positive, continuous distributions (Tijms, [13]). Lee and Lin [12] also demonstrated that a uniform distribution, a mixture of two gamma distributions, a generalised Pareto distribution, and the lognormal distribution can be approximated by a mixture of Erlang distributions. Finally, Lee and Lin [12] fitted the mixture of Erlang distributions to the US catastrophic loss data. Miljkovic and Grün [14] stated that a drawback of using the mixture of Erlangs with the same scale parameter could be that more components may be required to obtain an adequate fit that could have otherwise been attained without this restriction. Next, Miljkovic and Grün [14] proposed mixtures of non-Gaussian distributions with no restrictions on the parameters. Their best three models for the Danish data based on minimum BIC were the two-component Burr mixture, the three-component inverse Burr mixture, and the five-component lognormal mixture. In addition, Miljkovic and Grün [14] further added that these three models have lower negative log-likelihood (NLL), Akaike Information Criterion (AIC), and BIC in comparison to the composite Weibull-Burr, composite Weibull-Loglogistic, and the Weibull-Inverse paralogistic distributions which were considered to be the best three composite models in [7]. Abu Bakar et al. [15] proposed six two-component mixture models for fitting three real datasets—the Danish, AON Re Belgium, and Norwegian fire loss datasets. The two-component Burr mixture and the two-component lognormal were the first- and second-best models for the three datasets, respectively. However, the two-component exponential mixture was the worst for the Belgian and Danish fire loss datasets, while the two-component Pareto mixture was the worst for the Norwegian fire loss data. Abu Bakar and Nadarajah [16] proposed two-component mixture models based on the inverse transformed gamma and the transformed beta families, where their fit was illustrated using the Danish fire loss data. Furthermore, Abu Bakar and Nadarajah [16] stated that these families are appropriate for modelling loss data because of the high degree of skewness present in the tails of the distributions. This resulted in seventeen two-component mixtures, all with the inverse transformed gamma as the first component distribution and they found that these models have a better fit for the Danish data based on the BIC than all the composite models and mixture models that had been considered for the Danish dataset in the past.

In the spirit of modelling insurance claims data, many other authors have studied loss distributions. Asgharzadeh et al. [17] introduced the generalised inverse Lindley distribution for the Danish data and found it to be better than most of the classical heavy-tailed distributions, but not as good as the composite models. Next, Punzo et al. [18] introduced nine compound models using three real-life datasets (namely, the US indemnity losses, automobile insurance claims, and Norwegian fire claims). These models were said to have more flexibility than the unimodal two-parameter lognormal, inverse Gaussian, and gamma distributions due to the additional parameters. Bhati and Ravi [19] proposed the use of the generalised log-Moyal distribution and fitted it to the Danish and Norwegian fire loss datasets. Motivated by the research work of [18,19], Li et al. [20] proposed the use of the three-parameter gamma mixture of the generalised log-Moyal distribution, and it was shown to be a special case of the four-parameter generalised beta of the second kind. Zhao et al. [21] and Ahmad et al. [22] introduced additional new heavy-tailed distributions for use in insurance data analysis. While the above review is by no means comprehensive, it provides an overall view of the current literature for heavy-tailed (insurance/claims) data analysis.

This paper is motivated by the recent work by: (i) Grün and Miljkovic [8] where 256 composite models were evaluated for the Danish fire loss data (note though, the corresponding 256 mixture models have not all been considered before); and (ii) Maphalla et al. [23] where the standard loss distributions with the best goodness of fit for the South African taxi claims data were found to be the lognormal and the Pareto, and the potential future research idea of modelling the South African taxi claims data using mixture models were suggested. One should note that due to the flexibility of two-component composite and mixture models, overfitting may easily occur. Thus, care needs to be exercised when fitting these models, especially when using mixture models with more than two components, as this may easily lead to overfitting and greatly violate the principle of parsimony.

In this paper, we consider a thorough comparison of 256 composite models and 256 mixture models for curve-fitting which are derived from 16 popular parametric distributions listed in Table A1 in Appendix A. This study focuses on the following objectives: (i) To discover composite models that have not been studied previously for the South African taxi claims data; (ii) To discover mixture models that have not been studied previously for the South African taxi claims data and Danish fire loss data; and (iii) To assess the implications of the different composite models and mixture models using risk measures, such as Value-at-Risk (VaR) and Tail Value-at-Risk (TVaR).

This paper is structured as follows: Section 2 provides the methodology, which includes model specification, risk measures, and model selection criteria. Section 3 provides the analysis, wherein all the results for the top 20 composite models and mixture models that yield the best goodness-of-fit to the Danish fire loss data and the South African taxi claims data are discussed. Different information criteria and risk measures are computed and presented for models studied in this paper, with additional results provided in Appendix A. Finally, Section 4 provides the concluding remarks.

2. Methodology

2.1. The Composite Model

2.1.1. Model Specification

The probability density function (pdf) of a composite model which was introduced in [7] and adapted by [8] is given by

$$f(\vartheta_1, \vartheta_2, \theta, \phi) = \begin{cases} \frac{1}{1+\phi} f_1^*(x|\vartheta_1, \theta), & \text{if } 0 < x \leq \theta, \\ \frac{\phi}{1+\phi} f_2^*(x|\vartheta_2, \theta), & \text{if } \theta < x < \infty. \end{cases} \qquad (1)$$

The continuity condition and the differentiability conditions are imposed at the threshold θ such that,

$$f(\vartheta_1, \vartheta_2, \theta, \phi) = f(\vartheta_1, \vartheta_2, \theta, \phi) \qquad (2)$$

$$f'(\vartheta_1, \vartheta_2, \theta, \phi) = f'(\vartheta_1, \vartheta_2, \theta, \phi) \qquad (3)$$

where ϑ_1 and ϑ_2 are the parameter sets associated with the pdfs on the disjoint intervals, $(0, \theta]$ and (θ, ∞), respectively. The continuity and differentiability conditions ensure that the threshold parameter θ and the weight parameter $\phi > 0$ are defined as functions of the other parameters, ϑ_1 and ϑ_2. In addition, $\frac{1}{1+\phi}$ and $\frac{\phi}{1+\phi}$ are referred to as mixing weights, see [4]. Moreover, the continuity condition at threshold θ ensures that the weight parameter ϕ is expressed as a function of the other parameters ϑ_1, ϑ_2, θ, and cumulative distribution function (cdf) in closed form as,

$$\phi = -\frac{\frac{d \ln F_1(\theta|\vartheta_1)}{d\theta}}{\frac{d \ln[1-F_2(\theta|\vartheta_2)]}{d\theta}} = \frac{\frac{f_1(\theta|\vartheta_1)}{F_1(\theta|\vartheta_1)}}{\frac{f_2(x|\vartheta_2)}{1-F_2(\theta|\vartheta_2)}}. \qquad (4)$$

Substituting the expression for ϕ obtained in Equation (4) into the differentiability condition in Equation (3) gives the following condition for the threshold θ, which simplifies to,

$$\frac{d}{d\theta} \ln\left[\frac{f_1(\theta|\vartheta_1)}{f_2(\theta|\vartheta_2)}\right] = 0$$
$$\frac{f_1'(\theta|\vartheta_1)}{f_1(\theta|\vartheta_1)} = \frac{f_2'(\theta|\vartheta_2)}{f_2(\theta|\vartheta_2)}. \qquad (5)$$

Lastly, $f_1^*(x|\vartheta_1, \theta)$ and $f_2^*(x|\vartheta_2, \theta)$ are truncated pdfs which are defined in terms of their corresponding pdfs and cdfs are

$$f_1^*(x|\vartheta_1, \theta) = \frac{f_1(x|\vartheta_1)}{F_1(\theta|\vartheta_1)}, \qquad (6)$$

$$f_2^*(x|\vartheta_2, \theta) = \frac{f_2(x|\vartheta_2)}{1 - F_2(\theta|\vartheta_2)}, \qquad (7)$$

and also,

$$F(\vartheta_1, \vartheta_2, \theta, \phi) = \begin{cases} \frac{1}{1+\phi} \frac{F_1(x|\vartheta_1)}{F_1(\theta|\vartheta_1)}, & \text{if } 0 < x \leq \theta, \\ \frac{1}{1+\phi}\left[1 + \phi \frac{F_2(x|\vartheta_2) - F_2(\theta|\vartheta_2)}{1 - F_2(x|\vartheta_2)}\right], & \text{if } \theta < x < \infty. \end{cases} \quad (8)$$

The kth raw moment of the composite model is given in Grün and Miljkovic [8] as

$$\mathbb{E}\left[X^k\right] = \frac{1}{1+\phi}\mathbb{E}\left[X_1^k\right]\frac{F_1^{(k)}(\theta|\vartheta_1)}{F_1(\theta|\vartheta_1)} + \frac{\phi}{1+\phi}\mathbb{E}\left[X_2^k\right]\frac{1 - F_2^{(k)}(\theta|\vartheta_2)}{1 - F_2(\theta|\vartheta_2)}, \quad (9)$$

where X_i is the random variable associated with the ith component and $F_i^{(k)}$ is the kth incomplete moment distribution of the ith component distribution. For a random sample $x = \{x_1, x_2, \ldots, x_n\}$, the log-likelihood function which was introduced in Grün and Miljkovic [8] is given by

$$\ell(\vartheta_1, \vartheta_2|x) = \sum_{i=1}^n \ln(f(x_i|\vartheta_1, \vartheta_2)). \quad (10)$$

2.1.2. Risk Measures

Abu Bakar et al. [7] and Grün and Miljkovic [8] defined the theoretical estimate for the VaR of X as

$$\text{VaR}_p(X) = \begin{cases} F_1^{-1}(p(1+\phi)F_1(\theta)), & \text{if } 0 < p \leq \frac{1}{1+\phi}, \\ F_2^{-1}(F_2(\theta) + (p(1+\phi) - 1)(1 - F_2(\theta))/\phi), & \text{if } \frac{1}{1+\phi} < p < 1. \end{cases} \quad (11)$$

The theoretical estimates for the TVaR of X are defined in [8] as

$$\text{TVaR}_p(X) = \begin{cases} \frac{1}{1-p}\left[\frac{\int_{\pi_p}^\theta xf_1(x)dx}{F_1(\theta)} + \frac{\int_\theta^\infty xf_2(x)dx}{1-F_2(\theta)}\right], & \text{if } 0 < p \leq \frac{1}{1+\phi}, \\ \frac{1}{1-p}\frac{1}{1-F_2(\theta)}\left[\int_{\pi_p}^\infty xf_2(x)dx\right], & \text{if } \frac{1}{1+\phi} < p < 1. \end{cases} \quad (12)$$

Finite values of Equation (12) can only be obtained if the first moment of the tail distribution exists (Grün and Miljkovic, [8]).

2.2. The Mixture Model
2.2.1. Model Specification

The pdf of a two-component mixture model is given by

$$f(x|\vartheta_1, \vartheta_2, \phi) = \frac{1}{1+\phi}f_1(x|\vartheta_1) + \frac{\phi}{1+\phi}f_2(x|\vartheta_2), \quad x > 0, \quad (13)$$

where $\phi > 0$ is the weight parameter, and ϑ_1 and ϑ_2 are the parameter sets associated with the first and second component distributions, respectively, where f_1 and f_2 are the corresponding pdfs. The component distributions are both defined on R^+. Therefore, the set of parameters of the mixture model is $\{\phi, \vartheta_1, \vartheta_2\}$. Unlike the composite model, the weight parameter, ϕ, is not a function of the other parameters. Rather, the weight parameter is also a model parameter, which is estimated by the maximum likelihood method in a similar fashion as the other model parameters. The coefficients of f_1 and f_2 are called mixing weights and for $\phi > 0$, it is clear that $\frac{1}{1+\phi} + \frac{\phi}{1+\phi} = 1$. For $\phi = 1$, the component distributions have equal mixing weights of 0.5, i.e., $\frac{1}{1+\phi} = \frac{\phi}{1+\phi} = 0.5$. For $\phi < 1$, the first component distribution has a greater weight to the mixture model than the second component distribution. For $\phi > 1$, the second component distribution has a greater weight to the mixture model than the first component distribution.

The corresponding cdf is given by

$$F(x|\vartheta_1, \vartheta_2, \phi) = \frac{1}{1+\phi}F_1(x|\vartheta_1) + \frac{\phi}{1+\phi}F_2(x|\vartheta_2), x > 0, \tag{14}$$

where F_1 and F_2 are the cdfs of the first and second components, respectively. The kth raw moment of a two-component mixture model is given by

$$E\left[X^k\right] = \frac{1}{1+\phi}\mathbb{E}\left[X_1^k\right] + \frac{\phi}{1+\phi}\mathbb{E}\left[X_2^k\right], x > 0, \tag{15}$$

where $\mathbb{E}\left[X_1^k\right]$ and $\mathbb{E}\left[X_2^k\right]$ are the kth raw moments of the first and second components, respectively, given that they exist. The moment-generating function (mgf) of a two-component mixture model is given by

$$M_X(t) = \frac{1}{1+\phi}\mathbb{M}_{X_1}(t) + \frac{\phi}{1+\phi}\mathbb{M}_{X_2}(t), \tag{16}$$

where $\mathbb{M}_{X_1}(t)$ and $\mathbb{M}_{X_2}(t)$ are the mgfs of the first and second components, respectively, given that they exist.

For a random sample $x = \{x_1, x_2, \ldots, x_n\}$, the log-likelihood function was introduced in Abu Bakar and Nadarajah [16] as

$$l(\vartheta_1, \vartheta_2, \phi|x) = -n\ln(1+\phi) + \sum_{i=1}^{n}\{\ln[f_1(x_i|\vartheta_1) + \phi f_2(x_i|\vartheta_2)]\}. \tag{17}$$

2.2.2. Flexibility for Unimodal and Multimodal Data

Abu Bakar and Nadarajah [16] illustrated the flexibility of two-component mixture models by their adaptability to unimodal and bimodal density functions. In this section, we extend the demonstration of flexibility (adaptability to unimodality and bimodality) by illustrating graphically with two additional two-component mixture models. The first model has different parametric distributions (i.e., inverse transformed gamma and transformed beta distributions) and the second has the same parametric distribution (i.e., Burr distributions) with different parameters. Varying the parameter estimates as indicated in Table 1 leads to different shapes of the pdfs in Figure 1 to illustrate how the mixture of inverse transformed gamma and transformed beta tends to account for unimodality and bimodality.

Table 1. Parameter estimates corresponding to the models in Figure 1.

	ϕ	ϑ_1	ϑ_2
Model A	$\phi = 1$	$\alpha = 0.5, \tau = 2, \theta = 1$	$\alpha = 0.1, \gamma = 0.5, \tau = 2, \theta = 1$
Model B	$\phi = 2.5$	$\alpha = 0.1, \tau = 0.5, \theta = 1$	$\alpha = 0.9, \gamma = 1.5, \tau = 5, \theta = 1$
Model C	$\phi = 5$	$\alpha = 10, \tau = 3, \theta = 10$	$\alpha = 0.2, \gamma = 5, \tau = 2, \theta = 0.5$

Similarly, varying the parameter estimates as indicated in Table 2 leads to different shapes of the pdfs in Figure 2 to illustrate how the mixture of two Burr distributions tends to account for unimodality and bimodality. Note that similar patterns can be illustrated for other types of mixture distributions.

Table 2. Parameter estimates corresponding to the models in Figure 2.

	ϕ	ϑ_1	ϑ_2
Model A	$\phi = 1$	$\alpha = 0.1,\ \gamma = 2,\ \theta = 1$	$\alpha = 2,\ \gamma = 5,\ \theta = 1$
Model B	$\phi = 2$	$\alpha = 2,\ \gamma = 0.6,\ \theta = 1$	$\alpha = 3,\ \gamma = 0.2,\ \theta = 1$
Model C	$\phi = 1$	$\alpha = 10,\ \gamma = 10,\ \theta = 3$	$\alpha = 2.5,\ \gamma = 8,\ \theta = 8$

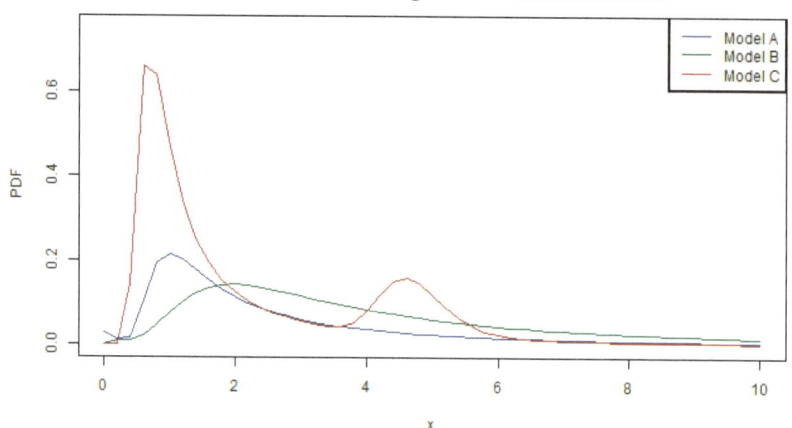

Figure 1. The pdfs of the two-component inverse transformed gamma and transformed beta mixture for different parameters.

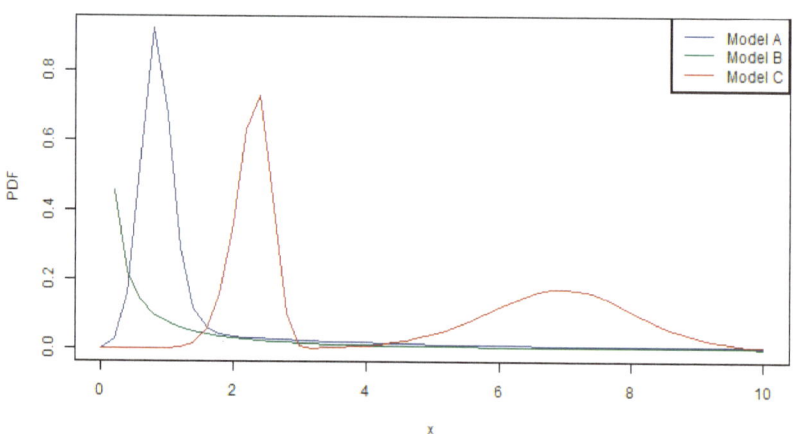

Figure 2. The pdfs of the two-component Burr mixture for different parameters.

2.2.3. Risk Measures

The theoretical estimate for the VaR of the mixture model does not have a closed-form solution and requires a numerical solution of

$$F_X(\text{VaR}_p(X)) = p \tag{18}$$

which can be evaluated using software as stated in [14]. However, for the mixture model, the theoretical estimate for the TVaR of X can be simplified by the "linearity" property stated in [14] to give the weighted sum of the $\text{TVaR}_p(X)$ of each of the component distributions.

2.3. Model Selection Criteria

This section discusses some commonly used model selection criteria that appear in the area of loss distributions. Three information criteria are considered: NLL, AIC, and BIC; see Abu Bakar et al. [7]. The BIC is also known as the Schwarz's Bayesian Criterion (SBC). For all three criteria, a lower value implies that the theoretical model provides a better fit to the data. The NLL is appropriate only when comparing models with the same number of parameters; however, the AIC and BIC are more appropriate for comparing models with a different number of parameters. Let $l(\theta)$ denote the maximised log-likelihood function of a model, then the NLL is defined as

$$\text{NLL} = -l(\theta). \tag{19}$$

The AIC was introduced by Akaike [24] and is defined as

$$\text{AIC} = 2\text{NLL} + 2p, \tag{20}$$

where p is the number of free parameters or degrees of freedom. The BIC was introduced by Schwarz [25] and it is defined as

$$\text{BIC} = 2\text{NLL} + p\log(n), \tag{21}$$

where n is the number of observations. An analysis of the results is given with an emphasis on the BIC.

3. Empirical Analysis

In this section, the statistical computations were performed in R (R Core Team, [26]). Two real-life datasets are considered—the South African taxi claims data and the Danish fire insurance loss data. The taxi (or minibus) industry in South Africa, well known for its taxi turf wars, provides the most commonly used mode of public transport, especially for lower-income communities (which account for a larger proportion of the population due to South Africa's high level of inequality in income levels and high unemployment rates). The types of disasters that this industry faces include and are not limited to road accidents (due to potholes, tyre bursts, improper road infrastructure, vehicle malfunctioning, and drunk driving), hijacking, theft of taxi parts, and damage or fires due to public protests because of poor service delivery by elected officials. However, the Danish fire loss data, which was collected by Copenhagen Reinsurance, covers losses from fire due to buildings, contents, and profits. The Danish data is from Denmark, which is in Europe (Northern Hemisphere), a developed first-world country, whereas the South African taxi claims data is from South Africa, which is in the southernmost part of Africa in the Southern Hemisphere, a developing third-world country. Considering that Denmark is a well-developed country, in the case of a fire hazard, the fire can be extinguished quickly because of Denmark's well-developed social service delivery. On the other hand, with the many hazards that can occur in the taxi industry in South Africa, they may not all be avoidable because of the many underdevelopments. When it comes to economic development, a large portion of the South African population is impoverished and burdened with unemployment, whereas only a small portion of the population in Denmark is lacking.

The South African taxi claims data, which was kindly provided for our study by [23] (this data has been made available in the Supplementary Materials of this paper), consists of 48,043 observations and was divided by 100 for computational ease. The Danish fire loss data, however, is very popular and has a long history of applications. It consists of 2492 observations which were adjusted for inflation to reflect 1985 values. Most of the composite models in actuarial literature have used the Danish data as an application. The Danish dataset is available in the *SMPracticals* package Version 1.4-3 in R, Davidson [27]. The full R code used for the analysis in this paper has been made available in the Supplementary Materials of this paper. Tables 3 and 4 provide the summary of the descriptive statistics for the South African taxi claims data and Danish fire loss data, respectively.

Table 3. Descriptive statistics of the South African taxi claims data.

Minimum	Quantiles	Mean	Maximum	Standard Deviation	Coefficient of Variation	Skewness	Kurtosis
0.1	(20.8, 45, 120.8)	132.3	4803.3	284.1563	2.15	6.474	63.64

Table 4. Descriptive statistics of the Danish fire loss data.

Minimum	Quantiles	Mean	Maximum	Standard Deviation	Coefficient of Variation	Skewness	Kurtosis
0.3134	(1.1572, 1.6339, 2.6455)	3.0627	263.2504	7.976703	2.60	19.896	549.5736

Figure 3 provides the boxplots for the South African taxi claims data and the Danish fire loss data, respectively. The dotted vertical line represents the mean value for the datasets. For both datasets, it is clear that the data are skewed to the right.

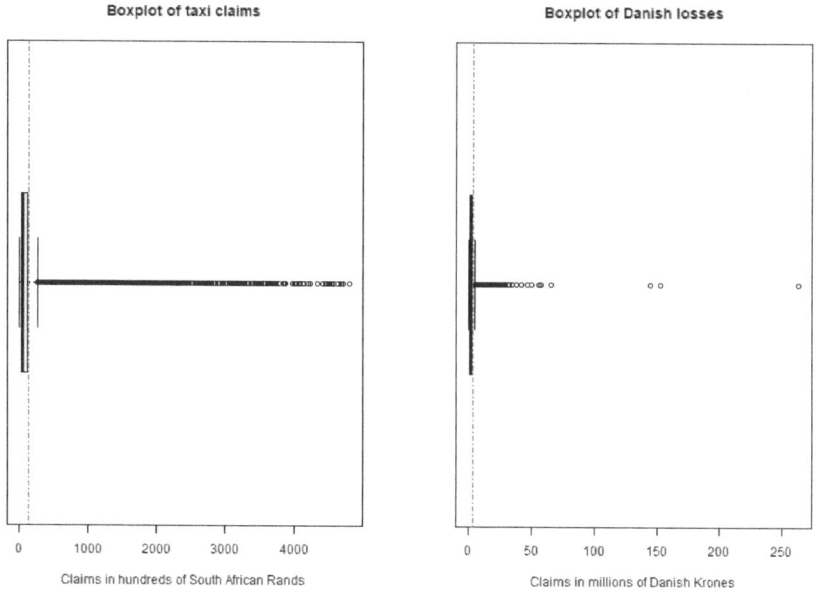

Figure 3. Boxplots of the taxi claims and the Danish loss data.

Figure 4 displays the histograms of the taxi claims and the Danish fire loss data, respectively. By visual inspection of the histograms, the claims data are positive (or at least nonnegative), unimodal and hump-shaped, skewed to the right with long upper tails, and the smaller claims occur with more frequency whereas the larger claims are less frequent.

Figure 5 displays the mean excess plots of the taxi claims on the left and the Danish fire losses data on the right. The mean excess plot for the taxi claims data is initially ultimately increasing, then ultimately constant, and then ultimately decreasing for the remainder of the plot. Therefore, the underlying distribution of the taxi claims data can be said to be heavy-tailed for the lower (left) tail and light-tailed for the upper (right) tail. The mean excess plot for the Danish losses is ultimately increasing (with two observations being the exception). Therefore, the underlying distribution of the Danish data can be said to be heavy-tailed throughout, apart from two observations.

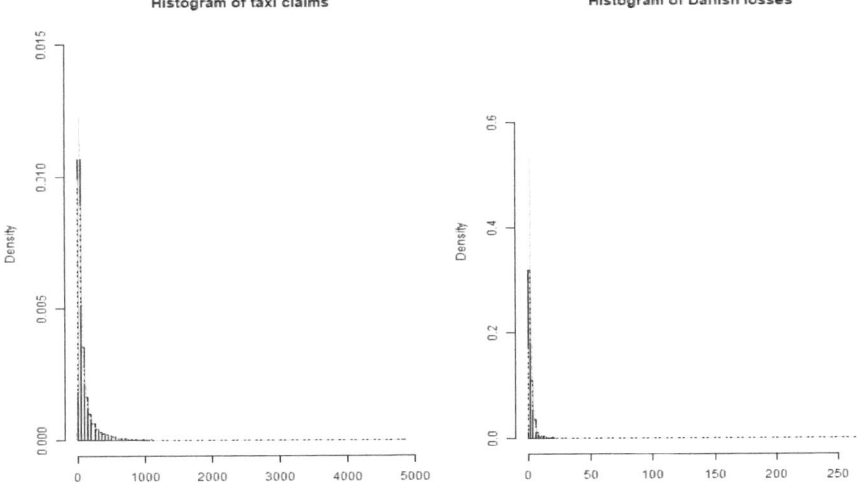

Figure 4. Histograms of the taxi claims and the Danish loss data.

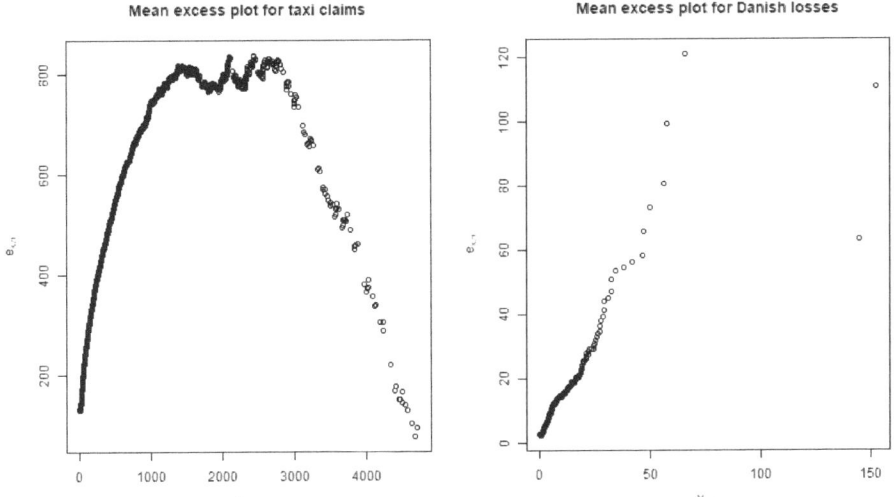

Figure 5. Mean excess plots of the taxi claims and the Danish loss data.

It is important to note that the approach used in this paper is based on the 'single best model' as performed in well-known studies like [7,8,14,16]. That is, using the BIC (as the main model selection criteria), we extract the top 20 best goodness-of-fit models, calculate the corresponding risk metrics, and select the model with the VaR and/or TVaR closer to the corresponding empirical values. Note though that Blostein and Miljkovic [28] as well as Miljkovic and Grün [29] suggested two alternative methods to select the optimal model, which consider both the goodness-of-fit measures and the risk metrics. It is worth mentioning that [29] stated that the 'single best model' approach is still the most used one because although *"model averaging has been recognized in the actuarial field, it has not yet been embraced as a standard practice neither in risk management nor the regulatory capital environment"*, but researchers and practitioners are starting to realise the importance of applying the 'model averaging approach'.

Fitting composite models to the taxi claims data

Using the 16 loss distributions outlined in Appendix A's Table A1, it is observed in Table 5 that using distributions such as the gamma, loglogistic, paralogistic, and inverse paralogistic in the head is found to be ideal for modelling the small and moderate size claims of the taxi claims data. However, the tail distributions such as the Weibull, inverse Gaussian, Burr, Pareto, or generalised Pareto and lognormal seem to be the best choices for modelling the upper tail of taxi claims. In an effort to conserve writing space, the corresponding parameter estimates of the top 20 models in Table 5 are provided in Table A2 in Appendix A.

Table 5. Summary of the information criteria of the top 20 composite models for taxi claims (based on the BIC).

Head	Tail	p	NLL	AIC	BIC
Gamma	Weibull	4	270,197.8	540,403.6	540,438.8
Paralogistic	Inverse Gaussian	4	270,198.1	540,404.2	540,439.3
Loglogistic	Inverse Gaussian	4	270,200.7	540,409.5	540,444.6
Paralogistic	Weibull	4	270,201.2	540,410.5	540,445.6
Inverse paralogistic	Inverse Gaussian	4	270,217.0	540,442.1	540,447.2
Weibull	Weibull	4	270,202.4	540,412.8	540,447.9
Gamma	Burr	5	270,197.8	540,405.6	540,449.5
Loglogistic	Weibull	4	270,204.4	540,416.8	540,452.0
Paralogistic	Burr	5	270,201.1	540,412.5	540,456.4
Weibull	Burr	5	270,202.4	540,414.8	540,458.7
Inverse Burr	Weibull	5	270,202.5	540,415.1	540,459.0
Loglogistic	Burr	5	270,204.4	540,418.8	540,462.7
Inverse Burr	Burr	6	270,201.2	540,417.1	540,469.8
Inverse paralogistic	Weibull	4	270,223.4	540,454.7	540,489.9
Inverse paralogistic	Burr	5	270,223.4	540,456.8	540,500.7
Burr	Pareto	5	270,246.1	540,502.3	540,546.2
Weibull	Lognormal	4	270,259.9	540,527.8	540,562.9
Gamma	Lognormal	4	270,260	540,528.8	540,563.9
Gamma	Generalised Pareto	5	270,257.0	540,524.0	540,567.9
Paralogistic	Lognormal	4	270,262.9	540,533.9	540,569.0

Table 6 reports the empirical risk estimates, the estimated risk measures for the top 20 composite models for taxi claims, and the percentage deviation in parenthesis of each estimated risk measure with respect to the empirical risk estimates. The risk estimates obtained from using the top 20 composite distributions closely match the empirical risk estimates. However, using the lognormal or the generalised Pareto distributions as a tail distribution leads to much higher estimates for the TVaR than when using the Burr distribution, the Pareto distribution, or the Weibull distribution. Using the inverse Gaussian distribution as the tail distribution leads to TVaR estimates that are much lower than the empirical estimates.

Table 6. Summary of the empirical risk estimates, risk measures of the top 20 composite models for taxi claims data, and the percentage deviation with respect to the empirical risk estimates in parenthesis.

		$VaR_{0.95}$	$VaR_{0.99}$	$TVaR_{0.95}$	$TVaR_{0.99}$
Empirical Estimates		525.1509	1396.901	1085.583	2206.203
Parametric					
Head	Tail				
Gamma	Weibull	521.76 (−0.6%)	1396.27 (0.0%)	1125.77 (3.7%)	2422.04 (9.8%)
Paralogistic	Inverse Gaussian	547.96 (4.3%)	1361.21 (−2.6%)	1068.6 (−1.6%)	2055.10 (−6.8%)
Loglogistic	Inverse Gaussian	547.84 (4.3%)	1361.62 (−2.5%)	1068.80 (−1.5%)	2056.09 (−6.8%)
Paralogistic	Weibull	521.92 (−0.6%)	1394.70 (−0.2%)	1124.46 (3.6%)	2416.23 (9.5%)
Inverse paralogistic	Inverse Gaussian	547.38 (4.2%)	1363.79 (−2.4%)	1070.18 (−1.4%)	2061.91 (−6.5%)
Weibull	Weibull	522.21 (−0.6%)	1389.41 (−0.5%)	1119.77 (3.1%)	2396.8 (8.6%)
Gamma	Burr	521.78 (−0.6%)	1396.5 (0.0%)	1126.1 (3.7%)	2423.17 (9.8%)
Loglogistic	Weibull	521.62 (−0.7%)	1393.86 (−0.2%)	1123.85 (3.5%)	2415.2 (9.5%)
Paralogistic	Burr	521.92 (−0.6%)	1394.71 (−0.2%)	1124.46 (3.6%)	2416.26 (9.5%)
Weibull	Burr	-	-	-	-
Inverse Burr	Weibull	522.15 (−0.6%)	1395.0 (−0.1%)	1124.66 (3.6%)	2416.12 (9.5%)
Loglogistic	Burr	-	-	-	-
Inverse Burr	Burr	521.91 (−0.6%)	1394.08 (−0.2%)	1123.92 (3.5%)	2414.26 (9.4%)
Inverse paralogistic	Weibull	520.81 (−0.8%)	1393.196 (−0.3%)	1123.73 (3.5%)	2418.417 (9.6%)
Inverse paralogistic	Burr	521.06 (−0.8%)	1394.12 (−0.2%)	1124.48 (3.6%)	2420.34 (9.7%)
Burr	Pareto	532.21 (1.3%)	1334.65 (−4.5%)	1112.35 (2.5%)	2391.68 (8.4%)
Weibull	Lognormal	516.02 (−1.7%)	1513.17 (8.3%)	1270.78 (17.1%)	3067.21 (39.0%)
Gamma	Lognormal	516.54 (−1.6%)	1524.43 (9.1%)	1286.50 (18.5%)	3133.02 (42.0%)
Gamma	Generalised Pareto	585.11 (11.4%)	1587.63 (13.7%)	1396.13 (28.6%)	3404.88 (54.3%)
Paralogistic	Lognormal	513.46 (−2.2%)	1511.45 (8.2%)	1274.91 (17.4%)	3098.64 (40.5%)

Fitting mixture models to the taxi claims data

Using the 16 loss distributions outlined in Appendix A's Table A1, it is observed from the results in Table 7 that the lognormal distribution seems to be an ideal component distribution for most of the best-fitting mixture models. It seems that the conclusion by Maphalla et al. [23] that the lognormal distribution is the best for taxi claims data is supported by the top mixture models with a lognormal distribution component. In an effort to conserve writing space, the corresponding parameter estimates of the top 20 models in Table 7 are provided in Table A3 in Appendix A.

Table 7. Summary of the information criteria of the top 20 mixture models for taxi claims data (based on the BIC).

First Component	Second Component	p	NLL	AIC	BIC
Inverse gamma	Lognormal	5	270,142.25	**540,294.5**	**540,338.4**
Inverse Gaussian	Lognormal	5	270,142.45	540,294.91	540,338.81
Generalised Pareto	Lognormal	6	**270,142.17**	540,296.35	540,349.03
Inverse paralogistic	Lognormal	5	270,148.17	540,306.35	540,350.25
Inverse Weibull	Lognormal	5	270,148.29	540,306.58	540,350.48
Inverse Burr	Lognormal	6	270,146.59	540,305.17	540,357.85
Loglogistic	Lognormal	5	270,158.197	540,326.39	540,370.29
Burr	Lognormal	6	270,155.86	540,323.72	540,376.4
Gamma	Lognormal	5	270,164.59	540,339.19	540,383.09
Paralogistic	Lognormal	5	270,167.79	540,345.59	540,389.49
Lognormal	Weibull	5	270,186.3	540,382.6	540,426.5
Loglogistic	Generalised Pareto	6	270,225.71	540,463.43	540,516.11
Generalised Pareto	Paralogistic	6	270,227.97	540,467.94	540,520.62
Loglogistic	Paralogistic	5	270,239.59	540,489.17	540,533.07
Burr	Loglogistic	6	270,236.85	540,485.70	540,538.38
Paralogistic	Paralogistic	5	270,247.06	540,504.11	540,548.01
Burr	Burr	7	270,236.71	540,487.43	540,548.89
Inverse gamma	Paralogistic	5	270,248.70	540,507.41	540,551.31
Inverse gamma	Generalised Pareto	6	270,243.88	540,499.75	540,552.43
Paralogistic	Burr	6	270,246.83	540,505.67	540,558.35

For the taxi claims data, the two-component Burr mixture also performs better than the two-component gamma mixture, the two-component Pareto mixture, the two-component Weibull mixture, and the two-component exponential mixture—this is similar to the results observed in Abu Bakar et al. [15] for the Danish, Belgian, and Norwegian loss datasets. Additionally, for the taxi claims data, the two-component paralogistic mixture performs better than the two-component Burr based on the BIC. In fact, the two-component gamma mixture, the two-component exponential mixture, and the two-component Weibull mixture did not converge for the taxi claims data. Other components such as the paralogistic distribution, the Burr distribution, the generalised Pareto distribution, the loglogistic distribution, and the inverse gamma distribution also seem to be optimal component distributions for the mixture models for the taxi claims data.

The mixture models considered provide fair estimates for the VaR at both 95% and 99% security levels, although the VaR at a 95% security level is underestimated by all the models (see Table 8). The TVaR at both 95% and 99% security levels is not underestimated for any of the models, which provides a bit of comfort since the TvaR is a coherent risk measure and more attractive than the VaR.

Table 8. Summary of the empirical risk estimates, risk measures of the top 20 mixture models for the taxi claims data, and the percentage deviation with respect to the empirical risk estimates in parenthesis.

		$VaR_{0.95}$	$VaR_{0.99}$	$TVaR_{0.95}$	$TVaR_{0.99}$
Empirical Estimates		525.1509	1396.901	1085.583	2206.203
		Parametric			
First Component	Second Component				
Inverse gamma	Lognormal	513.98 (−2.1%)	1382.65 (−1.0%)	1142.3 (5.2%)	2558.96 (16.0%)
Inverse Gaussian	Lognormal	514.55 (−2.0%)	1373.99 (−1.6%)	1134.85 (4.5%)	2529.36 (14.6%)
Generalised Pareto	Lognormal	514.63 (−2.0%)	1383.4 (−1.0%)	1142.86 (5.3%)	2558.81 (16.0%)
Inverse paralogistic	Lognormal	515.07 (−1.9%)	1390.59 (−0.5%)	1149.02 (5.8%)	2580.52 (17.0%)
Inverse Weibull	Lognormal	510.55 (−2.8%)	1378.04 (−1.4%)	1139.31 (4.9%)	2560.72 (16.1%)
Inverse Burr	Lognormal	513.78 (−2.2%)	1389.68 (−0.5%)	1148.59 (5.8%)	2583.77 (17.1%)
Loglogistic	Lognormal	517.67 (−1.4%)	1391.51 (−0.4%)	1149.05 (5.8%)	2570.78 (16.5%)
Burr	Lognormal	516.78 (−1.6%)	1400.85 (0.3%)	1157.8 (6.7%)	2607.82 (18.2%)
Gamma	Lognormal	513.25 (−2.3%)	1360.23 (−2.6%)	1123.05 (3.5%)	2489.42 (12.8%)
Paralogistic	Lognormal	516.78 (−1.6%)	1373.68 (−1.7%)	1133.63 (4.4%)	2515.69 (14.0%)
Lognormal	Weibull	512.52 (−2.4%)	1342.93 (−3.9%)	1107.41 (2.0%)	2431.57 (10.2%)
Loglogistic	Generalised Pareto	518.2 (−1.3%)	1349.01 (−3.4%)	1157.27 (6.6%)	2664.54 (20.8%)
Generalised Pareto	Paralogistic	511.48 (−2.6%)	1363.65 (−2.4%)	1190.45 (9.7%)	2846.36 (29.0%)
Loglogistic	Paralogistic	513.22 (−2.3%)	1372.25 (−1.8%)	1223.63 (12.7%)	3002.93 (36.1%)
Burr	Loglogistic	515.45 (−1.8%)	1351.42 (−3.3%)	1181.42 (8.8%)	2799.21 (26.9%)
Paralogistic	Paralogistic	507.18 (−3.4%)	1384.18 (−0.9%)	1255.56 (15.7%)	3176.84 (44.0%)
Burr	Burr	516.17 (−1.7%)	1351.72 (−3.2%)	1181.84 (8.9%)	2798.08 (26.8%)
Inverse gamma	Paralogistic	504.42 (−3.9%)	1436.60 (2.8%)	1327.5 (22.3%)	3496.71 (58.5%)
Inverse gamma	Generalised Pareto	507.28 (−3.4%)	1402.21 (0.4%)	1268.23 (16.8%)	3220.81 (46.0%)
Paralogistic	Burr	507.44 (−3.4%)	1376.18 (−1.5%)	1240.94 (14.3%)	3109.60 (40.9%)

Fitting composite models to the Danish data

Using the 16 loss distributions outlined in Appendix A's Table A1, it is observed that having distributions such as the Weibull, paralogistic, and inverse Burr in the head is found to be ideal for modelling the small and moderate size claims of Danish fire losses (Grün and Miljkovic, [8]). The tail distributions such as inverse Weibull, inverse paralogistic, loglogistic, Burr, inverse gamma, and paralogistic seem to be the best choices for modelling the long tail of Danish fire losses (Grün and Miljkovic, [8]). In an effort to conserve writing space, the corresponding parameter estimates of the top 20 models in Table 9 are provided in Table A4 in Appendix A.

Table 9. Summary of the information criteria of the top 20 composite models for Danish fire loss data (based on the BIC)—these results are similar to those reported in Grün and Miljkovic [8].

Head	Tail	p	NLL	AIC	BIC
Weibull	Inverse Weibull	4	3820.01	7648.02	**7671.30**
Paralogistic	Inverse Weibull	4	3820.14	7648.28	7671.56
Inverse Burr	Inverse Weibull	5	3816.34	7642.68	7671.79
Weibull	Inverse paralogistic	4	3820.93	7649.87	7673.15
Inverse Burr	Inverse paralogistic	5	3817.07	7644.14	7673.25
Paralogistic	Inverse paralogistic	4	3821.04	7650.08	7673.36
Weibull	Loglogistic	4	3821.23	7650.46	7673.74
Inverse Burr	Loglogistic	5	3817.37	7644.74	7673.85
Paralogistic	Loglogistic	4	3821.32	7650.65	7673.93
Loglogistic	Inverse Weibull	4	3821.38	7650.76	7674.04
Weibull	Burr	5	3817.57	7645.14	7674.24
Paralogistic	Burr	5	3817.72	7645.43	7674.54
Inverse Burr	Burr	6	**3814.00**	**7639.99**	7674.92
Loglogistic	Inverse paralogistic	4	3822.15	7652.31	7675.59
Inverse Burr	Inverse gamma	5	3818.30	7646.61	7675.71
Paralogistic	Inverse gamma	4	3822.22	7652.43	7675.72
Loglogistic	Loglogistic	4	3822.41	7652.82	7676.10
Weibull	Paralogistic	4	3822.44	7652.88	7676.17
Paralogistic	Paralogistic	4	3822.53	7653.05	7676.34
Inverse Burr	Paralogistic	5	3818.68	7647.37	7676.47

Although the composite inverse Burr-Burr model has the lowest NLL and AIC among the other models in Table 9 (and the 256 considered), there is no strong evidence that it provides a better fit than the other models—its BIC is not at least 10 units less than the BIC of the other models in Table 9 (see Abu Bakar et al. [7]). Additionally, the composite inverse Burr-Burr model has six parameters, and the principle of parsimony does not favour it. Rather, a simpler four-parameter composite model is more favourable here.

Table 10 reports the empirical risk estimates, the estimated risk measures for the top 20 composite models for Danish fire loss data and the percentage deviation in parenthesis of each estimated risk measure with respect to the empirical risk estimates. Most of the risk estimates in Table 10 obtained from using the top 20 composite distributions closely match the empirical risk estimates (except those with the Burr as the tail distribution, in terms of the TVaR). The top 20 composite models provide fair estimates for the VaR at both 95% and 99% security levels, although the VaR at a 95% security level is underestimated by all the models (see Table 10). Using the Burr distribution as a tail distribution leads to much higher estimates for the TVaR than using the inverse Weibull distribution, the inverse paralogistic distribution, or the loglogistic distribution. Using the inverse gamma or the paralogistic distribution as the tail distribution leads to TVaR estimates that are lower than the empirical estimates at a 95% security level.

Table 10. Summary of the empirical risk estimates, the risk measures of the top 20 composite models for Danish fire loss data (reported in Grün and Miljkovic [8]) and the percentage deviation with respect to the empirical risk estimates in parenthesis.

		$VaR_{0.95}$	$VaR_{0.99}$	$TVaR_{0.95}$	$TVaR_{0.99}$
Empirical Estimates		8.406298	24.61378	22.15509	54.60396
Parametric					
Head	Tail				
Weibull	Inverse Weibull	8.02 (−4.6%)	22.77 (−7.5%)	22.64 (2.2%)	63.86 (17.0%)
Paralogistic	Inverse Weibull	8.02 (−4.6%)	22.79 (−7.4%)	22.67 (2.3%)	64.00 (17.2%)
Inverse Burr	Inverse Weibull	8.01 (−4.7%)	22.73 (−7.7%)	22.59 (2.0%)	63.67 (16.6%)
Weibull	Inverse paralogistic	8.03 (−4.5%)	22.64 (−8.0%)	22.38 (1.0%)	62.65 (14.7%)
Inverse Burr	Inverse paralogistic	8.03 (−4.5%)	22.65 (−8.0%)	22.39 (1.1%)	62.69 (14.8%)
Paralogistic	Inverse paralogistic	8.03 (−4.5%)	22.68 (−7.9%)	22.44 (1.3%)	62.89 (15.2%)
Weibull	Loglogistic	8.05 (−4.2%)	22.7 (−7.8%)	22.43 (1.2%)	62.8 (15.0%)
Inverse Burr	Loglogistic	8.04 (−4.4%)	22.64 (−8.0%)	22.35 (0.9%)	62.46 (14.4%)
Paralogistic	Loglogistic	8.05 (−4.2%)	22.71 (−7.7%)	22.46 (1.4%)	62.89 (15.2%)
Loglogistic	Inverse Weibull	8.05 (−4.2%)	22.96 (−6.7%)	22.93 (3.5%)	65.02 (19.1%)
Weibull	Burr	8.22 (−2.2%)	25.18 (2.3%)	26.98 (21.8%)	82.59 (51.3%)
Paralogistic	Burr	8.22 (−2.2%)	25.18 (2.3%)	26.98 (21.8%)	82.61 (51.3%)
Inverse Burr	Burr	8.22 (−2.2%)	25.13 (2.1%)	26.88 (21.3%)	82.15 (50.4%)
Loglogistic	Inverse paralogistic	8.05 (−4.2%)	22.79 (−7.4%)	22.6 (2.0%)	63.55 (16.4%)
Inverse Burr	Inverse gamma	8.1 (−3.6%)	22.33 (−9.3%)	21.42 (−3.3%)	57.83 (5.9%)
Paralogistic	Inverse gamma	8.11 (−3.5%)	22.44 (−8.8%)	21.57 (−2.6%)	58.48 (7.1%)
Loglogistic	Loglogistic	8.06 (−4.1%)	22.82 (−7.3%)	22.61 (2.1%)	63.52 (16.3%)
Weibull	Paralogistic	8.11 (−3.5%)	22.6 (−8.2%)	21.98 (−0.8%)	60.35 (10.5%)
Paralogistic	Paralogistic	8.11 (−3.5%)	22.62 (−8.1%)	21.99 (−0.7%)	60.41 (10.6%)
Inverse Burr	Paralogistic	8.1 (−3.6%)	22.47 (−8.7%)	21.78 (−1.7%)	59.52 (9.0%)

Fitting mixture models to the Danish data

Using the 16 loss distributions outlined in Appendix A's Table A1, it is observed from the results in Table 11 that the Burr distribution seems to be an ideal component distribution for most of the best mixture models. For the Danish fire loss data, the two-component Burr mixture performs better than the two-component gamma mixture, the two-component Pareto mixture, the two-component Weibull mixture, and the two-component exponential mixture as also concluded in Abu Bakar et al. [15] for the three fire loss datasets considered (i.e., Danish, Belgian and Norwegian). In an effort to conserve writing space, the corresponding parameter estimates of the top 20 models in Table 11 are provided in Table A5 in Appendix A.

Table 11. Summary of the information criteria of the top 20 mixture models for Danish fire loss data (based on the BIC).

First Component	Second Component	p	NLL	AIC	BIC
Burr	Burr	7	3786.47	7586.95	7627.69
Inverse Weibull	Burr	6	3790.61	7593.22	7628.15
Loglogistic	Burr	6	3791.60	7595.20	7630.13
Inverse paralogistic	Burr	6	3792.02	7596.03	7630.96
Paralogistic	Burr	6	3794.36	7600.72	7635.64
Inverse Burr	Burr	7	3790.73	7595.46	7636.21
Gamma	Burr	6	3798.004	7608.01	7642.93
Lognormal	Burr	6	3799.06	7610.12	7645.05
Generalised Pareto	Burr	7	3797.91	7609.83	7650.57
Inverse Gaussian	Burr	6	3801.97	7615.94	7650.86
Inverse gamma	Burr	6	3803.79	7619.57	7654.5
Inverse exponential	Burr	5	3810.03	7630.06	7659.17
Exponential	Burr	5	3811.32	7632.63	7661.74
Inverse Pareto	Burr	6	3810.06	7632.12	7667.04
Weibull	Burr	6	3810.82	7633.65	7668.57
Pareto	Burr	6	3811.33	7634.65	7669.58
Inverse Weibull	Inverse Burr	6	3833.76	7679.53	7714.45
Inverse paralogistic	Inverse Weibull	5	3840.24	7690.49	7719.59
Inverse Weibull	Inverse gamma	5	3840.53	7691.07	7720.17
Inverse Burr	Inverse Burr	7	3833.79	7681.57	7722.32

Other than the Burr distribution, the inverse Weibull distribution and the inverse gamma distribution also seem like optimal component distributions for the mixture models of the Danish fire loss data.

Table 12 reports the empirical risk estimates, the estimated risk measures for the top 20 mixture models for the Danish fire loss data, and the percentage deviation in parenthesis of each estimated risk measure with respect to the empirical risk estimates. The mixture distributions considered provide fair estimates for the VaR at both 95% and 99% security levels. Most of the mixture models have TVaR estimates much higher than the empirical estimates. For the Danish data, the mixture models proposed, especially the ones with the Burr component, do not adequately capture the area under the tail.

Table 12. Summary of the empirical risk estimates, the risk measures of the top 20 mixture models for Danish fire loss data and the percentage deviation with respect to the empirical risk estimates in parenthesis.

		$VaR_{0.95}$	$VaR_{0.99}$	$TVaR_{0.95}$	$TVaR_{0.99}$
Empirical Estimates		8.406298	24.61378	22.15509	54.60396
		Parametric			
First Component	Second Component				
Burr	Burr	8.26 (−1.7%)	27.0 (9.7%)	34.74 (56.8%)	119.72 (119.3%)
Inverse Weibull	Burr	8.19 (−2.6%)	25.23 (2.5%)	27.21 (22.8%)	83.83 (53.5%)
Loglogistic	Burr	9.245 (10.0%)	36.07 (46.5%)	60.00 (170.8%)	234.13 (328.8%)
Inverse paralogistic	Burr	8.2025 (−2.4%)	25.32 (2.9%)	27.37 (23.5%)	84.498 (54.7%)
Inverse Burr	Burr	8.1902 (−2.6%)	25.23 (2.5%)	27.22 (22.9%)	83.87 (53.6%)
Gamma	Burr	9.395 (11.8%)	36.48 (48.2%)	59.80 (169.9%)	232.22 (325.3%)
Inverse Gaussian	Burr	8.59 (2.2%)	31.88 (29.5%)	46.53 (110.0%)	172.91 (216.7%)
Lognormal	Burr	8.71 (3.6%)	32.67 (32.7%)	48.82 (120.4%)	183.18 (235.5%)
Generalised Pareto	Burr	8.81 (4.8%)	33.31 (35.3%)	50.72 (128.9%)	191.77 (251.2%)
Inverse gamma	Burr	9.14 (8.7%)	32.64 (32.6%)	43.71 (97.3%)	156.09 (185.9%)
Inverse exponential	Burr	9.61 (14.3%)	34.555 (40.4%)	-	-
Exponential	Burr	9.51 (13.1%)	33.85 (37.5%)	45.41 (105.0%)	162.22 (197.1%)
Inverse Pareto	Burr	9.61 (14.3%)	34.55 (40.4%)	-	-
Paralogistic	Burr	8.43 (0.3%)	26.85 (9.1%)	30.07 (35.7%)	95.73 (75.3%)
Weibull	Burr	8.44 (0.4%)	26.89 (9.2%)	30.14 (36.0%)	96.04 (75.9%)
Pareto	Burr	9.51 (13.1%)	33.85 (37.5%)	45.39 (104.9%)	162.15 (197.0%)
Inverse Weibull	Inverse Burr	8.15 (−3.0%)	21.51 (−12.6%)	20.11 (−9.2%)	51.88 (−5.0%)
Inverse paralogistic	Inverse Weibull	8.13 (−3.3%)	19.92 (−19.1%)	17.89 (−19.3%)	42.34 (−22.5%)
Inverse Weibull	Inverse gamma	8.53 (1.5%)	22.16 (−10.0%)	19.84 (−10.4%)	48.35 (−11.5%)
Inverse Burr	Inverse Burr	8.15 (−3.0%)	21.49 (−12.7%)	20.096 (−9.3%)	51.83 (−5.1%)

Finally, although the results in Tables 5, 7, 9 and 11 are sorted in terms of the BIC (in the last column), the boldfaced value in each column provides the best goodness of fit with respect to the minimum model selection criterion (NLL, AIC, BIC).

4. Conclusions

For the composite models, it seems that the composite Paralogistic-Burr, composite Weibull-Burr, and composite Inverse Burr-Burr are optimal models for both datasets as they both appear in the top 20 composite models. However, for the mixture models, it seems that the two-component Burr mixture, the two-component paralogistic and Burr mixture, and the two-component lognormal and Burr mixture are optimal models for both datasets as they also both appear in the top 20 mixture models. In general, the composite models provide better risk estimates for both of the datasets. The mixture models seem to not adequately capture the area under the tail, especially when using the Burr distribution as a component distribution for the Danish data. Finally, model selection criteria (NLL, AIC, BIC) evaluate the quality of fit of the entire model and not just the tail, so both the model selection criteria and risk estimates are important for deciding which model is optimal.

As it can be observed here, there is no single universal composite or mixture model that is better than the others. Stated differently, the best model depends on the underlying data being used to fit the model and the corresponding risk metrics. Finally, care needs to be taken when interpreting the risk metrics because a model with an excessively large risk metric as compared to the empirical estimate implies that more funds need to be kept in reserve rather than being invested elsewhere, which leads to less profits.

For future research, composite and mixture models (with more than two components) can be fitted to the taxi claims data to evaluate their suitability and other appropriate risk metrics. More importantly, it would be of interest to investigate what would be the best possible back-testing technique that is appropriate for the considered models and risk metrics. While a lot of composite and mixture distributions were considered in this paper, a reader can extend this list by considering the distributions that are discussed in [30]. Next, if data are time-dependent, readers are advised to also investigate analytical methods that involve hidden Markov models. Given that this paper used the 'single best model' approach, it would be interesting to investigate the 'grid map' and 'model averaging' methods discussed in [28,29,31,32] using the datasets and models discussed in this paper. There is also a need for academics to engage with the private sector so that they can be granted access to large datasets and be able to use more advanced and accurate machine learning techniques where data can be split into training, validation, and test sets, as well as for back-testing purposes. However, considering that the data from private companies are usually under many proprietary laws, this is a major limitation when it comes to the analysis of real-life insurance data.

Supplementary Materials: The following supporting information can be downloaded at: https://www.mdpi.com/article/10.3390/math12020335/s1. Supplementary File: R Code for Mathematics. Table S1. Taxi Claims Data.

Author Contributions: Methodology, W.A.M.; Software, W.A.M.; Validation, W.A.M. and S.C.S.; Formal analysis, W.A.M. and S.C.S.; Investigation, W.A.M. and S.C.S.; Resources, S.C.S.; Data curation, W.A.M. and S.C.S.; Writing—original draft, W.A.M.; Writing—review & editing, S.C.S. All authors have read and agreed to the published version of the manuscript.

Funding: This research received no external funding. The authors would like to acknowledge the Department of Mathematical Statistics and Actuarial Science, and the Open Access Publication Fund at University of the Free State for assistance with APC.

Data Availability Statement: The data and R codes used in this paper are provided as Supplementary Materials.

Conflicts of Interest: The authors declare no conflict of interest.

Appendix A

Table A1. Sixteen distributions that are considered as head and/or tail in the composite model, or first and second components in the mixture model.

Distribution	Parameters	PDF	CDF	$E[X^k]$
Burr	$\alpha > 0, \gamma > 0, \theta > 0$	$\frac{\alpha\gamma\left(\frac{x}{\theta}\right)^{\gamma}}{x\left[1+\left(\frac{x}{\theta}\right)^{\gamma}\right]^{\alpha+1}}$	$1 - u^{\alpha}, \quad u = \frac{1}{1+\left(\frac{x}{\theta}\right)^{\gamma}}$	$\frac{\theta^{k}\Gamma\left(1+\frac{k}{\gamma}\right)\Gamma\left(\alpha-\frac{k}{\gamma}\right)}{\Gamma(\alpha)}, -\gamma < k < \alpha\gamma$
Exponential	$\theta > 0$	$\frac{e^{-\frac{x}{\theta}}}{\theta}$	$1 - e^{-\frac{x}{\theta}}$	$\begin{cases} \theta^{k}\Gamma(k+1), & k > -1 \\ \theta^{k}k!, & \text{if } k \text{ is a positive integer} \end{cases}$
Gamma	$\alpha > 0, \theta > 0$	$\frac{\left(\frac{x}{\theta}\right)^{\alpha}e^{-\frac{x}{\theta}}}{x\Gamma(\alpha)}$	$\Gamma\left(\alpha;\frac{x}{\theta}\right)$	$\begin{cases} \frac{\theta^{k}\Gamma(\alpha+k)}{\Gamma(\alpha)}, & k > -\alpha \\ \theta^{k}(\alpha+k-1)\cdots\alpha, & \text{if } k \text{ is a positive integer} \end{cases}$

Table A1. Cont.

Distribution	Parameters	PDF	CDF	$E[X^k]$
Generalised Pareto	$\alpha > 0, \tau > 0, \theta > 0$	$\frac{\Gamma(\alpha+\tau)}{\Gamma(\alpha)\Gamma(\tau)} \frac{\theta^\alpha x^{\tau-1}}{(x+\theta)^{\alpha+\tau}}$	$\beta(\tau, \alpha; u), \ u = \frac{x}{x+\theta}$	$\begin{cases} \frac{\theta^k \Gamma(\tau+k)\Gamma(\alpha-k)}{\Gamma(\alpha)\Gamma(\tau)}, & -\tau < k < \alpha \\ \frac{\theta^k \tau(\tau+1)\cdots(\tau+k-1)}{(\alpha-1)\cdots(\alpha-k)}, & \text{if } k \text{ is a positive integer} \end{cases}$
Inverse Burr	$\tau > 0, \gamma > 0, \theta > 0$	$\frac{\tau\gamma\left(\frac{x}{\theta}\right)^{\gamma\tau}}{x\left[1+\left(\frac{x}{\theta}\right)^\gamma\right]^{\tau+1}}$	$u^\tau, \ u = \frac{\left(\frac{x}{\theta}\right)^\gamma}{1+\left(\frac{x}{\theta}\right)^\gamma}$	$\frac{\theta^k \Gamma\left(\tau+\frac{k}{\gamma}\right)\Gamma\left(1-\frac{k}{\gamma}\right)}{\Gamma(\tau)}, -\tau\gamma < k < \gamma$
Inverse Exponential	$\theta > 0$	$\frac{\theta e^{-\frac{\theta}{x}}}{x^2}$	$e^{-\frac{\theta}{x}}$	$\theta^k \Gamma(1-k), \ k < 1$
Inverse Gamma	$\alpha > 0, \theta > 0$	$\frac{\left(\frac{\theta}{x}\right)^\alpha e^{-\frac{\theta}{x}}}{x\Gamma(\alpha)}$	$1 - \Gamma(\alpha; \frac{\theta}{x})$	$\begin{cases} \frac{\theta^k \Gamma(\alpha-k)}{\Gamma(\alpha)}, & \text{if } k < \alpha \\ \frac{\theta^k}{(\alpha-1)\cdots(\alpha-k)}, & \text{if } k \text{ is a positive integer} \end{cases}$
Inverse Gaussian	$\mu > 0, \theta > 0$	$\left(\frac{\theta}{2\pi x^3}\right)^{\frac{1}{2}} e^{-\frac{\theta z^2}{2x}}$, $z = \frac{x-\mu}{\mu}$	$\Phi\left[z\left(\frac{\theta}{x}\right)^{\frac{1}{2}}\right] + e^{\frac{2\theta}{\mu}}\Phi\left[-y\left(\frac{\theta}{x}\right)^{\frac{1}{2}}\right]$, $z = \frac{x-\mu}{\mu}$	$\sum_{n=0}^{k-1}\frac{(k+n-1)!}{(k-n-1)!n!}\frac{\mu^{n+k}}{(2\theta)^n}, \ k = 1,2,\ldots$
Inverse Paralogistic	$\tau > 0, \theta > 0$	$\frac{\tau^2\left(\frac{x}{\theta}\right)^{\tau^2}}{x\left[1+\left(\frac{x}{\theta}\right)^\tau\right]^{\tau+1}}$	$u^\tau, \ u = \frac{\left(\frac{x}{\theta}\right)^\tau}{1+\left(\frac{x}{\theta}\right)^\tau}$	$\frac{\theta^k \Gamma\left(\tau+\frac{k}{\tau}\right)\Gamma\left(1-\frac{k}{\tau}\right)}{\Gamma(\tau)}, -\tau^2 < k < \tau$
Inverse Pareto	$\tau > 0, \theta > 0$	$\frac{\tau\theta x^{\tau-1}}{(x+\theta)^{\tau+1}}$	$\left(\frac{x}{x+\theta}\right)^\tau$	$\begin{cases} \frac{\theta^k \Gamma(\tau+k)\Gamma(1-k)}{\Gamma(\tau)}, & -\tau < k < 1 \\ \frac{\theta^k (-k)!}{(\tau-1)\cdots(\tau+k)}, & \text{if } k \text{ is a negative integer} \end{cases}$
Inverse Weibull	$\tau > 0, \theta > 0$	$\frac{\tau\left(\frac{\theta}{x}\right)^\tau e^{-\left(\frac{\theta}{x}\right)^\tau}}{x}$	$e^{-\left(\frac{\theta}{x}\right)^\tau}$	$\theta^k \Gamma\left(1-\frac{k}{\tau}\right), \ k < \tau$
Loglogistic	$\gamma > 0, \theta > 0$	$\frac{\gamma\left(\frac{x}{\theta}\right)^\gamma}{x\left[1+\left(\frac{x}{\theta}\right)^\gamma\right]^2}$	$\frac{\left(\frac{x}{\theta}\right)^\gamma}{1+\left(\frac{x}{\theta}\right)^\gamma}$	$\theta^k \Gamma\left(1+\frac{k}{\gamma}\right)\Gamma\left(1-\frac{k}{\gamma}\right), -\gamma < k < \gamma$
Lognormal	$\mu > 0, \sigma > 0$	$\frac{1}{x\sigma\sqrt{2\pi}}e^{-\frac{z^2}{2}} = \frac{\phi(z)}{\sigma x}$, $z = \frac{\ln x - \mu}{\sigma}$	$\Phi(z)$	$e^{k\mu + \frac{1}{2}k^2\sigma^2}$
Paralogistic	$\alpha > 0, \theta > 0$	$\frac{\alpha^2\left(\frac{x}{\theta}\right)^\alpha}{x\left[1+\left(\frac{x}{\theta}\right)^\alpha\right]^{\alpha+1}}$	$1 - u^\alpha, \ u = \frac{1}{1+\left(\frac{x}{\theta}\right)^\alpha}$	$\frac{\theta^k \Gamma\left(1+\frac{k}{\alpha}\right)\Gamma\left(\alpha-\frac{k}{\alpha}\right)}{\Gamma(\alpha)}, -\alpha < k < \alpha^2$
Pareto	$\alpha > 0, \theta > 0$	$\frac{\alpha\theta^\alpha}{(x+\theta)^{\alpha+1}}$	$1 - \left(\frac{\theta}{x+\theta}\right)^\alpha$	$\begin{cases} \frac{\theta^k \Gamma(k+1)\Gamma(\alpha-k)}{\Gamma(\alpha)}, & -1 < k < \alpha \\ \frac{\theta^k k!}{(\alpha-1)\cdots(\alpha-k)}, & \text{if } k \text{ is a positive integer} \end{cases}$
Weibull	$\tau > 0, \theta > 0$	$\frac{\tau\left(\frac{x}{\theta}\right)^\tau e^{-\left(\frac{x}{\theta}\right)^\tau}}{x}$	$1 - e^{-\left(\frac{x}{\theta}\right)^\tau}$	$\theta^k \Gamma\left(1+\frac{k}{\tau}\right), \ k > -\tau$

Table A2. Parameter estimates and standard errors in parenthesis of the top 20 composite models for the taxi claims data.

Head	Tail	ϑ_1	ϑ_2	θ	ϕ
Gamma	Weibull	$\alpha = 1.8955 \ (0.0229)$, $\frac{1}{\theta} = 0.0599 (0.0014)$	$\tau = 0.3325 \ (0.0058)$, $\theta = 7.1857 (0.72095)$	35.5261	1.3957
Paralogistic	Inverse Gaussian	$\alpha = 1.6984 (101.315)$, $\frac{1}{\theta} = 0.0231 (17.4617)$	$\mu = 101.31497 \ (3.7430)$, $\theta = 17.46168 \ (1.4671)$	37.1193	1.2982
Loglogistic	Inverse Gaussian	$\gamma = 1.7352 \ (0.0155)$, $\theta = 32.5366 \ (0.4623)$	$\mu = 99.64726 \ (4.2422)$, $\theta = 16.8705 \ (1.6193)$	40.5153	1.1594
Paralogistic	Weibull	$\alpha = 1.7090 \ (0.0154)$, $\frac{1}{\theta} = 0.02368 \ (0.0003)$	$\tau = 0.3338 \ (0.0058)$, $\theta = 7.3820 \ (0.735)$	34.7893	1.4319
Inverse paralogistic	Inverse Gaussian	$\tau = 1.3895 \ (0.0072)$, $\frac{1}{\theta} = 0.0364 \ (0.0005)$	$\mu = 91.2859 \ (6.0897)$, $\theta = 14.0619 \ (2.0543)$	51.2797	0.8695
Weibull	Weibull	$\tau = 1.6529 \ (0.0166)$, $\theta = 28.8585 \ (0.4634)$	$\tau = 0.3376205 \ (0.0058)$, $\theta = 7.9324 \ (0.76697)$	30.8894	1.6791
Gamma	Burr	$\alpha = 1.8953 (0.0236)$, $\frac{1}{\theta} = 0.05999 \ (0.0015)$	$\alpha = 9290451 (2166981)$, $\gamma = 0.3324 \ (0.0061)$, $\frac{1}{\theta} = 1.519089 \times 10^{-22}$ $(1.195389 \times 10^{-23})$	35.5336	1.3953

Table A2. Cont.

Head	Tail	ϑ_1	ϑ_2	θ	ϕ
Loglogistic	Weibull	$\gamma = 1.7465\ (0.016)$, $\theta = 31.722669$	$\tau = 0.3335(0.0058)$, $\theta = 7.3185(0.7314)$	37.162	1.314
Paralogistic	Burr	$\alpha = 1.7090\ (0.0152)$, $\theta = 0.0237(0.0003)$	$\alpha = 132231.5,\ \gamma = 0.3338$, $\frac{1}{\theta} = 6.186885 \times 10^{-17}$	34.7898	1.4319
Weibull	Burr	$\tau = 1.6528\ (0.0166)$, $\theta = 28.8594\ (0.4602)$	$\alpha = 873910.9$, $\gamma = 0.3376(0.0057)$, $\frac{1}{\theta} = 3.173873 \times 10^{-19}$	30.891	1.6789
Inverse Burr	Weibull	$\tau = 0.8631\ (0.0595)$, $\gamma = 1.9451\ (0.1031)$, $\frac{1}{\theta} = 0.0301(0.0007)$	$\tau = 0.3341\ (0.0058)$, $\theta = 7.4239(0.7363)$	35.0194	1.4186
Loglogistic	Burr	$\gamma = 1.7465\ (0.0156)$, $\theta = 31.7227\ (0.41667)$	$\alpha = 31127.62,\ \gamma = 0.3335$, $\frac{1}{\theta} = 4.603593 \times 10^{-15}$	37.1616	1.3140
Inverse Burr	Burr	$\tau = 0.8632(0.0599)$, $\gamma = 1.9450\ (0.1035)$, $\frac{1}{\theta} = 0.03012(0.00066)$	$\alpha = 18358.92$, $\gamma = 0.334237$, $\frac{1}{\theta} = 2.354395 \times 10^{-14}$	35.01608	1.4188
Inverse paralogistic	Weibull	$\tau = 1.3941(0.0077)$, $\frac{1}{\theta} = 0.0372\ (0.0006)$	$\tau = 0.3313\ (0.0062)$, $\theta = 6.9658\ (0.7622)$	44.1095	1.0634
Inverse paralogistic	Burr	$\tau = 1.3941(0.0077)$, $\frac{1}{\theta} = 0.0372\ (0.0006)$	$\alpha = 29397.29$, $\gamma = 0.3313(0.0006)$, $\frac{1}{\theta} = 4.6688 \times 10^{-15}$ $\left(1.0827 \times 10^{-17}\right)$	44.1084	1.0634
Burr	Pareto	$\alpha = 0.3761\ (0.0112)$, $\gamma = 1.8324\ (0.0187)$, $\frac{1}{\theta} = 0.0487\ (0.0010)$	$\alpha = 2.6827\ (0.164)$, $\theta = 443.996\ (61.2433)$	371.0913	0.0883
Weibull	Lognormal	$\tau = 1.68\ (0.0180)$, $\theta = 27.3341\ (0.5253)$	$\mu = 3.2995\ (0.0398)$, $\sigma = 1.6521\ (0.0180)$	27.3694	2.0122
Gamma	Lognormal	$\alpha = 1.9054\ (0.0247)$, $\frac{1}{\theta} = 0.0614\ (0.0017)$	$\mu = 3.2496\ (0.0443)$, $\sigma = 1.6686\ (0.0188)$	32.3335	1.6029
Gamma	Generalised Pareto	$\alpha = 1.9193\ (0.0243)$, $\frac{1}{\theta} = 0.063\ (0.0016)$	$\alpha = 2.00855\ (0.0590)$, $\tau = 0.00000008$ $\left(1.4945 \times 10^{-15}\right)$, $\frac{1}{\theta} = 0.00263\ (0.00015)$	33.0272	1.5616
Paralogistic	Lognormal	$\alpha = 1.7207(0.0264)$, $\frac{1}{\theta} = 0.0243(0.0005)$	$\mu = 3.2653(0.0550)$, $\sigma = 1.6637\ (0.0264)$	30.9552	1.6977

Table A3. Parameter estimates of the top 20 mixture models for the taxi claims data.

First Component	Second Component	ϑ_1	ϑ_2	ϕ
Inverse gamma	Lognormal	$\alpha = 3.8170,\ \theta = 93.6396$	$\mu = 4.074,\ \sigma = 1.4041$	4.4276
Inverse Gaussian	Lognormal	$\mu = 29.8338,\ \theta = 107.9798$	$\mu = 4.0794,\ \sigma = 1.3961$	4.7267
Generalised Pareto	Lognormal	$\alpha = 4.1712,\ \tau = 45.299$, $\theta = 2.2854$	$\mu = 4.0765,\ \sigma = 1.4031$	4.4435
Inverse paralogistic	Lognormal	$\tau = 2.4166,\ \theta = 17.1673$	$\mu = 4.0813,\ \sigma = 1.4069$	4.0799
Inverse Weibull	Lognormal	$\tau = 2.0034\theta = 23.086563$	$\mu = 4.060686,\ \sigma = 1.4098$	4.1863
Inverse Burr	Lognormal	$\tau = 4.5381,\ \gamma = 2.1940$, $\theta = 12.0926$	$\mu = 4.0736,\ \sigma = 1.4101$	4.0411
Loglogistic	Lognormal	$\gamma = 2.9332,\ \theta = 26.4047$	$\mu = 4.1085,\ \sigma = 1.3974$	3.9066

Table A3. Cont.

First Component	Second Component	ϑ_1	ϑ_2	ϕ
Burr	Lognormal	$\alpha = 0.5817, \gamma = 3.4263, \theta = 21.4809$	$\mu = 4.0956802, \sigma = 1.4086$	3.6253
Gamma	Lognormal	$\alpha = 4.2428, \theta = 6.5837$	$\mu = 4.091793, \sigma = 1.3857$	4.769
Paralogistic	Lognormal	$\alpha = 2.403, \theta = 40.2599$	$\mu = 4.126997, \sigma = 1.3822$	4.002
Lognormal	Weibull	$\mu = 4.1531, \sigma = 1.3612$	$\tau = 1.9636, \theta = 29.5290$	0.2536
Loglogistic	Generalised Pareto	$\gamma = 1.7759, \theta = 29.8996$	$\alpha = 2.2800, \tau = 1.5141, \theta = 247.543$	0.52375
Generalised Pareto	Paralogistic	$\alpha = 2.0570, \tau = 1.614, \theta = 168.3791$	$\alpha = 1.7326, \theta = 39.1302$	1.1726
Loglogistic	Paralogistic	$\gamma = 1.7641, \theta = 30.5826$	$\alpha = 1.3814, \theta = 228.3339$	0.4576
Burr	Loglogistic	$\alpha = 1.6342, \gamma = 1.2558, \theta = 256.4782$	$\gamma = 1.7836, \theta = 30.326$	1.9518
Paralogistic	Paralogistic	$\alpha = 1.7332, \theta = 40.1747$	$\alpha = 1.3464, \theta = 180.1915$	0.7779
Burr	Burr	$\alpha = 0.95765, \gamma = 1.7884, \theta = 29.6828$	$\alpha = 1.663, \gamma = 1.2498, \theta = 264.2923$	0.4903
Inverse gamma	Paralogistic	$\alpha = 1.7495, \theta = 294.6358$	$\alpha = 1.6836, \theta = 43.7961$	2.4313
Inverse gamma	Generalised Pareto	$\alpha = 1.8386, \theta = 334.996$	$\alpha = 6.2739, \tau = 1.9704, \theta = 107.74715$	2.2563
Paralogistic	Burr	$\alpha = 1.7378, \theta = 40.1805$	$\alpha = 1.3951, \tau = 1.3194, \theta = 184.542$	0.797

Table A4. Parameter estimates and standard errors in parenthesis of the top 20 composite models for the Danish fire loss data.

Head	Tail	ϑ_1	ϑ_2	θ	ϕ
Weibull	Inverse Weibull	$\tau = 16.094(1.554), \theta = 0.955(0.0149)$	$\tau = 1.555(0.0505), \frac{1}{\theta} = 1.102(0.0994)$	0.955	9.854
Paralogistic	Inverse Weibull	$\alpha = 16.088(1.581), \frac{1}{\theta} = 0.879(0.0232)$	$\tau = 1.554(0.0507), \frac{1}{\theta} = 1.105(0.101)$	0.957	9.688
Inverse Burr	Inverse Weibull	$\tau = 0.204(0.235), \gamma = 68.745(73.872), \frac{1}{\theta} = 1.046(0.0176)$	$\tau = 1.557(0.0493), \frac{1}{\theta} = 1.096(0.0919)$	0.934	12.609
Weibull	Inverse paralogistic	$\tau = 15.806(1.728), \theta = 0.960(0.0176)$	$\tau = 1.567(0.0565), \frac{1}{\theta} = 1.777(0.213)$	0.961	9.256
Inverse Burr	Inverse paralogistic	$\tau = 0.000383(0.00001), \gamma = 35290(1049), \frac{1}{\theta} = 1.078(0.00005)$	$\tau = 1.567(0.053), \frac{1}{\theta} = 1.775(0.181)$	0.928	14.086
Paralogistic	Inverse paralogistic	$\alpha = 15.745(1.968), \frac{1}{\theta} = 0.872(0.0301)$	$\tau = 1.566(0.057), \frac{1}{\theta} = 1.787(0.221)$	0.964	9.054
Weibull	Loglogistic	$\tau = 15.652(1.939), \theta = 0.962(0.0206)$	$\gamma = 1.568(0.0593), \theta = 0.680(0.0979)$	0.964	9.030
Inverse Burr	Loglogistic	$\tau = 0.000395(0.00002), \gamma = 34285(1.517), \frac{1}{\theta} = 1.078(0.00005)$	$\gamma = 1.570(0.0278), \theta = 0.688(0.00275)$	0.928	14.020

Table A4. Cont.

Head	Tail	ϑ_1	ϑ_2	θ	ϕ
Paralogistic	Loglogistic	$\alpha = 15.683(1.205)$, $\frac{1}{\theta} = 0.871(0.0174)$	$\gamma = 1.567(0.0569)$, $\theta = 0.678(0.0888)$	0.965	8.906
Loglogistic	Inverse Weibull	$\gamma = 16.267(1.264)$, $\theta = 0.975(0.0127)$	$\alpha = 1.547(0.0502)$, $\frac{1}{\theta} = 1.130(0.105)$	0.976	8.216
Weibull	Burr	$\tau = 16.267(1.264)$, $\theta = 0.949(0.00107)$	$\alpha = 0.395(0.104)$, $\gamma = 3.646(0.880)$, $\frac{1}{\theta} = 1.182(0.0693)$	0.947	11.13
Paralogistic	Burr	$\alpha = 16.278(1.257)$, $\frac{1}{\theta} = 0.887(0.887)$	$\alpha = 0.394(0.104)$, $\gamma = 3.649(0.884)$, $\frac{1}{\theta} = 1.182(0.0697)$	0.947	11.043
Inverse Burr	Burr	$\tau = 0.262(0.108)$, $\gamma = 53.766(19.976)$, $\frac{1}{\theta} = 1.046(0.0099)$	$\alpha = 0.406(0.107)$, $\gamma = 3.549(0.853)$, $\frac{1}{\theta} = 1.190(0.0713)$	0.932	13.251
Loglogistic	Inverse paralogistic	$\gamma = 16.197(1.358)$, $\theta = 0.977(0.0141)$	$\tau = 1.561(0.0554)$, $\frac{1}{\theta} = 1.819(0.216)$	0.980	7.876
Inverse Burr	Inverse gamma	$\tau = 4.4300(0.000000225)$, $\gamma = 30761(0.885)$, $\frac{1}{\theta} = 1.078(0.0000315)$	$\alpha = 1.641(0.0399)$, $\frac{1}{\theta} = 1.148(0.0263)$	0.928	13.945
Paralogistic	Inverse gamma	$\alpha = 15.635(1.285)$, $\frac{1}{\theta} = 0.869(0.0186)$	$\alpha = 1.635(0.0733)$, $\frac{1}{\theta} = 1.119(0.188)$	0.967	8.753
Loglogistic	Loglogistic	$\gamma = 16.153(1.387)$, $\theta = 0.978(0.0145)$	$\gamma = 1.562(0.0573)$, $\theta = 0.666(0.0911)$	0.981	7.761
Weibull	Paralogistic	$\tau = 15.511(1.314)$, $\theta = 0.965(0.00129)$	$\alpha = 1.267(0.0273)$, $\frac{1}{\theta} = 1.607(0.265)$	0.968	8.660
Paralogistic	Paralogistic	$\alpha = 15.557(1.3555)$, $\frac{1}{\theta} = 0.867(0.020)$	$\alpha = 1.266(0.0273)$, $\frac{1}{\theta} = 1.611(0.267)$	0.969	8.551
Inverse Burr	Paralogistic	$\tau = 0.000929(0.0000014)$, $\gamma = 14718(2.036)$, $\frac{1}{\theta} = 1.077(0.0000152)$	$\alpha = 1.270(0.0136)$, $\frac{1}{\theta} = 1.559(0.0307)$	0.928	13.775

Table A5. Parameter estimates of the top 20 mixture models for the Danish fire loss data.

First Component	Second Component	ϑ_1	ϑ_2	ϕ
Burr	Burr	$\alpha = 0.2706$, $\gamma = 6.6542$, $\theta = 1.2869$	$\alpha = 0.0257$, $\gamma = 49.3079$, $\theta = 0.8573$	2.1565
Inverse Weibull	Burr	$\tau = 10.5701$, $\theta = 0.9465$	$\alpha = 0.1577$, $\gamma = 9.0711$, $\theta = 1.1658$	4.3468
Loglogistic	Burr	$\gamma = 5.1028$, $\theta = 1.7185$	$\alpha = 0.028$, $\gamma = 41.8835$, $\theta = 0.8645$	4.6562
Inverse paralogistic	Burr	$\gamma = 11.85025$, $\theta = 0.7666$	$\alpha = 0.1541$, $\gamma = 9.2683$, $\theta = 1.1497$	4.7785
Inverse Burr	Burr	$\tau = 136.2773$, $\gamma = 10.6678$, $\theta = 0.5971$	$\alpha = 0.1574$, $\gamma = 9.0858$, $\theta = 1.1646$	4.3782
Gamma	Burr	$\alpha = 8.7777$, $\theta = 0.2029$	$\alpha = 0.0298$, $\gamma = 39.7623$, $\theta = 0.8677$	5.40768

Table A5. Cont.

First Component	Second Component	ϑ_1	ϑ_2	ϕ
Inverse Gaussian	Burr	$\mu = 2.1500,\ \theta = 8.79$	$\alpha = 0.0315,\ \gamma = 38.8700,\ \theta = 0.8699$	4.7799
Lognormal	Burr	$\mu = 0.6349,\ \sigma = 0.4417$	$\alpha = 0.0308,\ \gamma = 39.48825,\ \theta = 0.8686$	4.7594
Generalised Pareto	Burr	$\alpha = 12.2499,\ \tau = 10.915,\ \theta = 2.0900$	$\alpha = 0.0303,\ \gamma = 39.960,\ \theta = 0.8679$	4.7499
Inverse gamma	Burr	$\alpha = 3.6500,\ \theta = 6.61997$	$\alpha = 0.0404,\ \gamma = 31.330,\ \theta = 0.877$	4.2999
Inverse exponential	Burr	$\theta = 0.98115$	$\alpha = 0.04097,\ \gamma = 25.728,\ \theta = 0.8903$	134.9374
Exponential	Burr	$\frac{1}{\theta} = 0.3686$	$\alpha = 0.04999,\ \gamma = 25.2807,\ \theta = 0.8908$	119.5614
Inverse Pareto	Burr	$\tau = 66.423,\ \theta = 0.015575$	$\alpha = 0.0491,\ \gamma = 25.7105,\ \theta = 0.8903$	134.9823
Paralogistic	Burr	$\alpha = 20.0916,\ \theta = 1.1077$	$\alpha = 0.1294,\ \gamma = 10.7409,\ \theta = 1.0445$	10.19554
Weibull	Burr	$\tau = 19.9054,\ \theta = 0.95505$	$\alpha = 0.1291,\ \gamma = 10.7569,\ \theta = 1.0427$	10.3812
Pareto	Burr	$\alpha = 37.699,\ \theta = 100.9595$	$\alpha = 0.0501,\ \gamma = 25.2500,\ \theta = 0.8908$	121.669
Inverse Weibull	Inverse Burr	$\tau = 3.9523,\ \theta = 1.1564$	$\tau = 5.8476,\ \gamma = 1.714638,\ \theta = 0.866$	0.7465
Inverse paralogistic	Inverse Weibull	$\tau = 1.9012,\ \theta = 2.3835$	$\tau = 3.4688,\ \theta = 1.2099$	2.2879
Inverse gamma	Inverse Weibull	$\alpha = 3.6924,\ \theta = 1.183$	$\tau = 1.91129,\ \theta = 4.8491$	0.6334
Inverse Burr	Inverse Burr	$\tau = 5.8468,\ \gamma = 1.7148,\ \theta = 0.866$	$\tau = 4119.4997,\ \gamma = 3.9533,\ \theta = 0.1408$	1.33905

References

1. Cooray, K.; Ananda, M.M.A. Modeling actuarial data with a composite lognormal-Pareto model. *Scand. Actuar. J.* **2007**, *2005*, 321–334. [CrossRef]
2. Scollnik, D.P.M. On composite lognormal-Pareto models. *Scand. Actuar. J.* **2007**, *2007*, 20–33. [CrossRef]
3. Pigeon, M.; Denuit, M. Composite Lognormal-Pareto model with random threshold. *Scand. Actuar. J.* **2010**, *2011*, 177–192. [CrossRef]
4. Nadarajah, S.; Bakar, S.A.A. New composite models for the Danish fire insurance data. *Scand. Actuar. J.* **2012**, *2014*, 180–187. [CrossRef]
5. Ciumara, R. An actuarial model based on the composite Weibull-Pareto distribution. *Math. Rep.* **2006**, *8*, 401–414.
6. Scollnik, D.P.M.; Sun, C. Modeling with Weibull-Pareto models. *N. Am. Actuar. J.* **2012**, *16*, 260–272. [CrossRef]
7. Abu Bakar, S.A.; Hamzah, N.A.; Maghsoudi, M.; Nadarajah, S. Modeling loss data using composite models. *Insur. Math. Econ.* **2015**, *61*, 146–154. [CrossRef]
8. Grün, B.; Miljkovic, T. Extending composite loss models using a general framework of advanced computational tools. *Scand. Actuar. J.* **2019**, *2019*, 642–660. [CrossRef]
9. Calderin-Ojeda, E.; Kwok, C.F. Modeling claims data with composite Stoppa models. *Scand. Actuar. J.* **2015**, *2016*, 817–836. [CrossRef]
10. Keatinge, C.L. Modeling losses with the mixed exponential distribution. *Proc. Casualty Actuar. Soc.* **1999**, *LXXXVI*, 654–698.
11. Klugman, S.; Rioux, J. Toward a Unified Approach to Fitting Loss Models. *N. Am. Actuar. J.* **2006**, *10*, 63–83. [CrossRef]
12. Lee, S.C.K.; Lin, S.X. Modeling and Evaluating Insurance Losses Via mixtures of Erlang Distributions. *N. Am. Actuar. J.* **2010**, *14*, 107–130. [CrossRef]
13. Tijms, H. *Stochastic Models: An Algorithm Approach*; Wiley: Hoboken, NJ, USA, 1994; ISBN 0-471-95123-4.
14. Miljkovic, T.; Grün, B. Modeling loss data using mixtures of distributions. *Insur. Math. Econ.* **2016**, *70*, 387–396. [CrossRef]
15. Abu Bakar, S.A.; Nadarajah, S.; Adzhar, Z.A.A.K. Loss modeling using Burr mixtures. *Empir. Econ.* **2017**, *54*, 1503–1516. [CrossRef]

16. Abu Bakar, S.A.; Nadarajah, S. Risk measure estimation under two component mixture models with trimmed data. *J. Appl. Stat.* **2019**, *46*, 835–852. [CrossRef]
17. Asgharzadeh, A.; Nadarajah, S.; Sharafi, F. Generalized inverse Lindley distribution with application to Danish fire insurance data. *Commun. Stat. Theory Methods* **2017**, *46*, 5000–5021. [CrossRef]
18. Punzo, A.; Bagnato, L.; Maruotti, A. Compound unimodal distributions for insurance losses. *Insur. Math. Econ.* **2018**, *81*, 95–107. [CrossRef]
19. Bhati, D.; Ravi, S. On generalized log-Moyal distribution: A new heavy tailed size distribution. *Insur. Math. Econ.* **2018**, *79*, 247–259. [CrossRef]
20. Li, Z.; Beirlant, J.; Meng, S. Generalizing the log-Moyal distribution and regression models for heavy-tailed loss data. *ASTIN Bull. J. IAA* **2021**, *51*, 57–99. [CrossRef]
21. Zhao, J.; Ahmad, Z.; Mahmoudi, E.; Hafez, E.H.; Mohie El-Din, M.M. A new class of heavy-tailed distributions: Modeling and simulating actuarial measures. *Complexity* **2021**, *2021*, 5580228. [CrossRef]
22. Ahmad, Z.; Mahmoudi, E.; Dey, S. A new family of heavy tailed distributions with an application to the heavy tailed insurance loss data. *Commun. Stat.-Simul. Comput.* **2022**, *51*, 4372–4395. [CrossRef]
23. Maphalla, R.; Mokhoabane, M.; Ndou, M.; Shongwe, S.C. Quantifying risk using loss distributions. In *Applied Probability Theory—New Perspectives, Recent Advances and Trends*; Jaoude, A.A., Ed.; IntechOpen: London, UK, 2023; pp. 139–161. [CrossRef]
24. Akaike, H. A new look at the statistical model identification. *IEEE Trans. Autom. Control* **1974**, *19*, 716–723. [CrossRef]
25. Schwarz, G. Estimating the Dimension of a Model. *Ann. Stat.* **1978**, *6*, 461–464. [CrossRef]
26. R Core Team. R: A Language and Environment for Statistical Computing. 2022. Available online: https://www.R-project.org/ (accessed on 1 April 2023).
27. Davison, A. SMPracticals: Practicals for Use with Davison (2003) Statistical Models. 2019. R Package Version 1.4-3. Available online: https://CRAN.R-project.org/package=SMPracticals (accessed on 1 May 2023).
28. Blostein, M.; Miljkovic, T. On modeling left-truncated loss data using mixtures of distributions. *Insur. Math. Econ.* **2019**, *85*, 35–46. [CrossRef]
29. Miljkovic, T.; Grün, B. Using Model Averaging to Determine Suitable Risk Measure Estimates. *N. Am. Actuar. J.* **2021**, *25*, 562–579. [CrossRef]
30. Rigby, R.A.; Stasinopoulos, M.D.; Heller, G.Z.; De Bastiani, F. *Distributions for Modeling Location, Scale, and Shape: Using GAMLSS in R*; CRC Press: Boca Raton, FL, USA, 2019. [CrossRef]
31. Zou, J.; Wang, W.; Zhang, X.; Zou, G. Optimal model averaging for divergent-dimensional Poisson regressions. *Econom. Rev.* **2022**, *41*, 775–805. [CrossRef]
32. Zou, J.; Yuan, C.; Zhang, X.; Zou, G.; Wan, A.T.K. Model averaging for support vector classifier by cross-validation. *Stat. Comput.* **2023**, *33*, 117. [CrossRef]

Disclaimer/Publisher's Note: The statements, opinions and data contained in all publications are solely those of the individual author(s) and contributor(s) and not of MDPI and/or the editor(s). MDPI and/or the editor(s) disclaim responsibility for any injury to people or property resulting from any ideas, methods, instructions or products referred to in the content.

Article

Addressing Concerns about Single Path Analysis in Business Cycle Turning Points: The Case of Learning Vector Quantization

David Enck [1], Mario Beruvides [1], Víctor G. Tercero-Gómez [2] and Alvaro E. Cordero-Franco [3,*]

1. Department of Industrial Manufacturing & Systems Engineering, Texas Tech University, Lubbock, TX 79409, USA; david.enck@ttu.edu (D.E.); mario.beruvides@ttu.edu (M.B.)
2. School of Engineering and Sciences, Tecnologico de Monterrey, Monterrey 64849, Mexico; victor.tercero@tec.mx
3. Facultad de Ciencias Físico Matemáticas, Universidad Autónoma de Nuevo León, San Nicolás de los Garza 66451, Mexico
* Correspondence: alvaro.corderofr@uanl.edu.mx

Citation: Enck, D.; Beruvides, M.; Tercero-Gómez, V.G.; Cordero-Franco, A.E. Addressing Concerns about Single Path Analysis in Business Cycle Turning Points: The Case of Learning Vector Quantization. *Mathematics* 2024, 12, 678. https://doi.org/10.3390/math12050678

Academic Editors: Arne Johannssen and Nataliya Chukhrova

Received: 29 December 2023
Revised: 5 February 2024
Accepted: 6 February 2024
Published: 26 February 2024

Copyright: © 2024 by the authors. Licensee MDPI, Basel, Switzerland. This article is an open access article distributed under the terms and conditions of the Creative Commons Attribution (CC BY) license (https://creativecommons.org/licenses/by/4.0/).

Abstract: Data-driven approaches in machine learning are increasingly applied in economic analysis, particularly for identifying business cycle (BC) turning points. However, temporal dependence in BCs is often overlooked, leading to what we term single path analysis (SPA). SPA neglects the diverse potential routes of a temporal data structure. It hinders the evaluation and calibration of algorithms. This study emphasizes the significance of acknowledging temporal dependence in BC analysis and illustrates the problem of SPA using learning vector quantization (LVQ) as a case study. LVQ was previously adapted to use economic indicators to determine the current BC phase, exhibiting flexibility in adapting to evolving patterns. To address temporal complexities, we employed a multivariate Monte Carlo simulation incorporating a specified number of change-points, autocorrelation, and cross-correlations, from a second-order vector autoregressive model. Calibrated with varying levels of observed economic leading indicators, our approach offers a deeper understanding of LVQ's uncertainties. Our results demonstrate the inadequacy of SPA, unveiling diverse risks and worst-case protection strategies. By encouraging researchers to consider temporal dependence, this study contributes to enhancing the robustness of data-driven approaches in financial and economic analyses, offering a comprehensive framework for addressing SPA concerns.

Keywords: data-driven methods; temporal dependence; Monte Carlo simulation; robustness; multivariate analysis; economic indicators

MSC: 62M10; 62H30; 9110

1. Introduction

Data driven approaches in machine learning offer powerful alternatives for statistical classification, optimal decision-making, and pattern recognition. Its increasing use in economic analysis does not come as a surprise, where the application to identify business cycle (BC) turning points (TPs) has gained great popularity. A BC TP refers to the transition between different phases of an economic system. The recurring pattern of economic expansion and contraction consists of four main phases: expansion, peak, contraction (or recession), and trough. A TP marks the shift from one phase to another (NBER [1]). Identification of TPs allows economists to determine the impact of mitigating policies that were implemented, and timely identification helps countries, businesses, and citizens make financial decisions (Hamilton [2]). As BCs are related to time series data, they often exhibit autocorrelation, meaning that the observations are correlated with their past values. In such cases, the temporal structure of the data becomes crucial, and the performance of machine learning algorithms needs to address this reality. Related literature on BC analysis often fails to acknowledge this situation, and algorithms are evaluated with a given historical series that represents a single path of all possible potential routes a data structure can take. Approaches that fail to acknowledge this situation fall into the trap of what we call single path analysis (SPA). We encourage researchers doing BC analysis to be mindful of the temporal dependence when evaluating the performance of an algorithm to shed light on what to expect of an approach when applied to a new sequence of observations.

To illustrate the issue, we focus our attention on the learning vector quantization (LVQ) algorithm. LVQ is particularly useful for classification tasks where the decision boundaries between classes are well-defined. It has been applied in various domains, including pattern recognition, image classification, and the monitoring of BCs (Giusto and Piger [3]). In the context of BC monitoring, LVQ is used to classify economic indicators and determine the current phase of the BC based on historical indicators. The algorithm's ability to adapt to changing patterns and different levels of information makes it suitable for tasks where the characteristics of different cycles may evolve over time.

To address the temporal structure of the data, we use a multivariate Monte Carlo simulation with a structure with a specific number of change-points, autocorrelation and cross-correlations. A second-order vector autoregressive (VAR) model, calibrated with different levels of economic performance observed in the economic leading indicators (LIs) was used in the analysis. Results show the uncertainties of the LVQ algorithm, revealing different levels of risks and strategies for worst-case protections that were not possible to assess with single path analysis.

A common practice when evaluating BC TP algorithms is to use a single historical time series to select an algorithm's design parameters and to compare the ability of different algorithms to detect BC TPs. Given the defined design parameters, such as smoothing time periods and probability thresholds, an algorithm decides whether a time point is in a recession or growth period. Authors frequently evaluate the time to signal (TTS) for an algorithm's decisions against the economic status, and TTS is defined by the National Bureau of Economic Research (NBER) Business Cycle Dating Committee (BCDC) in Cambridge, MA, USA.

The NBER declares the TPs for peaks and troughs that are used by economists to define BCs. The NBER looks for multiple economic indicators to show a significantly broad decrease or increase in economic activity that lasts more than a few months to declare a BC TP. Declarations of peaks and troughs are important, as they impact behaviors of financial institutions, governments, and citizens (Hamilton [2]). There is no stated model for the analysis; however, the NBER offers guidance on variables and patterns they look for when making decisions (NBER [1,4]; The White House [5]). Due to the lack of a reproducible numerical approach, authors have developed algorithms to try and estimate the declared BC TPs while providing transparency and a shorter TTS. An important part of the method selection and algorithm optimization involves making decisions on the design parameters that adjust the decisions made by the algorithms.

Hamilton [6] wrote a foundational paper on how to effectively predict BC TPs using Markov Switching Models. Hamilton based his decisions regarding the smoothing periods and probability threshold on observed performance of the algorithm for the dates of study. Goodwin [7] explored Hamilton's model on eight developed market economies and made smoothing period and probability threshold decisions based only on the data being evaluated in his study. Some authors explored extensions or alternatives to Hamilton's Markov switching approach, including Birchenhall et al. [8], Chauvet and Piger [9], Giusto and Piger [3], Soybilgen [10], and others. These authors may use different criteria for selecting design parameters; however, they all based their selections on the single set of data they used in their study.

Depending on the algorithm, the design parameters can include the number of time periods to smooth over, number of time periods in a forecast, the probability decision threshold, and/or the number of months the threshold should be exceeded. Design parameters that provide the best statistics for TTS and which incur the fewest false positives are selected, for that algorithm for that historical dataset. As an example, Chauvet and

Piger [9] evaluated a dynamic factor Markov-switching (DFMS) model of Chauvet [11] using a two-step approach to convert recession probability estimates into a claim on whether the economy is in recession or expansion. The two-step approach used 0.8 and 0.5 as probability thresholds with a 2-period lag. The parameters worked well, but the peak in 2001 was declared after the NBER announcement and declared to be 2 months prior to the NBER date of March 2001. Two other peaks found matched the NBER announcements, and one peak was only one month ahead. The DFMS model also identified the October 1982 trough 38 days ahead of the NBER announcement when the other three trough announcements studied had a range from 189 to 420 days ahead. Recessions and growth periods were clearly behaving differently relative to the specific changes in the economic indicators being monitored. This in-sample evaluation represents what we call the SPA problem. Algorithms calibrated using SPA usually show great performance in terms of false positives and TTS; however, evaluations cannot extend beyond the observed data for both the depth of the recession (growth) period and the distributional pattern of the observations.

This paper re-assesses the LVQ implementation by Giusto and Piger [3] with a Monte Carlo approach, allowing researchers to simulate specific changes in a data generating process (DGP) and evaluate the probability distribution for the TTS. The approach can be adapted to approach any algorithm used to evaluate BC TPs and avoid the SPA problem. Our proposed simulation allows researchers to develop a better understanding of performance of a set of definable design parameters for a particular TP.

Section 2 provides a comprehensive review of the LVQ algorithm, while Section 3 proposes a multi-path analysis to counteract SPA through Monte Carlo simulation. In Section 4, we apply the Monte Carlo approach to evaluate BC TPs, re-assessing the work of Giusto and Piger [3]. Section 5 presents a discussion on the obtained results, and finally, Section 6 provides conclusions and suggestions for future research.

2. Learning Vector Quantization in Business Cycle Analysis

LVQ is a family of machine-learning algorithms used for real-time statistical classification based on the use of codebook vectors, which are used to represent a particular class or category in a dataset. These vectors are used to classify new data points based on their similarity to the codebook vectors. In the LVQ algorithm, the codebook vectors are adjusted based on their proximity to new data points, allowing them to adapt to the characteristics of the data and improve the classification process (see Kohonen [12]). LVQ classifier presents some advantages when comparing multilayers perceptron and support vector machine algorithms, in terms of computational cost and interpretability (Nova and Estévez [13]). LVQ classifiers have been studied by several authors, either in theoretical studies looking for improvement terms of convergence or robustness, as it is seen in Sato and Yamada [14] in their General LVQ, and Seo and Obermayer [15] with their Soft LVQ, among others; or using the LVQ, or its adaptations, in different applications in image and signal processing (see Nanopoulos et al. [16]), medicine (see Pesu et al. [17]), or finance (see Giusto and Piger [3]).

Let $x_i \in \mathbb{R}^n$, $i = 1, \ldots, N$ a vector associated to a class y_i, where $y_i \in \{c_1, c_2, \ldots, c_k\}$. Algorithm LVQ trains over the pairs $\{x_i, y_i\}$ to create a model M that predicts the class of a new vector x. The LVQ algorithm is a supervised learning algorithm based on codebook vectors, which are vectors used to represent the different classes. Let $M = \left\{m_j^1\right\}_{j=1}^{N}$ be a set of initial codevectors, where $m_j^1 \in \mathbb{R}^n$ and $N \in \mathbb{Z}$, $N \in [k, n]$. It should be noted that one class can have more than one codevector. The algorithm uses two parameters, α to iteratively adjust the codevectors, and G as the number of iterations. The codevectors are updated in a linear fashion according to the distances of these codevectors to the closets datapoint. Let m_j^g be the codevector at step g, $g = 1, \ldots, G$. The description of the algorithm is given next:

1. Initialize $g \leftarrow 1$, $i \leftarrow 1$.
2. Identify the codevector m_c^g closest to x_i, using a preferred metric distance:

$$c = \underset{j \in \{1,2,\ldots,N\}}{\operatorname{argmin}} \left\{ \|x_i - m_j^g\| \right\}$$

3. Update the codevector set:
   - If x_i and its closest codevectors are in the same class:

   $$m_c^{g+1} = m_c^g + \alpha^g \left(x_i - m_c^g \right)$$

   - If x_i and its closest codevectors are not in the same class:

   $$m_c^{g+1} = m_c^g - \alpha^g \left(x_i - m_c^g \right)$$

4. If $i + 1 \leq N$, let $i \leftarrow i + 1$ and return to step 2. If not, go to step 5.
5. If $g \leq G$, let $i \leftarrow 1$, $g \leftarrow g + 1$, and return to step 2. Otherwise stop.

The final placement of the codebook vectors in an LVQ algorithm is not invariant to the initialization, so practitioners need to run multi-start procedures to initialize the codebook vectors. The class prediction of a new data point x consists of the class of the closest codevector:

$$c = \underset{j \in \{1,2,\ldots,N\}}{\operatorname{argmin}} \left\{ \|x - m_j\| \right\}$$

Giusto and Piger [3] proposed using the LVQ algorithm to identify United States BC TPs using the available NBER chronology and the four leading indicators (LIs) favored by the NBER Business Cycle Dating Committee (BCDC): non-farm payroll employment (E), industrial production (I), real personal income excluding transfer receipts (P), and real manufacturing and trade sales (M), over a period from November 1976 to July 2013. Recession and expansion phases are persistent according to their estimators from this period of time. The LVQ algorithm added this persistency by classifying a vector with their current values and a lag 1 of the series of data. In their research, at time $T + 1$, an analyst is looking to determine if a TP has occurred in the recent past, $T - j$ months, where the NBER classification is known. The time-window j varies according to some assumptions: (1) the date of a new peak or trough is known; (2) 12 months is the maximum time between a new peak and its announcement by the NBER; and (3) 6 months is the minimum amount of time of each BC phase. The determination of a new TP in a prediction period occurs if three consecutive months are classified in a different phase than the prior one, which is consistent with the NBER definition of a TP and the assumption of persistent periods. This process is executed 100 times looking to avoid a false identification of new TPs; if the LVQ identified a new TP in at least 80% of the runs, then this is assumed as true. This threshold presents better results in terms of accuracy and avoids false positives and negatives when comparing values from 50% to 90%.

Giusto and Piger [3] presented excellent results in the accuracy of their proposed methodology. With respect to the TPs identified by the NBER, the average lag to a new peak and a new trough improves in 90 and 212 days, respectively. Also, the LVQ algorithm never produces any false positives and negatives with respect to the ones predicted by the NBER. Because the NBER corrects their prediction results historically, the LVQ algorithm was also compared with other statistical methods. For the period of the great recession, 2007–2009, Hamilton [2] reported that the methods in use during that time did not identify a BC peak, whereas the LVQ algorithm detected one peak on 6 June 2008. He also compared the LVQ algorithm with the results presented by Chauvet and Piger [9] of the dynamic factor Markov-switching (SDMF) model of Chauvet [11] for identifying TPs over the period of November 1976 to June 2006. LVQ improves the results of SDMF in the identification of both peaks and troughs. Economic models had such trouble with the great recession that Congress held hearings and invited five renowned economists to testify regarding the identification of BC TPs (U.S. Congress, [18]).

3. Using Monte Carlo Simulation for Multi-Path Analysis

3.1. A Change-Point Model to Operationalize Business Cycles

The algorithms that model BC TPs typically assume a bivariate response for economic performance and model the probability of a recession. The probability is translated into a prediction on whether the current time segment will be classified as a BC TP, which is a transition from recession to growth or growth to recession periods. Rather than relying just on the single data path observed during the period of the study, we propose that measures of economic performance, such as TTS, should account for stochastic variation of the LIs under study.

The distribution of the LIs can be used to reflect historical or future behaviors of interest, thereby allowing investigators the ability to explore an algorithm's performance across a broad range of simulated scenarios from historically observable recessions to potential future scenarios. This section describes a method for using Monte Carlo Simulation to select design parameters and compare algorithms by comparing distributions for the TTS across a range of changes in economic LIs. The description of the methodology assumes that multiple LIs were used to make the BC TP decision. The algorithm is provided at the end of this section.

Our proposed Monte Carlo simulation method uses a population model or DGP for a multi-variate change point process for a series of LIs. The performance of a range of algorithm design parameters is evaluated for changes in the DGP of varying severities. The change-point model:

$$Y_{ij} \sim \begin{cases} MV(\underline{\mu}_1, \Sigma_1), & i = 1, \ldots, \tau_1 - 1; j = 1, \ldots, m \\ MV(\underline{\mu}_2, \Sigma_2), & i = \tau_1, \ldots, \tau_2 - 1; j = 1, \ldots, m \\ \vdots \\ MV(\underline{\mu}_k, \Sigma_k), & i = \tau_{k-1}, \ldots, \tau_k - 1; j = 1, \ldots, m \end{cases} \qquad (1)$$

represents the DGP from time point 1 to τ_k, where Y_{ij} follows a multivariate distribution for a set of LIs of interest with a mean vector of $\underline{\mu}_i$ and a variance–covariance matrix Σ_i. Parameter τ_n represents the nth change point, and m the number of LIs that make up Y_{ij}.

The model provides the flexibility to test changes in the population mean and variance–covariance structure. Not all changes specified in Equation (1) might be large enough to be declared as a BC TP. Equation (1) represents a model of economic behavior, while the objective of the NBER BCDC is to identify the changes in economic performance significant enough to declare a BC TP. An algorithm's sensitivity should be adjustable to detect this range of changes based on what is important for the user. Detection of smaller changes may be of interest to an analyst while deeper changes would be the focus for researchers predicting BC TPs. This population model allows for the assessment of a spectrum of TPs, both historically observed and unobserved. Researchers can select design parameters for their algorithm such that the distribution for the TTS shows a high probability of detecting the shallowest recession the BCDC has ever declared and a low probability of declaring a BC TP when a change point in the DGP is shallower than the shallowest recession—a false alarm. Our reassessment of the LVQ algorithm in the following section provides an example of this type of analysis.

3.2. Evaluating Practical Significance

The idea of monitoring a time series and declaring a change for fluctuations that are considered significant is not new. The concept of identifying significant changes in a system over time has been addressed in the process monitoring literature. Ewan and Kemp [19], Freund [20], Woodall [21,22], Box et al. [23,24] and Yashchin [25] partitioned a response over time into: good, undefined, and bad zones.

A basic application of a practical significance analysis is the 2-zone approach by Woodall and Faltin [26], which combines the good and indifferent zones into a "Dead-Zone" where an algorithm should have a low likelihood of detecting a change and an out of control or "Bad-Zone" with a high likelihood of detecting a change. For the BC TP problem, the TP would be the transition between zones. The 2-zone model gives context to language and measures in BC literature. When authors describe results that include false alarms, they are referring to a situation where the LIs are in the Dead-Zone, which means no changes occurred or the changes in the mean vector of the LIs are not large enough for the NBER to declare a BC TP recession (growth), and yet the algorithm declares a change. The Bad-Zone is represented by the space of LI distributional characteristics where the NBER flagged BC TPs. The simulation approach that we propose allows researchers to account for how the TTS distribution changes across the Dead-Zone into the Bad-Zone. It also allows for the tunning of an algorithm's design parameters based on these probabilities.

3.3. A Fitting a Model for Monte Carlo Simulation

The Monte Carlo simulation approach has the potential of helping researchers decide on design parameters by adding a simulated multivariate time series to the end of a historical series while evaluating the TTS distribution. This approach creates the opportunity to explore the impact of (i) typical variation within a growth or recession period based on the DGP; as well as (ii) changes in the DGP or distributions of the LIs on the TTS distribution for different design parameters. Researchers can study any construction of linear and non-linear distributional changes between growth and recession periods. Since the point of the paper is to propose the assessment of multi-path behaviors, we will keep the assumptions simple and use a Multivariate Normal distribution for the LIs, with a defined shift in the mean and a constant variance-covariance matrix.

The first step of the algorithm is to fit the recessionary distribution that is to be detected. Figure 1 shows the recession behavior for two important LIs between 1965 and 2013. Mfg. and Trade Sales is in a state of dynamic equilibrium across the 7 recessions displayed, while the Non-Farm Payroll Employment can change in a more unpredictable pattern over a recession. Because recessions are typically declared within 7 months of a TP (Giusto and Piger [9], p. 180), researchers should consider estimating the mean using the first 7 months of the LIs they plan to use for BC TP identification.

The simulated means for the LIs for a recession, in the case study, are estimated by the average of the months available from the start of the identified BC TP to the month the TP was flagged.

The Dead- and Bad-Zones should be determined through evaluation of the means of the 4 LIs. The means span a range from the deepest recession period to the shallowest. The deepest recession would be the recession where the average growth for each of the LIs was furthest from 0. "Shallow" recessions have average growth for each of the LIs which are closest to 0. The radar chart in Figure 2 shows the shallowest and deepest recessions analyzed by Giusto and Piger [3]. The shallowest recession, which started in March 2001 took the longest to detect at 7 months. The deepest recession started in January 1980 and was detected in 3 months by the LVQ algorithm.

Given the defined recessions of interest, the second step requires the selection of the final growth period. The simulated recession added to this final period of growth. Multiple final growth periods could be used to evaluate the TTS for given changes between the population means for the final growth and recession periods per the specified design parameters.

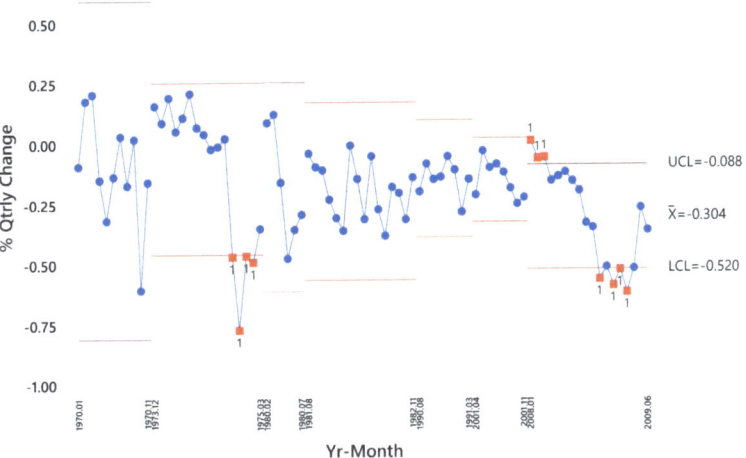

Figure 1. The graphs depict a concatenation of the last 7 recessions for real mfg trade sales and non-farm payroll employment. The upper and lower lines for each recession represent a rudimentary assessment of typical variation and the middle line represents the mean. The Red points represent values in a recession which are flagged as unusual. Understanding the type of variation a leading indicator (LI) experiences along with an understanding of when the variation occurs helps the researcher determine how best to model the behavior of the LI during the period that BC TP signals are typically observed, within 7 months. Performances of the two LIs depicted in Figure 1 reveal different behaviors between and within LIs for the 7 recessions. Real manufacturing trade and sales values less variation, however there are fewer values flagged as unusual than the non-farm payroll employment values. Given the flagged values tend to occur after 7 quarters we are less concerned about special patterns that occur within 2 years of the onset of a recession. This simplifies modeling recession behaviors for these two LIs.

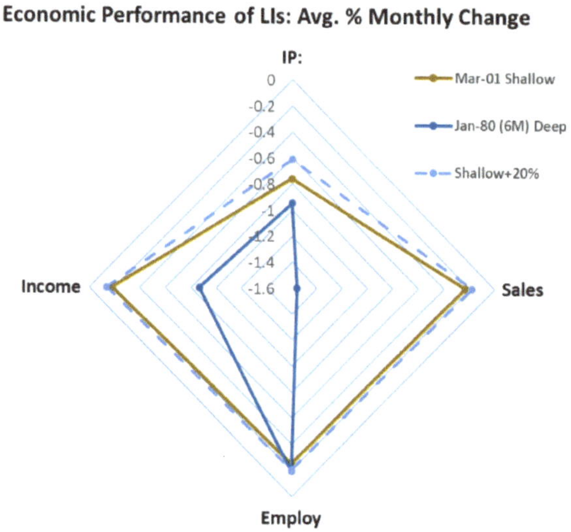

Figure 2. Depiction of 3 levels of economic performance in 2 zones: the Dead-Zone is represented by the Shallow Recession + 20%, and the Bad-Zone is represented by the Shallow Recession period starting in March 2001 and deep recession starting in January 1980.

The third step of the Monte Carlo simulation involves simulating the recession(s) of interest using the population model specified in Equation (1). Variables for the 4 LIs of a simulated recession stage are defined using the VAR model:

$$\underline{y}_{it} = \underline{b}_0 + B_t * \underline{y}_{it-1} + \underline{\varepsilon}_{it}, \tag{2}$$

where $\underline{y}_{it}$ is a $p \times 1$ vector, $\varepsilon_{it} \sim N(\underline{0}, \Sigma)$, and Σ and B_t are $(p+1) \times (p+1)$ matrices with $i = 1, \ldots, 4$. We use lag 1 for simplicity in Equation (2). We use information from the case study to fit the model described, which results in:

$$\hat{y}_{I,t} = -0.085 + 0.163 I_{t-1} + 0.113 E_{t-1} + 0.338 P_{t-1} + 0.063 M_{t-1} - 0.257 I_{t-2} \\ + 0.050 E_{t-2} + 0.095 P_{t-2} + 0.033 M_{t-2}, \tag{3}$$

$$\hat{y}_{E,t} = -0.001 + 0.126 I_{t-1} + 0.067 E_{t-1} - 0.101 P_{t-1} - 0.050 M_{t-1} - 0.156 I_{t-2} \\ - 0.451 E_{t-2} - 0.212 P_{t-2} - 0.056 M_{t-2}, \tag{4}$$

$$\hat{y}_{P,t} = 0.006 - 0.714 I_{t-1} - 0.921 E_{t-1} - 0.548 P_{t-1} + 0.160 Sales_{t-1} - 0.123 I_{t-2} \\ + 1.957 E_{t-2} - 0.239 P_{t-2} + 0.175 Sales_{t-2}, \tag{5}$$

$$\hat{y}_{M,t} = -0.094 + 1.557 I_{t-1} - 8.748 E_{t-1} - 0.113 P_{t-1} - 1.235 M_{t-1} - 0.586 I_{t-2} \\ - 0.792 E_{t-2} - 0.254 P_{t-2} - 0.963 M_{t-2}, \tag{6}$$

for a second-order VAR model. The model was fitted with R (R Core Team, 2014) [27] using the function VAR from the R package vars (Pfaff, [28]). The equations show how the autocorrelation and cross-correlations are accounted for within a VAR model. The approach shown is used for the Monte Carlo simulation when reassessing the LVQ algorithm. The evaluation was set with lag 2 with a different subset of the data. The models in Equations (3)–(6) were generated using a combined dataset from the July 1990 and March 2001 recessions, starting the month after the peak and ending the month of the next trough. When reviewing the LIs for the recessions, we found that they showed qualitatively similar performance. The LIs within each recession were standardized to a zero mean.

The final step of the algorithm would be to append the simulated values onto the end of the defined dataset and repeat the process with a new simulated set of values each iteration.

3.4. Multi-Path Assessment Algorithm with Monte Carlo Simulation

The algorithm summarized here is used in Section 4 to show how a statistical assessment can be made regarding the design parameters for the LVQ approach utilized by Giusto and Piger [3]. It follows:

1. Fit the recessionary distribution of the LIs $N\left(\underline{\mu}_r, \Sigma_r\right)$ that you wish to evaluate:
   a. Where Σ_r is estimated from the last growth period or a combination of recession periods to provide a large enough sample size, and
   b. $\underline{\mu}_r$ represents the mean vector for the LIs of the recession under study.
2. Select the final growth period for the historical data under study.
3. Simulate N observations using an assumed distribution of the recession:
   a. Data is standardized to 0 growth,
   b. Desired mean vector, $\underline{\mu}_r$, is added to standardized data,
   c. Estimate a VAR model of order w,
   d. Select stopping point for N based on max. expected time to detect a signal. N should be ≤24, as TP should be detected within 24 months.
   e. Generate simulated values for the N time intervals using the VAR model.
4. Add the simulated values to the end of the historical dataset (ending in a growth period) and iteratively apply the test for whether a recession has occurred. Define test criteria:
   a. Decision level: Probability that the current time-period is part of a recession period,
   b. Decision threshold: the limit that a decision level must exceed a set number of times to declare a recession (growth) period,
   c. Define the number of time periods that the decision level must be greater than (less than) the decision threshold to declare a recession (growth) period.

4. Re-Assessment of Giusto and Piger's LVQ Implementation

We used the proposed Monte Carlo simulation method from Section 3.4 to provide statistical assessments of the design parameters for the LVQ algorithm as presented by Giusto and Piger [3]. Given the goal of the case study is to re-approach the LVQ in a BC TP context with a stochastic analysis, we provide a direct comparison with their published selection of design parameters. The TTS for shallow to deep recessions is evaluated using the 4 LIs described in Section 3. The LVQ algorithm of Giusto and Piger [3] is used to determine an estimated BC TP and TTS for the known change.

The four LIs favored by the BCDC are: (1) nonfarm payroll employment; (2) industrial production; (3) real manufacturing and trade sales; and (4) real personal income excluding transfer payments. We use monthly data as available to users from November 1976 to September 2000. The measure available for each LI at each month is the % monthly growth. For consistency, we used the vintage data available for download from Giusto and Piger [3]. Vintage datasets are also available from the Federal Reserve Bank of St. Louis (St. Louis Federal Reserve [29]).

Our Monte Carlo simulation measured the time, in months, required for the LVQ algorithm to signal a peak BC TP after a change was made to the underlying DGP. Key design elements of the simulation algorithm are provided below:

- Baseline period was March 1967 to September 2000.
- The simulated data were added after September 2000, which is 6 months prior to the March 2001 recession to avoid any end effects close to the start of a recession.
- A VAR process for the % monthly change of the 4 LIs was modeled using the March 1991 to March 2001 growth period. The Var–Cov matrix, Σ_g, was used when simulating the recession data, as the recession periods were short.

- The simulated shifts for data starting after the growth period spanned a range of means representing the shallowest and deepest recessions observed across the dataset studied by Giusto and Piger [3]. The shallowest recession was the recession with the longest TTS for the LVQ algorithm (started in March 2001), while the deepest recession (started in January 1980) had the shortest.
- The means for simulated recessions were estimated using the months from the recession start to the month the recession was flagged.
- The mean was calculated with the vintage data available the month the BC TP was flagged.
- The probability or decision thresholds studied ranged from 0.6 to 0.9, and the months required to be at or below the decision threshold were 2 and 3.
- The simulation was stopped after 24 months if the simulated change was not detected. To be useful, a peak BC TP should be flagged within 12 months. Using 24 months as the limit allowed more separation of the TTS distributions in the simulation.

The top four graphs of Figure 3 show actual data for the first 5 months of the LIs for the December 2007 recession added onto the growth period ending in September 2000. The observed data are overlaid on a single set of simulated observations to provide a relative reference for the location and spread of the two datasets. The specific simulated result shown provides a reasonable representation of the actual mean and standard deviation of the observed data. Simulating the 4 LIs 5000 times generates different paths with varying TTS results. The second set of four graphs in Figure 3 displays three data sequences appended onto the final growth period. Each of the three sequences represents a simulated result that could have occurred from the same DGP. The sequences were selected from 5000 multivariate Monte Carlo runs. All four LIs were considered together for a TTS determination. One of the sequences generated a TTS of 3 months. The 3-month sequence had the lowest values of the three sequences for Personal Income and Mfg/Trade Sales, while its values were similar to the other two sequences for Industrial Production and Payroll Employment. The differences were enough to cause a TTS of 3 months. The sequence of four LIs with a TTS of 7 months had averages for Personal Income and Mfg/Trade Sales that were higher than the averages for the sequence with a 3-month TTS and less than the averages of the sequence with a 24-month TTS. The averages were similar for Industrial Production and Payroll Employment. For the same DGP, these types of fluctuations can arise due to differences in small nuisance factors that affect how the LIs occur. A researcher can evaluate different amounts of variation caused by the nuisance factors using historical or expected future economic behavior.

For the same VAR model, the TTS can vary from 3 to 24 months depending on fluctuations in the cross and auto-correlated LI results. The important part of the simulation is the distribution of the TTS. This potential for variation becomes important when selecting design parameters such as the probability threshold and number of months at or above the threshold. Rather than using an average of a single path to select the best options for decision criteria, we select decision criteria that perform well probabilistically (i.e., over a collection of future possible paths).

Monte Carlo simulation provides a flexible approach to simulating behaviors of future recessions. Whether the mean is degrading over time or making a step change, the Monte Carlo simulation approach can provide a general prediction or understanding of the BC TP algorithm's performance with regards to TTS for a selection of design parameters.

Using the LVQ algorithm and SPA, Giusto and Piger [3] (p.181) found that for the 3-month criterion, the probability threshold of 0.9 and 0.8 required a similar amount of time to identify a peak, approximately 4.5 months, with no false alarms. The 0.8 threshold required 1 month more than the 0.7 threshold; however, the 0.7 threshold produced a false alarm. The 0.6 and 0.5 thresholds required the same average TTS, but the 0.6 threshold produced one false alarm and the 0.5 threshold produced three false alarms. The general implication of the SPA is that the 3-month, 0.8 threshold should be used to identify peak BC TPs. A limitation of this analysis is that there is no opportunity to study how adjusting the month and probability threshold criteria can be optimized for the severity of recession that the researcher may be interested in.

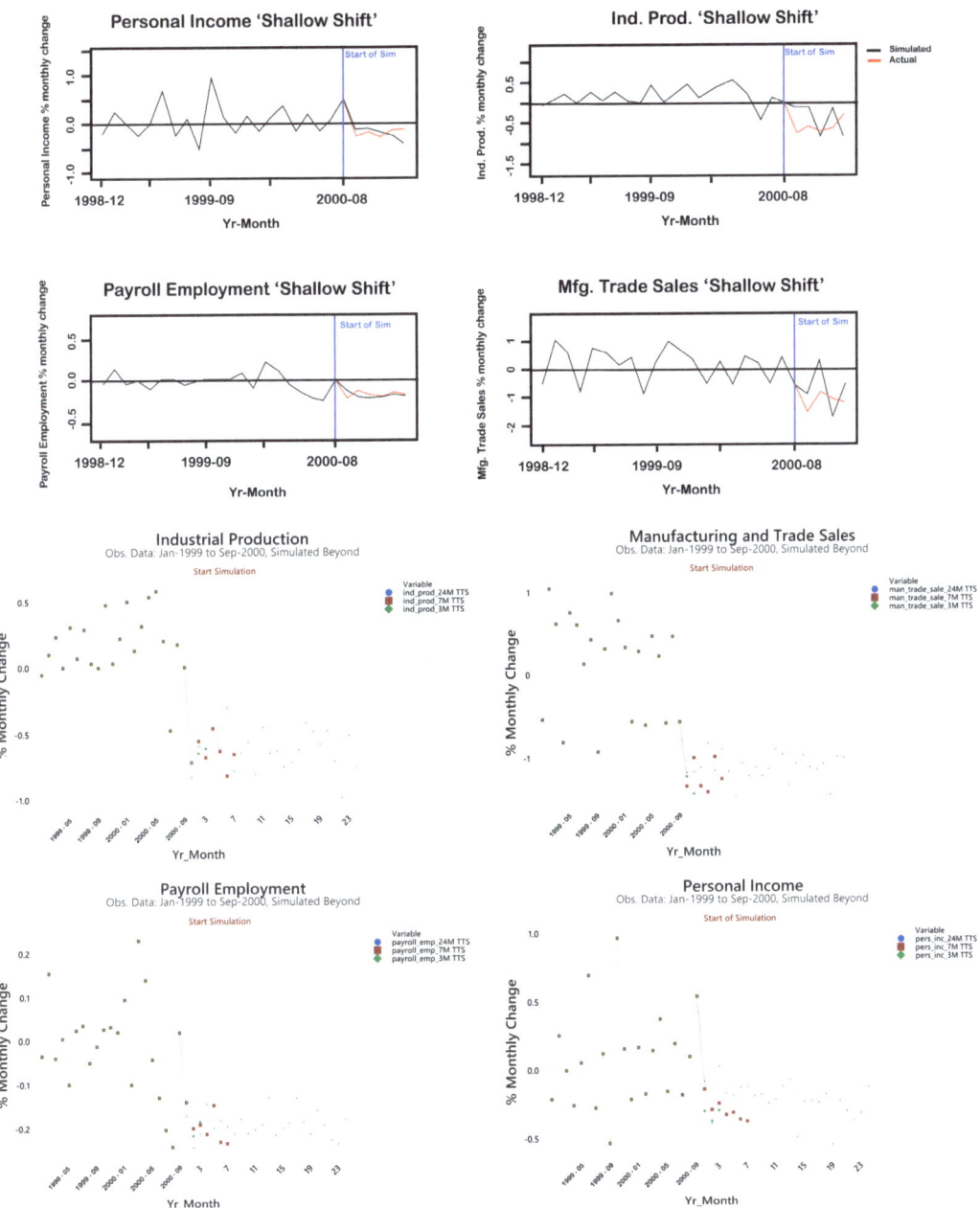

Figure 3. Top four graphs represent a simulation of December 2007 recession added on to the end of the March 1991 Growth period. Bottom four graphs show three simulated sequences, from the same Data Generating Process, that resulted in a 3, 7, and 24 month TTS. The TTS was determined through joint assessment of the four LIs.

We analyzed the data with the R program provided by Giusto and Piger [3] and did not find the false signal at the 0.7 threshold, which implies that the 3-month criterion with 0.7 threshold might be a good option as it identifies peak BC TPs 1 month earlier than the 0.8 threshold, and no false alarms occurred. However, this leaves the question of how much of a risk is there for a false alarm and how consistently can the 0.7 threshold identify peak BC TPs across a range of recession severities? And does the 0.8 threshold provide similar performance with less risk of a false alarm? These questions cannot be addressed by the SPA.

Monte Carlo simulation provides a more complete understanding of the decision criteria than the SPA. Following the Monte Carlo simulation approach in Section 3.4 and the key design elements from Section 3.3, we generated a distribution of 5000 TTS values for the decision thresholds of 0.6, 0.7, 0.8, and 0.9 for 2 months and 3 months with three severities of recession (Shallow + 20%, Shallow, Deep) using the 4 LIs previously mentioned. In Table 1, we show the TTS values for the Shallow + 20%, Shallow and Deep recessions with the 5th, 25th, 50th, 75th, and 95th percentiles for the simulated TTS values. The mean % monthly growth values for the Deep Recession (I, M, E, P) are: $-0.95, -1.56, -0.192, -0.869$, for the Shallow Recession: $-0.76, -0.23, -0.25, -0.18$, and for the Shallow Recession + 20% are: $-0.61, -0.18, -0.20, -0.14$.

Table 1. Percentiles (5th, 25th, 50th, 75th, and 95th) for varying decision level thresholds. The Dead-Zone represents a change that is real, albeit not significant enough for the BCDC to declare a recession. The means for the LIs in the Dead-Zone represent a 20% improvement over the Shallowest Recession and are still far from the means for the growth period (starting April 1991): 0.344, 0.359, 0.168, 0.326.

Monthly Criterion	Probability Threshold	Dead Zone: Shallow Rec + 20%	Shallow Recession (March 2001)	Deep Recession (January 1980)
$m = 2$	0.6	3, 5, 10, 24, 24	3, 3, 3, 3, 5	2, 2, 2, 2, 2
	0.7	3, 7, 19, 24, 24	3, 3, 3, 4, 6	2, 2, 2, 2, 2
	0.8	3, 9, 24, 24, 24	3, 3, 3, 4, 7	2, 2, 2, 2, 3
	0.9	4, 24, 24, 24, 24	3, 3, 3, 5, 9	2, 2, 2, 3, 3
$m = 3$	0.6	5, 24, 24, 24, 24	3, 4, 4, 5, 9	3, 3, 3, 3, 3
	0.7	7, 24, 24, 24, 24	4, 4, 4, 7, 11	3, 3, 3, 3, 3
	0.8	10, 24, 24, 24, 24	4, 4, 5, 8, 24	3, 3, 3, 3, 3
	0.9	12, 24, 24, 24, 24	4, 4, 7, 10, 24	3, 3, 3, 3, 3

There are several analyses that can be conducted. We compared the simulated results for the 3-month 0.8 and 0.7 thresholds with the SPA to show how the Monte Carlo simulation approach improves the understanding of performance. The analysis for the Monte Carlo simulation provides a more complete understanding of the impact on TTS for the design parameters due to the opportunity to understand the changes in distribution for different severities of recessions.

The analysis can be broken down into 3 steps that can be iterated: (1) select the recession depth; (2) pick the monthly criteria; and (3) compare the distribution for the TTS across different probability thresholds.

1. Select the shallow recession to start, as the deep recession is easily identified by all design parameter options.
2. Because the results of a SPA from Giusto and Piger [3] selected $m = 3$, start with 3 months for the number of months the probability threshold must be met.
3. Distribution of TTS for a peak:
   a. The 50th percentile of the TTS with probability threshold of 0.6 and 0.7 are both 4 months while 0.8 is 5 and 0.9 is 7 months. This aligns well with Giusto and Piger [3], who identified five peaks with a median of 4.1 and longest of

7.1 months. The months for the Monte Carlo simulation results are larger, as the SPA combines all recession severities and therefore biases the average TTS. The Shallow recession from Table 1 is the least severe recession within the Giusto and Piger [3] dataset, and the probability threshold of 0.7 provides a faster median TTS than 0.8.

    b. The 75th percentile of the TTS with probability threshold of 0.6 is 5 months, 0.7 is 7 months, while 0.8 is 8 and 0.9 is 10 months. This makes sense, as increasing the proportion of detected peaks out of 5000 will result in an increase in the number of months to detect; however, there is still a 1-month shift between the 0.7 and 0.8 probability thresholds.

    c. The 95th percentile of the TTS with the probability threshold of 0.6 is 9 months, 0.7 is 11 months, while 0.8 and 0.9 are greater than 24 months. This result was not possible to explain with the single path analysis, which addresses an average for all recessions. For a shallow recession, the selection of the probability threshold of 0.8 or 0.9 creates a 5% probability of taking longer than 24 months to detect a peak. The probability threshold of 0.7 provides much better protection against poor performance.

4. Iterate step 3 with a different severity of recession. The Monte Carlo simulation allows for the evaluation of changes in the LIs that have either not occurred or have not been declared as a recession by the BCDC. In our simulation, we created a Dead-Zone change in the LIs by adding 20% to the averages of the Shallowest recession declared by the BCDC for the dataset studied by Giusto and Piger [3]. This change is used to evaluate the probability of a false alarm.

    a. The 50th and 25th percentiles of the TTS, with probability threshold of 0.6, 0.7, 0.8, and 0.9, are all greater than 24 months (this is also true for the 75th and 95th percentiles).

    b. The 5th percentile of the TTS with probability threshold of 0.6 is 5 months and 0.7 is 7 months, while 0.8 is 10 and 0.9 is 12 months. There is a 5% probability of declaring this change a peak in 7 months if the probability threshold of 0.7 is used and in 10 months for 0.8. This translates to a low probability of a false alarm for 0.7 or 0.8.

5. The entire analysis can be repeated with the monthly threshold, m, set to 2. A review of Table 1 shows that the 0.7 and 0.8 probability thresholds improve the detection of a Shallow recession. The median TTS decreased from 4 and 5 months for $m = 3$ to 3 months for $m = 2$ while the 95th percentiles went from 11 and 24 to 6 and 7, respectively. However, when using $m = 2$, there is now a 25% probability of declaring the Dead-Zone change a BC TP in 7 months with a probability threshold of 0.7 and 9 months with a probability threshold of 0.8. This is an increase of 20% over the $m = 3$ probability of a false alarm. If a researcher is willing to be more aggressive, they have an option to use $m = 2$, and they would know the change in risk levels for different severities of recession.

5. Discussion

The ability to evaluate the probability of detecting differing severities of recessions provides a more detailed understanding of the impact of design parameter choices than SPA. The re-assessment of the LVQ showed that the options for design parameters were (i) $m = 3$, $p = 0.7$; (ii) $m = 3$, $p = 0.8$; and (iii) $m = 2$, $p = 0.7$ or 0.8 along with the risks and benefits of the different options. We also saw that $m = 3$, $p = 0.07$ provides a significant reduction in the worst-case probability of missing a BC TP. The worst-case protection is not possible to assess with the SPA.

Future research can evaluate the impact of changes in distributional forms, behaviors of the mean over time, and structure of the variance–covariance matrices. This paper focused on the identification of BC peaks which represent the onset of a recession. Similar analyses can be conducted for BC troughs to simulate capabilities for decision criteria in identifying the onset of growth periods.

6. Conclusions

We have shown that it is important to consider the statistical variation in the LIs when selecting design parameters for an algorithm that flags BC TPs to predict the state of the economy. Historical SPA provides an initial assessment; however, the results provide a narrowly focused analysis which is restricted to the observed data and is not capable of assessing all potential risks. The proposed Monte Carlo simulation methodology can be applied to any algorithm that predicts BC TPs. Monte Carlo simulation allows the researcher to avoid biases from specific single path sequences and provides a mechanism to make more informed decisions regarding the next BC TP flag. Future work needs to explore topics such as the impact of changing distributional forms, dynamic transition periods from peak to trough, and the structure of the variance–covariance matrices on the TTS distribution.

Author Contributions: D.E.: methodology, software, writing, investigation, validation, data curation, original draft preparation; M.B.: investigation, supervision, project administration, editing; V.G.T.-G.: conceptualization, supervision, review, editing; and A.E.C.-F.: review, formal analysis, editing. All authors have read and agreed to the published version of the manuscript.

Funding: This research received no external funding.

Data Availability Statement: No new data were created.

Conflicts of Interest: The authors declare no conflicts of interest.

References

1. NBER. The NBER's Recession Dating Procedure. 7 January 2008. Available online: http://www.nber.org/cycles/jan08bcdc_memo.html (accessed on 5 April 2022).
2. Hamilton, J.D. Calling recessions in real time. *Int. J. Forecast.* **2011**, *27*, 1006–1026. [CrossRef]
3. Giusto, A.; Piger, J. Identifying Business Cycle Turning Points in Real Time with Vector Quantization. *Int. J. Forecast.* **2017**, *33*, 174–184. [CrossRef]
4. NBER. The National Bureau of Economic Research. September 2010. Available online: http://www.nber.org/cycles/recessions.html (accessed on 30 October 2022).
5. The White House. How Do Economists Determine Whether the Economy Is in a Recession. 21 July 2022. Available online: https://www.whitehouse.gov/cea/written-materials/2022/07/21/how-do-economists-determine-whether-the-economy-is-in-a-recession/#:~:text=The%20National%20Bureau%20of%20Economic,committee%20typically%20tracks%20include%20real (accessed on 7 February 2024).
6. Hamilton, J.D. A New Approach to the Economic Analysis of Nonstationary Time Series and the Business Cycle. *Econometrica* **1989**, *57*, 357–384. [CrossRef]
7. Goodwin, T.H. Business-Cycle Analysis with a Markov-Switching Model. *J. Bus. Econ. Stat.* **1993**, *11*, 331–339.
8. Birchenhall, C.R.; Jessen, H.; Osborn, D.R.; Simpson, P. Predicting U.S. Business-Cycle Regimes. *J. Bus. Econ. Stat.* **1999**, *17*, 313–323.
9. Chauvet, M.; Piger, A. Comparison of the Real-Time Performance of Business Cycle Dating Methods. *J. Bus. Econ. Stat.* **2008**, *26*, 42–49. [CrossRef]
10. Soybilgen, B. Identifying US business cycle regimes using dynamic factors and neural network models. *J. Forecast.* **2020**, *39*, 827–840. [CrossRef]
11. Chauvet, M. An econometric characterization of business cycle dynamics with factor structure. *Int. Econ. Rev.* **1998**, *39*, 969–996. [CrossRef]
12. Kohonen, T. Learning vector quantization. In *Self-Organizing Maps*; Springer: Berlin/Heidelberg, Germany, 1997; pp. 175–189.
13. Nova, D.; Estévez, P. A review of learning vector quantization classifiers. *Neural Comput. Appl.* **2014**, *25*, 511–524. [CrossRef]
14. Sato, A.; Yamada, K. Generalized learning vector quantization. *Adv. Neural Inf. Process. Syst.* **1995**, *8*, 423–429.
15. Seo, S.; Obermayer, K. Soft learning vector quantization. *Neural Comput.* **2003**, *15*, 1589–1604. [CrossRef]
16. Nanopoulos, A.; Alcock, R.; Manolopoulos, Y. Feature-based classification of time-series data. *Int. J. Comput. Res.* **2001**, *10*, 49–61.
17. Pesu, L.; Helistö, P.; Ademovič, E.; Pesquet, J.; Saarinen, A.; Sovijärvi, A. Classification of respiratory sounds based on wavelet packet decomposition and learning vector quantization. *Technol. Health Care* **1998**, *6*, 65–74. [CrossRef] [PubMed]
18. US Congress; US House Committee on Science and Technology. *Building a Science of Economics for the Real World*; Serial Number 111-106; Government Printing Office: Washington, DC, USA, 2010. Available online: https://app.thestorygraph.com/books/8925845f-3a94-426f-aa55-c61747844938 (accessed on 10 February 2024).

19. Ewan, W.D.; Kemp, K.W. Sampling inspection of continuous processes with no autocorrelation between successive results. *Biometrika* **1960**, *47*, 363–380. [CrossRef]
20. Freund, J.E. Some Results on Recurrent Events. *Am. Math. Mon.* **1957**, *64*, 718–720. [CrossRef]
21. Woodall, W.H. Weaknesses of The Economic Design of Control Charts. *Technometrics* **1986**, *28*, 408–409. [CrossRef]
22. Woodall, W.H. The Statistical Design of Quality Control Charts. *J. R. Stat. Society. Ser. D* **1985**, *34*, 155–160. [CrossRef]
23. Box, G.; Bisgaard, S.; Graves, S.; Kulahci, M.; Marko, K.; James, J.; Van Gilder, J.; Ting, T.; Zatorski, H.; Wu, C. Performance Evaluation of dynamic monitoring systems: The waterfall chart. *Qual. Eng.* **2003**, *16*, 183–191. [CrossRef]
24. Box, G.; Graves, S.; Bisgaard, S.; Van Gilder, J.; Marko, K.; James, J.; Seifer, M.; Poublon, M.; Fodale, F. Detecting Malfunctions in Dynamic Systems. *SAE Trans.* **2000**, *109*, 469–479.
25. Yashchin, E. Statistical monitoring of multi-stage processes. In *Frontiers in Statistical Quality Control*; Springer: Berlin/Heidelberg, Germany, 2018; pp. 185–209.
26. Woodall, W.H.; Faltin, F.W. Rethinking control chart design and evaluation. *Qual. Eng.* **2019**, *31*, 596–605. [CrossRef]
27. R Core Team. *R: A Language and Environment for Statistical Computing*; R Foundation for Statistical Computing: Vienna, Austria, 2014.
28. R Core Team. vars: VAR Modelling. R Package Version 1.5-6. 2021. Available online: https://CRAN.R-project.org/package=vars (accessed on 10 February 2024).
29. St. Louis Federal Reserve. Archival Federal Reserve Economic Data (ALFRED). 12 April 2020. Available online: https://alfred.stlouisfed.org/ (accessed on 7 February 2024).

Disclaimer/Publisher's Note: The statements, opinions and data contained in all publications are solely those of the individual author(s) and contributor(s) and not of MDPI and/or the editor(s). MDPI and/or the editor(s) disclaim responsibility for any injury to people or property resulting from any ideas, methods, instructions or products referred to in the content.

Article

Process Capability Index for Simple Linear Profile in the Presence of Within- and Between-Profile Autocorrelation

Aylin Pakzad [1,*], Ali Yeganeh [2], Rassoul Noorossana [3] and Sandile Charles Shongwe [2]

1. Department of Industrial Engineering, Kosar University of Bojnord, Bojnord 9453155168, Iran
2. Department of Mathematical Statistics and Actuarial Science, Faculty of Natural and Agricultural Sciences, University of the Free State, Bloemfontein 9301, South Africa; yeganeh.ali369@gmail.com (A.Y.); shongwesc@ufs.ac.za (S.C.S.)
3. Information Systems and Operations Management Department, College of Business, University of Central Oklahoma, Edmond, OK 73034, USA; rnoorossana@uco.edu
* Correspondence: a.pakzad@kub.ac.ir; Tel.: +98-58-31220000

Abstract: In many situations, the quality of a process or product can be characterized by a functional relationship or profile. It is well-known that the independence assumptions of the error terms within or between profiles are not always valid and could be violated due to within or between profile autocorrelation. Since most of the process capability indices (PCIs) have been developed for simple linear profiles (SLPs) without considering autocorrelation, this paper provides some novel methods to analyze the capability of SLP under each of the two different autocorrelation effects separately, as well as the case where both autocorrelation effects are present. We assume that the first-order autoregressive AR(1) model explains the within- and between-profile autocorrelation in error terms. To evaluate the process capability, a new functional index called $C_p''' (Profile)$ is introduced for SLP with independent errors, and then it is modified to include the three possible cases of within, between, and simultaneous autocorrelation. The simulation results demonstrate that the proposed schemes outperform existing schemes regarding bias and mean square error (MSE) criteria. Moreover, bootstrap confidence intervals for the proposed index are obtained. Finally, an illustrative example in the chemical industry is used to demonstrate the applicability of the proposed method.

Keywords: autocorrelated simple linear profiles; functional approach; process capability analysis; statistical analysis; within- and between-profile autocorrelation

MSC: 62P30

Citation: Pakzad, A.; Yeganeh, A.; Noorossana, R.; Shongwe, S.C. Process Capability Index for Simple Linear Profile in the Presence of Within- and Between-Profile Autocorrelation. *Mathematics* 2024, 12, 2549. https://doi.org/10.3390/math12162549

Academic Editor: Jiancang Zhuang

Received: 17 July 2024
Revised: 14 August 2024
Accepted: 15 August 2024
Published: 18 August 2024

Copyright: © 2024 by the authors. Licensee MDPI, Basel, Switzerland. This article is an open access article distributed under the terms and conditions of the Creative Commons Attribution (CC BY) license (https://creativecommons.org/licenses/by/4.0/).

1. Introduction

A control chart, as a featured tool of statistical process control (SPC), can be used to monitor a single or a vector of quality characteristics with the aim of improving manufacturing and service processes. In some applications, a functional relationship between a response and one or more explanatory variables could represent the quality of such processes effectively. This functional relationship, which is referred to as a profile, can be linear or nonlinear in nature. In the simplest but most fundamental case, a simple linear profile (SLP) represents a response variable as a function of a single explanatory variable [1]. In the related literature, several methods have been developed for monitoring simple linear profiles in Phase I and Phase II of control charts. As a pioneering work, Kang and Albin [2] developed a method to monitor SLP in Phase II using Hoteling T^2 and exponentially weighted moving average (EWMA) control charts. They showed that EWMA was a proper choice for detecting small and moderate shifts, while Hoteling T^2 performed well in detecting large shifts. Due to the remarkable performance of the EWMA scheme, some other researchers, including Kim et al. [3] and Li and Wang [4] developed several other monitoring schemes based on the EWMA control chart. Moreover, other ideas, including the implementation of the generalized likelihood ratio method [5], deep learning structures [6], and robust estimators [7] have been proposed to improve the efficiency of control charts in SLP monitoring. For more details on profile monitoring, interested readers are referred to the literature review done by Maleki et al. [8].

In most of the methods developed in the literature for profile monitoring, it is assumed that profiles are independent over time. This assumption can be easily violated in some processes where quality characteristics of interest from different samples collected over time are autocorrelated. It is well known that when methods such as those discussed above are applied to processes where autocorrelation is an inherent part of the process, one should expect misleading results due to errors in parameter estimation. Generally, we consider two types of autocorrelations in profile monitoring, namely between profile autocorrelation (BPAC) and within profile autocorrelation (WPAC). Between profile autocorrelation (BPAC) refers to the case where observations on each profile are related to each other via a specific autocorrelation model over time. For example, Noorossana et al. [9] assumed a first order autoregressive model AR(1) for profile observations and extended T^2 and EWMA control charts to monitor SLP. In this approach, a proper transformation of the response variables was applied to reach an independent profile model. To improve the detection ability of the EWMA scheme, Wang and Huang [10] suggested a more efficient statistic to remove the autocorrelation from the residuals. Some other ideas, such as U statistics [11] and multivariate EWMA [12], were also proposed for the elimination of BPAC. On the other hand, the within profile autocorrelation (WPAC) was considered in the second group. It was followed by a model in which the observations of each profile depend on each other under an autocorrelation effect. As a pioneering work, Soleimani et al. [13] first reduced the dimension of profile samples, and then four control charts were proposed to monitor the responses and residuals. They compared T^2 and EWMA control charts under the WPAC, and they concluded that EWMA based methods could detect small shifts faster. Jensen et al. [14] and Zhang et al. [15] proposed robust control charts to counteract the autocorrelation effect. Moreover, other methods such as principal component analysis and support vector machines were also developed to monitor profiles with WPAC [16,17]. Recently, Nadi et al. [18] considered SLPs with both between and within autocorrelation effects. We use BAC abbreviation to refer to the case when BPAC and WPAC occur in a profile simultaneously. Considering the studies by Noorossana et al. [9] and Soleimani et al. [13], Nadi et al. [18] introduced and compared four control charts for monitoring SLP processes with BAC effects.

When a process is under statistical control, one needs to evaluate process performance via process capability analysis (PCA). Process capacity indices (PCIs) provide a quantitative measure using specification limits (SLs) to identify capable processes. Generally, larger PCI values indicate a more capable process. PCA has been conducted for both monitoring quality characteristics [19–22] and different profile types [23–25]. The traditional PCA is conducted under the assumption that the in-control process observations are independent. However, in practical applications, this assumption can be violated. For univariate processes, several investigations were conducted in order to deal with autocorrelated data [26–28].

Despite the efforts in profile monitoring with the autocorrelation effect, only a few studies have focused on PCIs for autocorrelated profiles. On the other hand, the majority of studies on analyzing the capability of different types of profiles have been performed under the assumption that the errors of profiles are independent. For example, Hosseinifard and Abbasi [29] estimated PCI for SLP using the proportion of nonconformance. Furthermore, five approaches to estimating PCIs for non-normal linear profiles were examined and contrasted by Hosseinifard and Abbasi [30]. Ebadi and Shahriari [31] proposed two methods to evaluate process capability for simple linear profiles. The first method is based on the average percentage of non-conforming parts at each level of the explanatory variable, and the second one is based on a multivariate process capability vector consisting of three components. Process yield, the percentage of products that are within SLs, has been a standard metric for measuring the capability and performance of manufacturing processes. Wang [32] and Wang [33] proposed a process yield index for SLPs and also developed two new indices for SLPs with one-sided specification limits. Process yield analysis for between and within profile autocorrelations in linear profiles was addressed by Wang and Tamirat [34] and Wang and Tamirat [35], respectively. Mehri et al. [36] proposed two robust PCIs for multiple linear profiles using the M-estimator and the Fast-τ-estimator. Abbasi Ganji and Sadeghpour Gildeh [37] extended a competence index denoted by C'''_{ppM} for SLPs. It was shown that the proposed index outperformed earlier indices discussed by Ebadi and Shahriari [31] and Wang [32] in terms of precision and accuracy. Additionally, for SLPs with symmetric and asymmetric tolerances, the loss-based functional capability index $C_{pm}(Profile)$ and incapability index $C''_{pp}(Profile)$ were also derived by Pakzad et al. [38] and Pakzad and Basiri [39], respectively.

According to the literature review, research on the evaluation of process capability for profiles in the presence of autocorrelation effects is limited, and only a few studies have extended the well-known adequate process-yield index S_{pkA} for each type of autocorrelation effect individually [34,35]. However, in practice, one can experience between- and within-profile autocorrelation simultaneously. Generally, there are three different correlation structures for the in-control SLPs, each of which should employ a different PCI. As it was shown in Soleimani et al. [13] and Nadi et al. [18], the autocorrelation in error terms leads to an underestimated variance of the error terms. Therefore, to avoid the overestimation of the PCIs for SLP and ultimately incorrect decisions about process capability, methods that eliminate the autocorrelation effects and a definition of a new PCI that considers both autocorrelation effects are necessary.

This paper aims to develop a method to address PCI in autocorrelated profiles using distinct approaches to estimate the parameters of SLP and to define PCI for SLP with a general error model. The contributions of this study are summarized as follows:

- A functional capability index, denoted by $C'''_p(Profile)$ in the absence of the autocorrelation effect, is introduced.
- In the presence of the BPAC effect, using the proposed method by Noorossana et al. [9], the index $C'''_p(Profile)$ is modified to eliminate the autocorrelation effect.
- In the presence of the WPAC effect, the transformation method of Soleimani et al. [13] is used to eliminate the WPAC effect, and the index $C'''_p(Profile)$ is modified accordingly to perform capability analysis for SLP.
- The performance of the modified and conventional PCIs [34,35] is compared under different simulation scenarios based on bias and mean squared error (MSE) criteria.
- As the final and major contribution of this study, a novel method for analyzing the capability of SLP in the presence of both within- and between-profile autocorrelation is developed using the transformation approach proposed by Nadi et al. [18]. Then, the proposed PCI $C'''_p(Profile)$ is modified to perform capability analysis for SLP.

This paper is structured as follows: Section 2 presents the preliminaries and theoretical foundations of the SLP model as well as a brief overview of the existing PCI for SLPs in the presence of autocorrelation. Section 3 introduces new PCIs for autocorrelated linear profiles. The performance of the proposed method is compared with other competitors [34,35] in terms of bias and MSE criteria in Section 4. In Section 5, two bootstrap confidence intervals for the proposed method are discussed, and a simulation study is conducted to assess their performance. In Section 6, an illustrative example from the chemical industry is used to show how the proposed method can be applied. Our concluding remarks are presented in the final section.

2. Preliminaries

In the following subsections, some preliminaries are discussed, which will be used throughout the manuscript.

2.1. SLPs with a General Error Model

A linear profile consists of m samples of size n in the form $(X_j, Y_{ij}); i = 1, 2, \ldots, n$ and $j = 1, 2, \ldots, m$ in which there is a single explanatory variable X and a response vector Y. The SLP with general error model introduced by Nadi et al. [18] that relates the explanatory variable to the response variable for an in-control process is given as

$$\begin{aligned} Y_{ij} &= A_0 + A_1 X_i + \varepsilon_{ij}, \\ \varepsilon_{ij} &= \rho \varepsilon_{(i-1)j} + a_{ij}, \\ a_{ij} &= \varphi \varepsilon_{i(j-1)} + u_{ij}, \\ i &= 1, 2, \ldots, n, \\ j &= 1, 2, \ldots, m. \\ |\rho| &< 1, \\ |\varphi| &< 1. \end{aligned} \quad (1)$$

where the intercept A_0 and slope A_1 are profile parameters, and X_i is the explanatory variable that is assumed to have fixed values for each sample. In addition, ε_{ij} and a_{ij} are the correlated error terms, and u_{ij} are the independently and identically normally distributed errors with a mean zero and variance σ^2. In Equation (1), the AR(1) model correlation coefficients φ and ρ represent the BPAC and WPAC, respectively. The SLP with the general error model in Equation (1) can represent the traditional SLP model with independent error terms when $\rho = \varphi = 0$, as well as the BPAC, WPAC, and BAC models associated with $\rho = 0, \varphi = 0$, and $\rho \neq 0, \varphi \neq 0$, respectively. Considering the traditional SLP defined in Equation (1) with $\rho = \varphi = 0$, the stable values of the parameters A_0 and A_1 are unknown and should be estimated in Phase I by

$$\begin{aligned} \hat{A}_0 &= a_0 = \frac{\sum_{j=1}^{m} a_{0j}}{m}, \\ \hat{A}_1 &= a_1 = \frac{\sum_{j=1}^{m} a_{1j}}{m}. \end{aligned} \quad (2)$$

where the least squares estimator (LSE) of profile parameters for the jth sample can be calculated as

$$a_{0j} = \overline{Y}_j - a_{1j}\overline{X}, \quad a_{1j} = \frac{S_{XY(j)}}{S_{XX}}. \quad (3)$$

where $\overline{Y}_j = \frac{\sum_{i=1}^{n} Y_{ij}}{n}, \overline{X} = \frac{\sum_{i=1}^{n} X_i}{n}, S_{XY(j)} = \sum_{i=1}^{n}(X_i - \overline{X})Y_{ij}, S_{XX} = \sum_{i=1}^{n}(X_i - \overline{X})^2$ [2]. Thus, $\hat{Y}_{ij} = a_{0j} + a_{1j}X_i, i = 1, 2, \ldots, n$, and $\hat{Y}_{ij}$ denotes the predicted value of the jth response variable for a given level of the explanatory variable. The process variance (σ^2) is estimated using the mean square error (MSE) computed by $MSE = \frac{\sum_{j=1}^{m} MSE_j}{m}$, where $MSE_j = \frac{\sum_{i=1}^{n} e_{ij}^2}{(n-2)}$ is the unbiased estimator of σ^2 for sample j and e_{ij} denotes the residuals, which are defined by $e_{ij} = Y_{ij} - \hat{Y}_{ij}$ [3].

2.2. The Capability Index C'''_{ppM} for Traditional SLP

Let LSL_i, USL_i, T_i and μ_i be the lower SL (LSL), upper SL (USL), target value, and mean of the response variable at the ith level of the explanatory variable, respectively. Considering n levels, Abbasi Ganji and Sadeghpour Gildeh [37] proposed a new index $C'''_{ppM} = \frac{\sum_{i=1}^{n} C'''_{ppi}}{n}$ for SLP. The general idea had been extracted from the univariate index $C'''_p(u,v)$ for asymmetric tolerance proposed by Abbasi Ganji and Sadeghpour Gildeh [40]. The $C'''_p(u,v)$ parameters are binary numbers, and by choosing big values for u and v, it is possible to obtain an index that is more sensitive to process shifts. Abbasi Ganji and

Sadeghpour Gildeh [37] calculated $\hat{C}_{ppi}'''$ for each explanatory level by setting $u = v = 1$ as the following formula:

$$\hat{C}_{ppi}''' = \frac{d_i^* - \hat{A}_i^*}{3\sqrt{\hat{\sigma}^2\left(1 + \frac{1}{mn} + \frac{(X_i - \overline{X})^2}{m\sum_{i=1}^n (X_i - \overline{X})^2}\right) + \hat{A}_i^2}}, i = 1, 2, \ldots, n, \qquad (4)$$

where $D_{l_i} = T_i - LSL_i$, $D_{u_i} = USL_i - T_i$, $d_i^* = \min\{D_{l_i}, D_{u_i}\}$, $d_i = \frac{USL_i - LSL_i}{2}$, and

$$\hat{A}_i^* = \begin{cases} \frac{(T_i - \hat{\mu}_i)^2}{D_{l_i}} & \text{if } \hat{\mu}_i \leq T_i \\ \frac{(\hat{\mu}_i - T_i)^2}{D_{u_i}} & \text{if } \hat{\mu}_i > T_i' \end{cases}$$

$$\hat{A}_i^2 = \begin{cases} \frac{d_i^2(T_i - \hat{\mu}_i)^2}{D_{l_i}^2} & \text{if } \hat{\mu}_i \leq T_i \\ \frac{d_i^2(\hat{\mu}_i - T_i)^2}{D_{u_i}^2} & \text{if } \hat{\mu}_i > T_i \end{cases} \qquad (5)$$

2.3. Existing Process Yield Indices for SLP with BPAC and WPAC

Considering the general index $S_{pk} = \frac{1}{3}\Phi^{-1}\left\{\frac{1}{2}\Phi\left(\frac{USL-\mu}{\sigma}\right) + \frac{1}{2}\Phi\left(\frac{\mu-LSL}{\sigma}\right)\right\}$ proposed by Boyles [41], where μ and σ are the process mean and standard deviation, and LSL and USL are the lower and upper SLs, respectively, and AR(1) structure, Wang and Tamirat [34] and Wang and Tamirat [35] directly estimated the process yield index $S_{pkA;AR(1)}$ for SLP with BPAC and WPAC, which are described in the following subsections:

2.3.1. Process Yield Index for SLP with BPAC

For a normally distributed process at the ith level of the explanatory variable, the index S_{pk_i} introduced to establish the relationship between the SLs and the actual process performance. Based on the sample data from the in-control SLP model in Equation (1) with $\rho = 0$, an estimator for S_{pk_i} is derived as

$$\hat{S}_{pk_i} = \frac{1}{3}\Phi^{-1}\left\{\frac{1}{2}\Phi\left(\frac{USL_i - \overline{y}_i}{s_i}\right) + \frac{1}{2}\Phi\left(\frac{\overline{y}_i - LSL_i}{s_i}\right)\right\}, \qquad (6)$$

where $\overline{y}_i$ and s_i are the sample mean and the sample standard deviation for the response variable at the ith level of the explanatory variable. In Equation (6), $\Phi(.)$ denotes the standard normal cumulative distribution function, and so $\Phi^{-1}(.)$ is its inverse. Then, based on Bothe's idea [42], the total process yield is calculated by

$$\hat{P} = \frac{1}{n}\sum_{i=1}^n P_i = \frac{1}{n}\sum_{i=1}^n \left[2\Phi\left(3\hat{S}_{pk_i}\right) - 1\right]. \qquad (7)$$

Finally, the overall process yield index for SLP with the BPAC is determined as

$$\hat{S}_{pkA;AR(1)} = \frac{1}{3}\Phi^{-1}\left[\frac{1}{2}(1 + \hat{P})\right]. \qquad (8)$$

2.3.2. Process Yield Index for SLP with WPAC

Considering SLP with the WPAC effect, which is set equal to $\varphi = 0$ in Equation (1), an estimator for S_{pk_i} is computed for each level of the explanatory variable profile in Equation (6). Then, the process yield for the ith level of the explanatory variable is estimated by $\hat{P}_i = 2\Phi\left(3\hat{S}_{pk_i}\right) - 1$ [35]. In the case of SLP with WPAC, all yields on each level of the explanatory variable are represented by the AR(1) model as follows:

$$(P_i - \mu) = \rho(P_{i-1} - \mu) + e_i, \qquad (9)$$

where e_i are independent random variables with a mean of zero and a variance of σ_e^2. Thus, the overall process yield (P) for a SLP with WPAC is estimated as $\hat{P} = \hat{\mu}_0$, where $\hat{P} = \frac{1}{n}\sum_{i=1}^{n} P_i$. Therefore, the estimator for $S_{pkA;AR(1)}$ can be obtained using Equation (8).

3. The Proposed PCIs for Autocorrelated Linear Profiles

This section presents a novel PCI for autocorrelated linear profiles based on a functional approach. We first discuss the general formulas for the proposed method and then modify it accordingly based on the three autocorrelation effects, or BPAC, WPAC, and BAC.

The functional approach to process capability analysis for SLP was initially developed by Nemati Keshteli et al. [43], which has the advantage of measuring process capability in the entire domain of the explanatory variables. This method presents a functional form of PCIs using a reference profile, functional natural tolerances, and functional SLs. The area bounded between SLs and natural tolerance limits was used to determine a single value for the functional PCI of SLP. As stated in Section 2.2, the process capability index C_{ppM}''' proposed by Abbasi Ganji and Sadeghpour Gildeh [37] is an extension of the univariate index $C_p'''(u,v)$ for SLP. Consequently, despite the fact that the index C_{ppM}''' accurately assesses the process capability based on both the percentage of non-conforming products and the distance between the process mean and the target value, this index is not functional and only assesses the process capability for only the n levels of the explanatory variable, ignoring other X-values in the profile. As a result, this section introduces the development of the capability index C_{ppM}''' based on the functional approach for the traditional SLP model defined in Equation (1) with $\rho = \varphi = 0$.

Let $LSL_Y(X) = A_{0l} + A_{1l}X$, $USL_Y(X) = A_{0u} + A_{1u}X$, and $T_Y(X) = A_{0T} + A_{1T}X$ be the functional LSL, USL, and target values for the response variable, respectively. The explanatory variable $X \in [x_l, x_u]$, where x_l and x_u are the minimum and maximum values of the explanatory variable and A_{0l} and A_{1l} are the intercept and slope of LSL, respectively. Similarly, A_{0u}, A_{1u}, A_{0T}, and A_{1T} are the intercepts and slopes of USL and the target line of Y, respectively. Therefore, the estimator of index $C_p'''(Profile)$ for SLP is obtained by

$$\hat{C}_p'''(Profile) = \frac{\int_{x_l}^{x_u}\left(d_Y^*(X) - \hat{A}_Y^*(X)\right)dX}{\int_{x_l}^{x_u}\left(3\sqrt{\hat{\sigma}^2 + \hat{A}_Y^2(X)}\right)dX}, \tag{10}$$

where

$$d^*{}_Y(X) = \min\{D_{lY}(X), D_{uY}(X)\} = \min\{(T_Y(X) - LSL_Y(X)), (USL_Y(X) - T_Y(X))\}, \tag{11}$$

$$\hat{A}_Y^*(X) = \begin{cases} \frac{(T_Y(X) - \hat{\mu}_Y(X))^2}{D_{lY}(X)} & if\ \hat{\mu}_Y(X) \leq T_Y(X) \\ \frac{(\hat{\mu}_Y(X) - T_Y(X))^2}{D_{uY}(X)} & if\ \hat{\mu}_Y(X) > T_Y(X) \end{cases}, \tag{12}$$

$$\hat{A}_Y(X) = \begin{cases} \frac{d_Y(X)(T_Y(X) - \hat{\mu}_Y(X))}{D_{lY}(X)} & if\ \hat{\mu}_Y(X) \leq T_Y(X) \\ \frac{d_Y(X)(\hat{\mu}_Y(X) - T_Y(X))}{D_{uY}(X)} & if\ \hat{\mu}_Y(X) > T_Y(X) \end{cases}, \tag{13}$$

where, $\hat{\mu}_Y(X) = a_0 + a_1 X_i$ and $\hat{\sigma}^2 = MSE$ are the sample mean and variance for the response variable, which can be estimated from the in-control profile using the LSE method. In addition, $USL_Y(X) = a_{0u} + a_{1u}X$ and $LSL_Y(X) = a_{0l} + a_{1l}X$, where a_{0u} and a_{1u} (a_{0l} and a_{1l}), are the LSE for the intercept and slope of USL (LSL), respectively. To get the index $C_p'''(Profile)$, we have to determine the location of $\hat{\mu}_Y(X)$ relative to the $T_Y(X)$. As discussed by Pakzad and Basiri [39], four situations can be defined about the intersection of $\hat{\mu}_Y(X)$ and $T_Y(X)$ within the range of the explanatory variables as follows:

(i) If $\hat{\mu}_Y(X)$ and $T_Y(X)$ do not intersect in $X \in [x_l, x_u]$ and $\hat{\mu}_Y(X) > T_Y(X)$, we have

$$\hat{C}_p'''(Profile) = \frac{\int_{x_l}^{x_u} \left(d_Y^*(X) - \frac{(\hat{\mu}_Y(X) - T_Y(X))^2}{D_{uY}(X)} \right) dX}{\int_{x_l}^{x_u} \left(3\sqrt{\hat{\sigma}^2 + \left(\frac{d_Y(X)(\hat{\mu}_Y(X) - T_Y(X))}{D_{uY}(X)} \right)^2} \right) dX}. \tag{14}$$

(ii) If $\hat{\mu}_Y(X)$ and $T_Y(X)$ do not intersect in $X \in [x_l, x_u]$ and $\hat{\mu}_Y(X) \leq T_Y(X)$, we have

$$\hat{C}_p'''(Profile) = \frac{\int_{x_l}^{x_u} \left(d_Y^*(X) - \frac{(T_Y(X) - \hat{\mu}_Y(X))^2}{D_{lY}(X)} \right) dX}{\int_{x_l}^{x_u} \left(3\sqrt{\hat{\sigma}^2 + \left(\frac{d_Y(X)(T_Y(X) - \hat{\mu}_Y(X))}{D_{lY}(X)} \right)^2} \right) dX}. \tag{15}$$

(iii) If $\hat{\mu}_Y(X)$ and $T_Y(X)$ intersect in $X \in [x_l, x_u]$ at the intersection point x_m and $\hat{\mu}_Y(X) > T_Y(X), X \in [x_l, x_m]$ and $\hat{\mu}_Y(X) \leq T_Y(X), X \in [x_m, x_u]$, we have

$$\hat{C}_p'''(Profile) = \frac{\int_{x_l}^{x_m} \left(d_Y^*(X) - \frac{(\hat{\mu}_Y(X) - T_Y(X))^2}{D_{uY}(X)} \right) dX + \int_{x_m}^{x_u} \left(d_Y^*(X) - \frac{(T_Y(X) - \hat{\mu}_Y(X))^2}{D_{lY}(X)} \right) dX}{\int_{x_l}^{x_m} \left(3\sqrt{\hat{\sigma}^2 + \left(\frac{d_Y(X)(\hat{\mu}_Y(X) - T_Y(X))}{D_{uY}(X)} \right)^2} \right) dX + \int_{x_m}^{x_u} \left(3\sqrt{\hat{\sigma}^2 + \left(\frac{d_Y(X)(T_Y(X) - \hat{\mu}_Y(X))}{D_{lY}(X)} \right)^2} \right) dX}. \tag{16}$$

(iv) If $\hat{\mu}_Y(X)$ and $T_Y(X)$ intersect in $X \in [x_l, x_u]$ at the intersection point x_m and $\hat{\mu}_Y(X) \leq T_Y(X), X \in [x_l, x_m]$ and $\hat{\mu}_Y(X) > T_Y(X), X \in [x_m, x_u]$, we have

$$\hat{C}_p'''(Profile) = \frac{\int_{x_l}^{x_m} \left(d_Y^*(X) - \frac{(T_Y(X) - \hat{\mu}_Y(X))^2}{D_{lY}(X)} \right) dX + \int_{x_m}^{x_u} \left(d_Y^*(X) - \frac{(\hat{\mu}_Y(X) - T_Y(X))^2}{D_{uY}(X)} \right) dX}{\int_{x_l}^{x_m} \left(3\sqrt{\hat{\sigma}^2 + \left(\frac{d_Y(X)(T_Y(X) - \hat{\mu}_Y(X))}{D_{lY}(X)} \right)^2} \right) dX + \int_{x_m}^{x_u} \left(3\sqrt{\hat{\sigma}^2 + \left(\frac{d_Y(X)(\hat{\mu}_Y(X) - T_Y(X))}{D_{uY}(X)} \right)^2} \right) dX}. \tag{17}$$

It should be noted that if $C_p'''(Profile) \geq 1$, a process is called "capable", and if $C_p'''(Profile) < 1$, it is "incapable".

It is well known that the estimation of PCIs for SLP in Equation (1) requires estimation of the process parameters, i.e., A_0, A_1, and σ^2. However, the LSE method, which is usually applied to estimate the profile parameters, performs well when there are profiles without any autocorrelation effect. In other words, we may have an unbiased LSE for regression coefficients, but the estimation of error variance is seriously underestimated in the presence of autocorrelation effects [13,18]. To remedy this challenge and have an accurate PCI, it is necessary to adjust our proposed index based on the elimination of autocorrelation effects. In the next subsections, we suggest three strategies for SLP with BPAC, WPAC, and BAC effects to deal with the issue of removing autocorrelation from the profile model.

3.1. The Proposed PCI for SLP with BPAC

In the case of BPAC, when the observations of successive profiles are correlated, we have the SLP model defined in Equation (1) with $\rho = 0$. We suggest applying the transformation to the response variables proposed by Noorossana et al. [9] to obtain an independent profile model. Therefore, using $\hat{Y}_{ij} = \varphi Y_{i(j-1)} + (A_0 + A_1 X_i)$, the residuals can be calculated by

$$e_{ij} = Y_{ij} - \hat{Y}_{ij} = Y_{ij} - \varphi Y_{i(j-1)} - (1 - \varphi)(A_0 + A_1 X_i), \tag{18}$$

and the MSE can be obtained by

$$MSE = \frac{\sum_{j=1}^{m} MSE_j}{m},\tag{19}$$

where $MSE_j = \frac{\sum_{i=1}^{n} e_{ij}^2}{n}$. Then, the estimation of index $C_p'''(Profile)$ can be obtained using Equation (10), where $\hat{\mu}_Y(X) = a_0 + a_1 X$ can be estimated from the in-control profile utilizing the LSE method and $\hat{\sigma}^2 = MSE$ obtained by Equation (19).

3.2. The Proposed PCI for SLP with WPAC

To measure the capability of linear profiles with the WPAC effect, we introduce a two-step methodology. In the first step, the transformation method proposed by Soleimani et al. [13] is applied to eliminate the WPAC effect, while in the second step, the index $C_p'''(Profile)$ is modified to evaluate the process capability. To do this, let $Y_{ij}' = Y_{ij} - \rho Y_{(i-1)j}$, and then the transformed form of the SLP model in Equation (1) with $\varphi = 0$ is obtained as

$$Y_{ij}' = A_0(1-\rho) + A_1(X_i - \rho X_{i-1}) + \left(\varepsilon_{ij} - \rho \varepsilon_{(i-1)j}\right) = A_0' + A_1' X_i' + u_{ij},\tag{20}$$
$$i = 1, 2, 3, \ldots, n,$$

where u_{ij} are independent random variables with a mean of zero and a variance of σ^2. Therefore, by considering X_i' and Y_{ij}', it is possible to use the LSE method to estimate the model parameters. As a second step, in calculating the estimator of the index $C_p'''(Profile)$, $\hat{\mu}_{Y'}(X') = a_0' + a_1' X'$, and $\hat{\sigma}'^2$ are the sample mean and variance, respectively, which can be obtained from the LSE of the transformed in-control profile. In addition, functional forms of the transformed SLs and target line need to be obtained. Since LSL_i, USL_i, and T_i are the LSL, USL, and target value of the response variable at the ith level of the explanatory variable, respectively, the same transformation as the response variable is applied to the SLs and target value. Therefore, following the equations $LSL_i' = LSL_i - \rho LSL_{i-1}$, $USL_i' = USL_i - \rho USL_{i-1}$, and $T_i' = T_i - \rho T_{i-1}$, regression lines are fitted to X_i' and the values of LSL_i', USL_i', and T_i' to obtain $LSL_{Y'}(X')$, $USL_{Y'}(X')$, and $T_{Y'}(X')$. Finally, the index $C_p'''(Profile)$ can be estimated based on Equation (10).

3.3. The Proposed PCI for SLP with BAC

To evaluate the capability of SLP with BAC effects, we suggest using a transformation recommended by Nadi et al. [18] to eliminate both the within- and between-profile autocorrelations first and then using the index $C_p'''(Profile)$ to measure the process capability. Therefore, use the following transformed observations:

$$Y_{ij}' = Y_{ij} - \rho Y_{(i-1)j} - \varphi Y_{i(j-1)} + \rho \varphi Y_{(i-1)(j-1)}.\tag{21}$$

By replacing Y_{ij} values in Equation (21) by their equivalents from the SLP model in Equation (1), a SLP with independent error terms is obtained as

$$Y_{ij}' = A_0(1 - \rho - \varphi + \rho\varphi) + (A_1(1-\varphi))(X_i - \rho X_{i-1}) + u_{ij} = A_0' + A_1' X_i' + u_{ij},\tag{22}$$
$$i = 1, 2, 3, \ldots, n, \; j = 1, 2, \ldots, m,$$

where u_{ij} are independent random variables with a mean of zero and a variance of σ^2. Therefore, by considering X_i' and Y_{ij}', it is possible to use the LSE method to estimate the model parameters. As a second step, the index $C_p'''(Profile)$ can be estimated similarly, as stated in Section 3.2.

It should be noted that to obtain the transformed functional SLs and target line for the transformed response variable, Y_{ij}', the functional SLs and target line are calculated first by fitting the regression lines to the values of LSL_i, USL_i, and T_i, and then the transformed functional SLs and target line are calculated based on Equation (22). Finally, the index $\hat{C}_p'''(Profile)$ can be estimated based on Equation (10).

4. Simulation Study

In this section, we carry out a simulation study in MATLAB to investigate and assess how well the suggested approach performs. This section consists of four subsections. The first two subsections compare the performance of the proposed PCIs for SLP in the presence of the BPAC and WPAC against the existing process yield indices. The third subsection examines the performance of the proposed method for SLP with BAC, and the simulation results of the first three subsections are summed up in the fourth subsection. For this purpose, we use two criteria in terms of bias (the difference between the estimated value and the true value) and mean squared error (the average squared difference between the estimated values and the true value). In the simulation study, the following in-control SLP with a general error model is considered to generate the necessary data:

$$\begin{aligned} Y_{ij} &= 3 + 2X_i + \varepsilon_{ij}, \\ \varepsilon_{ij} &= \rho\varepsilon_{(i-1)j} + a_{ij}, \\ a_{ij} &= \varphi\varepsilon_{i(j-1)} + u_{ij}. \end{aligned} \quad (23)$$

where $u_{ij} \sim N(0,1)$ and in-control correlation coefficients ρ and φ are set based on the previous works (Wang and Tamirat [34]; Wang and Tamirat [35]) as detailed in the next subsections. We consider 10 levels for the explanatory variable ($n = 10$), and the values for the explanatory variables, SLs, and target values for the response variable at each level of the explanatory variable are shown in Table 1.

Table 1. SLs for each level of the explanatory variable based on Wang and Tamirat [35].

i	1	2	3	4	5	6	7	8	9	10
X_i	1	2	3	4	5	6	7	8	9	10
LSL_i	0.08	2.5	4.64	6.85	9.2	11.25	13.76	16.25	18.32	19.5
USL_i	7.58	10	12.14	14.35	16.7	18.75	21.26	23.75	25.82	27
T_i	3.83	6.25	8.39	10.6	12.95	15	17.51	20	22.07	23.25

For better clarification, the Monte Carlo simulation procedure for computing $\hat{C}_p'''(Profile)$, bias, and MSE is in Pseudocode 1. For each autocorrelation structure, the proposed parameter estimation scheme is used. This procedure can also be applied to obtain an estimate for the process capability index C_{ppM}'''. The green lines are some comments about the codes. It should be noted that the results reported here for the competing methods were all obtained using numerical simulation.

Pseudocode 1. The procedure for computing $\hat{C}_p'''(Profile)$, bias, and MSE using Monte Carlo simulation

---

Consider the in-control profile model, SLs' function, sigma = 1, n, m, ρ, and φ.
errorpre = normrnd (0,sigma,[1,n]); % Normal preerror with size n
while (we do not reach the number of generated samples)
    aij = normrnd (0,sigma,[1,n]); % Normal variable with size n
    % Generate data

    for i = 1:n
        if (i == 1)
            epsilon (i) = φ × errorpre (i) + aij (i);
        else
            epsilon(i) = ρ × epsilon (i − 1) + φ × errorpre (i) − φ × ρ × errorpre (i − 1) + aij (i);
        endif

Pseudocode 1. Cont.

```
    endfor
        Y = A₀ + A₁X+ epsilon;  % Equation (1)
        errorpre = epsilon;  % Update the previous error
endwhile
% For the two autocorrelation structures, it is necessary to transfer the SLs as we apply the transformation of the in-control
profile model.
if (WPAC|BAC)
        % The details for computations were given in Section 3.2 and Section 3.3
        Apply the transformation on the SLs.
endif
```
Compute $C_p'''(Profile)$ based on Equation (10) and the in-control parameters.
% The Monte Carlo simulation loop.
% The computed $\hat{C}_p'''(Profile)$ in each iteration is denoted by $\hat{C}_p'''(Profile)_{rep}$.
for rep = 1:1:10,000
   Generate m profiles by the in-control model.
   % The parameter estimation is performed based on the autocorrelation structure.
   $(a_0, a_1, \hat{\sigma}^2)$ = Estimate the profile parameters including intercept, slope, and standard deviation based on the m profiles.
   Compute $\hat{\mu}_Y(X) = a_0 + a_1 X$ and $\hat{\sigma}^2$ based on the proposed method for each category.
   $\hat{C}_p'''(Profile)_{rep}$ = Estimate $\hat{C}_p'''(Profile)$ based on Equation (10) by $\hat{\mu}_Y(X)$ and $\hat{\sigma}^2$.
   Store the values $\hat{C}_p'''(Profile)_{rep}$
endfor
$\hat{C}_p'''(Profile) = \frac{\sum_{rep=1}^{10,000} \hat{C}_p'''(Profile)_{rep}}{10,000}$.

$MSE = \frac{\sum_{rep=1}^{10,000} \left(\hat{C}_p'''(Profile)_{rep} - C_p'''(Profile)\right)^2}{10,000}$.

bias = $\hat{C}_p'''(Profile) - C_p'''(Profile)$.

4.1. Simulation Study for SLP in the Presence of BPAC

In order to ensure a fair comparison between the competing PCIs, we follow the simulation setting used in Wang and Tamirat [34]. To this end, the in-control SLP model in Equation (23) in the presence of autocorrelation between errors in the successive profiles ($\rho = 0$, $\varphi \geq 0$) and four fixed X_i-values of 2, 4, 6, and 8 are considered. In the simulation study, we use four profile samples ($m \in \{25, 50, 100, 200\}$), six correlation coefficients ($\varphi \in \{0, 0.1, 0.25, 0.4, 0.5, 0.7\}$), and six error terms ($u_{ij} \sim N\left(0, (0.8)^2\right)$, $u_{ij} \sim N\left(0, (1.0)^2\right)$, $u_{ij} \sim N\left(0, (1.2)^2\right)$, $u_{ij} \sim N\left(0.1, (0.8)^2\right)$, $u_{ij} \sim N\left(0.1, (1.0)^2\right)$, $u_{ij} \sim N\left(0.1, (1.2)^2\right)$). The SLs and target values for the response variable at X_i-values of 2, 4, 6, and 8 are shown in Table 1. Based on the values in Table 1, functional SLs and target line are obtained as $LSL_Y(X) = -2.2 + 2.2825X$, $USL_Y(X) = 5.3 + 2.2825X$, and $T_Y(X) = 1.55 + 2.2825X$, respectively.

For each simulated case, we calculated $\hat{S}_{pkA;AR(1)}$ given by Equation (8) in Section 2.3.1, $\hat{C}_p'''(Profile)$ based on the method described in Section 3.1, and $\hat{C}_{ppM}'''$ using the same procedure as described in Section 3.1 for $C_p'''(Profile)$. Following 10,000 iterations of the simulations, the average values of $\hat{S}_{pkA;AR(1)}$, $\hat{C}_p'''(Profile)$, and $\hat{C}_{ppM}'''$ are obtained and reported. For each simulated case, the true values of PCIs are calculated using the actual model parameters and listed (for more information, see Appendix A). Moreover, the bias and MSE for the PCIs are calculated. The simulation results are presented in Tables 2–4 as well as depicted in Figures 1–3.

Table 2. A comparison among $S_{pkA;AR(1)}$, C'''_{ppM}, and $C'''_p (Profile)$ for SLP with BPAC when $u_{ij} \sim N\left(0, (0.8)^2\right)$.

φ	m	$S_{pkA;AR(1)}$		C'''_{ppM}		$C'''_p (Profile)$	
		True Value	Estimate (Bias, MSE)	True Value	Estimate (Bias, MSE)	True Value	Estimate (Bias, MSE)
0	25	1.2971	1.2589 (−0.0382, 0.0126)	1.2363	1.2184 (−0.0179, **0.0067**)	1.3160	1.3162 (**0.0002**, 0.0073)
	50		1.2780 (−0.0191, 0.0064)		1.2193 (−0.0170, **0.0035**)		1.3166 (**0.0006**, 0.0036)
	100		1.2867 (−0.0104, 0.0034)		1.2190 (−0.0173, 0.0019)		1.3158 (**−0.0002, 0.0018**)
	200		1.2917 (−0.0054, 0.0018)		1.2187 (−0.0176, 0.0011)		1.3154 (**−0.0006, 0.0009**)
0.1	25	1.2916	1.2576 (−0.0340, 0.0127)	1.2363	1.2180 (−0.0183, **0.0072**)	1.3160	1.3151 (**−0.0009**, 0.0078)
	50		1.2738 (−0.0178, 0.0067)		1.2190 (−0.0173, **0.0037**)		1.3158 (**−0.0002**, 0.0039)
	100		1.2826 (−0.0090, 0.0036)		1.2194 (−0.0169, **0.0020**)		1.3160 (**0.0000, 0.0020**)
	200		1.2880 (−0.0036, 0.0019)		1.2198 (−0.0165, 0.0011)		1.3163 (**0.0003, 0.0010**)
0.25	25	1.2624	1.2328 (−0.0296, 0.0129)	1.2363	1.2152 (−0.0211, **0.0082**)	1.3160	1.3107 (**0.0053**, 0.0087)
	50		1.2461 (−0.0163, 0.0069)		1.2182 (−0.0181, **0.0043**)		1.3141 (**0.0019**, 0.0044)
	100		1.2540 (−0.0084, 0.0038)		1.2184 (−0.0179, 0.0024)		1.3145 (**−0.0015, 0.0023**)
	200		1.2582 (−0.0042, 0.0019)		1.2190 (−0.0173, 0.0013)		1.3154 (**−0.0006, 0.0011**)
0.4	25	1.2058	1.1858 (−0.0200, 0.0135)	1.2363	1.2148 (−0.0215, **0.0099**)	1.3160	1.3077 (**0.0083**, 0.0107)
	50		1.1944 (−0.0114, 0.0074)		1.2177 (−0.0186, **0.0053**)		1.3120 (**0.0040**, 0.0055)
	100		1.1986 (−0.0072, 0.0041)		1.2186 (−0.0177, **0.0029**)		1.3139 (**−0.0021, 0.0029**)
	200		1.2019 (−0.0039, 0.0021)		1.2188 (−0.0175, 0.0016)		1.3148 (**−0.0012, 0.0014**)
0.5	25	1.1503	1.1391 (**−0.0112**, 0.0144)	1.2363	1.2130 (−0.0233, **0.0123**)	1.3160	1.3027 (−0.0133, 0.0135)
	50		1.1424 (−0.0079, 0.0080)		1.2178 (−0.0185, **0.0066**)		1.3104 (**−0.0056**, 0.0069)
	100		1.1450 (−0.0053, 0.0041)		1.2187 (−0.0176, **0.0034**)		1.3129 (**−0.0031, 0.0034**)
	200		1.1466 (−0.0037, 0.0022)		1.2187 (−0.0176, 0.0020)		1.3138 (**−0.0022, 0.0018**)

Table 2. Cont.

φ	m	$S_{pkA;AR(1)}$		C'''_{ppM}		$C'''_p (Profile)$	
		True Value	Estimate (Bias, MSE)	True Value	Estimate (Bias, MSE)	True Value	Estimate (Bias, MSE)
0.7	25	0.9802	0.9974 (0.0172, 0.0169)	1.2363	1.1995 (−0.0368, 0.0219)	1.3160	1.2742 (−0.0418, 0.0258)
	50		0.9840 (**0.0038, 0.0089**)		1.2095 (−0.0268, 0.0127)		1.2934 (−0.0226, 0.0139)
	100		0.9788 (**−0.0014, 0.0046**)		1.2139 (−0.0224, 0.0071)		1.3035 (−0.0125, 0.0074)
	200		0.9792 (**−0.0010, 0.0024**)		1.2174 (−0.0189, 0.0038)		1.3101 (−0.0059, 0.0037)

Note: Bold values indicate the best-performing index in terms of Bias and MSE for each simulated case.

Table 3. A comparison among $S_{pkA;AR(1)}$, C'''_{ppM}, and $C'''_p(Profile)$ for SLP with BPAC when $u_{ij} \sim N\left(0, (1.0)^2\right)$.

φ	m	$S_{pkA;AR(1)}$		C'''_{ppM}		$C'''_p (Profile)$	
		True Value	Estimated (Bias, MSE)	True Value	Estimated (Bias, MSE)	True Value	Estimated (Bias, MSE)
0	25	1.0771	1.0488 (−0.0283, 0.0071)	1.0446	1.0254 (−0.0192, **0.0053**)	1.1048	1.1042 (**−0.0006**, 0.0056)
	50		1.0629 (−0.0142, 0.0037)		1.0270 (−0.0176, **0.0028**)		1.1050 (**0.0002, 0.0028**)
	100		1.0696 (−0.0075, 0.0009)		1.0270 (−0.0176, **0.0007**)		1.1048 (**0.0000, 0.0008**)
	200		1.0738 (−0.0033, 0.0010)		1.0275 (−0.0171, 0.0009)		1.1050 (**0.0002, 0.0007**)
0.1	25	1.0726	1.0474 (−0.0252, 0.0075)	1.0446	1.0251 (−0.0195, **0.0058**)	1.1048	1.1032 (**0.0016**, 0.0061)
	50		1.0581 (−0.0145 0.0038)		1.0260 (−0.0186, 0.0030)		1.1039 (**−0.0009, 0.0030**)
	100		1.0655 (−0.0071, 0.0019)		1.0269 (−0.0177, 0.0017)		1.1045 (**−0.0003, 0.0015**)
	200		1.0686 (−0.0040, 0.0010)		1.0267 (−0.0179, 0.0010)		1.1042 (**−0.0006, 0.0008**)
0.25	25	1.0486	1.0270 (−0.0216, 0.0077)	1.0446	1.0230 (−0.0216, **0.0065**)	1.1048	1.0998 (**−0.0050**, 0.0067)
	50		1.0375 (−0.0111, 0.0041)		1.0261 (−0.0185, 0.0034)		1.1030 (**−0.0018, 0.0033**)
	100		1.0424 (−0.0062, 0.0022)		1.0265 (−0.0181, 0.0019)		1.1037 (**−0.0011, 0.0017**)
	200		1.0459 (−0.0027, 0.0011)		1.0273 (−0.0173, 0.0011)		1.1045 (**−0.0003, 0.0009**)

Table 3. Cont.

φ	m	$S_{pkA;AR(1)}$		C_{ppM}'''		C_p'' (Profile)	
		True Value	Estimated (Bias, MSE)	True Value	Estimated (Bias, MSE)	True Value	Estimated (Bias, MSE)
0.4	25	1.0019	0.9876 (−0.0143, 0.0085)	1.0446	1.0208 (−0.0238, **0.0082**)	1.1048	1.0950 (**0.0098**, 0.0083)
	50		0.9921 (−0.0098, 0.0044)		1.0239 (−0.0207, 0.0043)		1.0995 (**−0.0053, 0.0041**)
	100		0.9967 (−0.0052, 0.0023)		1.0257 (−0.0189, 0.0024)		1.1022 (**−0.0026, 0.0021**)
	200		0.9986 (−0.0033, 0.0012)		1.0261 (−0.0185, 0.0013)		1.1030 (**−0.0018, 0.0010**)
0.5	25	0.9558	0.9469 (**−0.0089, 0.0090**)	1.0446	1.0183 (−0.0263, 0.0096)	1.1048	1.0899 (−0.0149, 0.0098)
	50		0.9491 (**−0.0067, 0.0048**)		1.0226 (−0.0220, 0.0054)		1.0964 (−0.0084, 0.0050)
	100		0.9519 (**−0.0039, 0.0025**)		1.0250 (−0.0196, 0.0029)		1.1005 (−0.0043, **0.0025**)
	200		0.9531 (**−0.0027, 0.0013**)		1.0260 (−0.0186, 0.0016)		1.1023 (**−0.0025, 0.0013**)
0.7	25	0.8125	0.8247 (**0.0122, 0.0104**)	1.0446	0.9973 (−0.0473, 0.0170)	1.1048	1.0564 (−0.0484, 0.0189)
	50		0.8164 (**0.0039, 0.0054**)		1.0123 (−0.0232, 0.0097)		1.0781 (−0.0267, 0.0098)
	100		0.8135 (**0.0010, 0.0028**)		1.0203 (−0.0243, 0.0053)		1.0909 (−0.0139, 0.0049)
	200		0.8124 (**−0.0001, 0.0014**)		1.0244 (−0.0202, 0.0030)		1.0981 (−0.0067, 0.0025)

Note: Bold values indicate the best-performing index in terms of Bias and MSE for each simulated case.

Table 4. A comparison among $S_{pkA;AR(1)}$, C'''_{ppM}, and $C'''_p(Profile)$ for SLP with BPAC when $u_{ij} \sim N\left(0, (1.2)^2\right)$.

φ	m	$S_{pkA;AR(1)}$		C'''_{ppM}		$C'''_p(Profile)$	
		True Value	Estimated (Bias, MSE)	True Value	Estimated (Bias, MSE)	True Value	Estimated (Bias, MSE)
0	25	0.9255	0.9041 (−0.0214, 0.0046)	0.9025	0.8828 (−0.0197, **0.0043**)	0.9477	0.9465 (−**0.0012**, 0.0044)
	50		0.9149 (−0.0106, 0.0023)		0.9096 (−0.0178, 0.0023)		0.9475 (−**0.0002**, 0.0022)
	100		0.9199 (−0.0056, 0.0012)		0.8850 (−0.0175, 0.0013)		0.9474 (−**0.0003**, 0.0011)
	200		0.9231 (−0.0024, 0.0006)		0.8855 (−0.0170, 0.0008)		0.9478 (**0.0001, 0.0005**)
0.1	25	0.9217	0.9024 (−0.0193, 0.0050)	0.9025	0.8823 (−0.0202, **0.0048**)	0.9477	0.9453 (−0.0024, 0.0048)
	50		0.9108 (−0.0109, 0.0025)		0.8837 (−0.0188, 0.0025)		0.9465 (−**0.0012**, 0.0023)
	100		0.9167 (−0.0050, 0.0013)		0.8850 (−0.0175, 0.0014)		0.9473 (−**0.0004, 0.0011**)
	200		0.9193 (−0.0024, **0.0006**)		0.8854 (−0.0171, 0.0008)		0.9476 (−**0.0001, 0.0006**)

Table 4. Cont.

φ	m	$S_{pkA;AR(1)}$		C'''_{ppM}		$\hat{C}'''_p$ (Profile)	
		True Value	Estimated (Bias, MSE)	True Value	Estimated (Bias, MSE)	True Value	Estimated (Bias, MSE)
0.25	25	0.9009	0.8847 (−0.0162, **0.0051**)	0.9025	0.8802 (−0.0223, 0.0054)	0.9477	0.9420 (**−0.0057**, 0.0052)
	50		0.8916 (−0.0093, 0.0027)		0.8831 (−0.0194, 0.0028)		0.9449 (**−0.0028, 0.0026**)
	100		0.8962 (−0.0047, 0.0014)		0.8841 (−0.0184, 0.0016)		0.9462 (**−0.0015, 0.0013**)
	200		0.8981 (−0.0028, 0.0007)		0.8845 (−0.0180, 0.0010)		0.9466 (**−0.0011, 0.0006**)
0.4	25	0.8602	0.8504 (**−0.0098, 0.0055**)	0.9025	0.8774 (−0.0251, 0.0065)	0.9477	0.9370 (−0.0107, 0.0063)
	50		0.8534 (−0.0068, 0.0030)		0.8813 (−0.0212, 0.0035)		0.9419 (**−0.0058**, 0.0031)
	100		0.8570 (−0.0032, **0.0015**)		0.8837 (−0.0188, 0.0019)		0.9450 (**−0.0027, 0.0015**)
	200		0.8575 (−0.0027, **0.0008**)		0.8839 (−0.0186, 0.0011)		0.9457 (**−0.0020, 0.0008**)
0.5	25	0.8198	0.8148 (**−0.0050, 0.0062**)	0.9025	0.8729 (−0.0296, 0.0079)	0.9477	0.9304 (0.0173, 0.0076)
	50		0.8156 (−0.0042, 0.0031)		0.8794 (−0.0231, 0.0041)		0.9387 (−0.0090, 0.0036)
	100		0.8176 (**−0.0022, 0.0016**)		0.8826 (−0.0199, 0.0023)		0.9431 (−0.0046, 0.0018)
	200		0.8181 (**−0.0017, 0.0009**)		0.8836 (−0.0189, 0.0013)		0.9449 (−0.0028, **0.0009**)
0.7	25	0.6939	0.7061 (**0.0122, 0.0073**)	0.9025	0.8486 (−0.0539, 0.0143)	0.9477	0.8969 (−0.0508, 0.0151)
	50		0.6989 (**0.0050, 0.0038**)		0.8666 (−0.0359, 0.0080)		0.9197 (−0.0280, 0.0075)
	100		0.6964 (**0.0025, 0.0020**)		0.8750 (−0.0275, 0.0044)		0.9320 (−0.0157, 0.0037)
	200		0.6949 (**0.0010, 0.0010**)		0.8812 (−0.0213, 0.0024)		0.9403 (−0.0074, 0.0018)

Note: Bold values indicate the best-performing index in terms of Bias and MSE for each simulated case.

According to Tables 2–4, among the three PCIs ($S_{pkA;AR(1)}$, C'''_{ppM}, $\hat{C}'''_p$ (Profile)), the index $\hat{C}'''_p$ (Profile), which significantly has the least bias, indicates that its mean is the most accurate representation of the true value. Additionally, regarding the MSE values, the index $\hat{C}'''_p$ (Profile) outperforms the index $\hat{S}_{pkA;AR(1)}$, especially for correlation coefficient φ less than 0.7. Also, the differences in MSE values of the indices $\hat{C}'''_p$ (Profile) and C'''_{ppM} are negligible. The smallest bias and MSE values are highlighted in Tables 2–4. On the other hand, the number of profile samples m affects the estimates of all PCIs. As the number of profile samples m increases, the values of bias and MSE for all PCIs decrease. Figures 1–3 clearly depict the findings about the performance evaluations of PCIs based on the MSE criterion for different correlation coefficients φ. Similar results can be drawn when the mean of the error terms is 0.1 under different variances. Results are available from the authors upon request. Hence, regarding bias and MSE, the suggested index $\hat{C}'''_p$ (Profile) and the existing index $S_{pkA;AR(1)}$ work best, respectively, for cases up to $\varphi = 0.5$ and $\varphi > 0.5$.

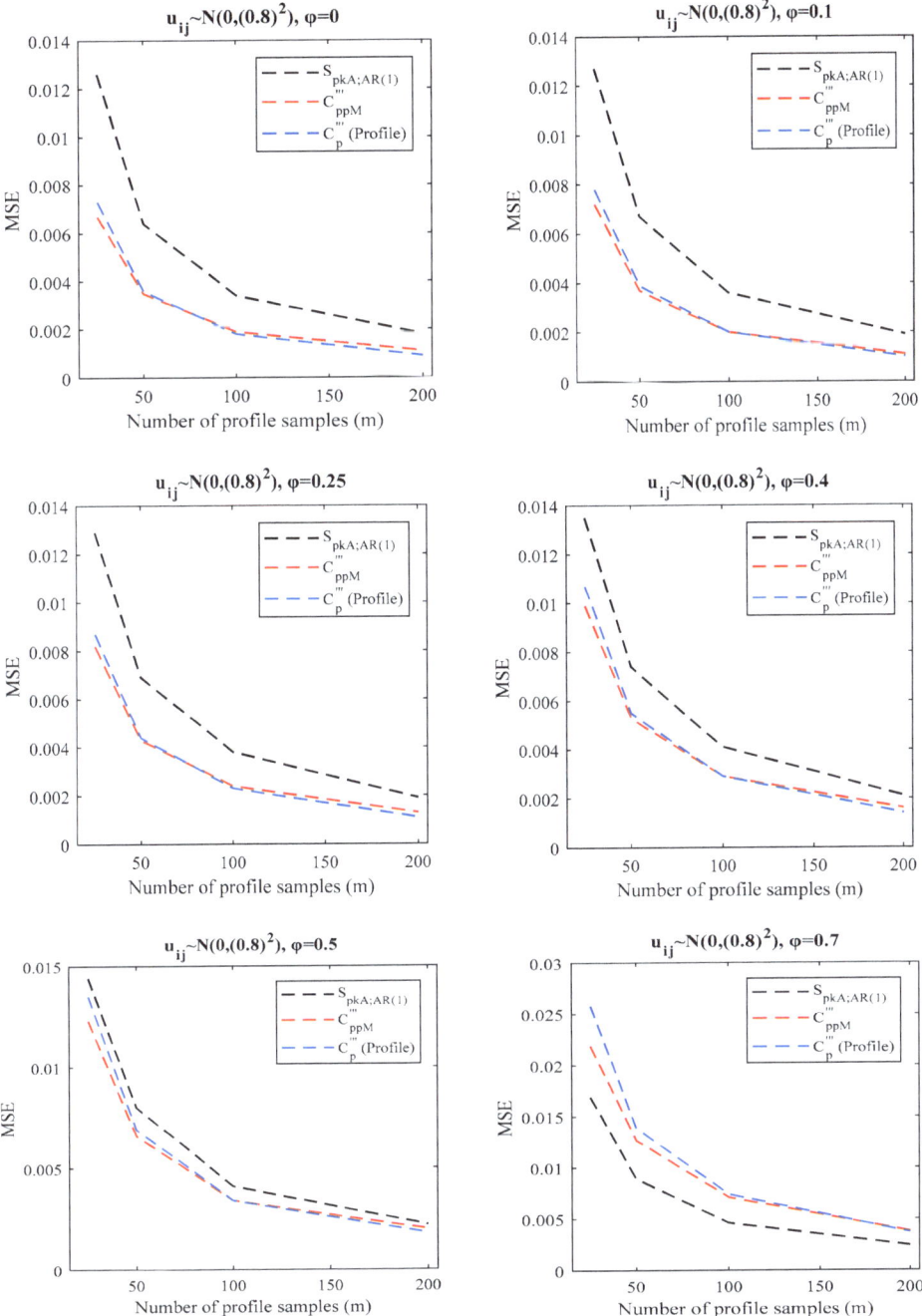

Figure 1. Comparison of various PCIs based on MSE criteria under different profile sample sizes m and parameters φ for SLP with BPAC when $u_{ij} \sim N\left(0, (0.8)^2\right)$.

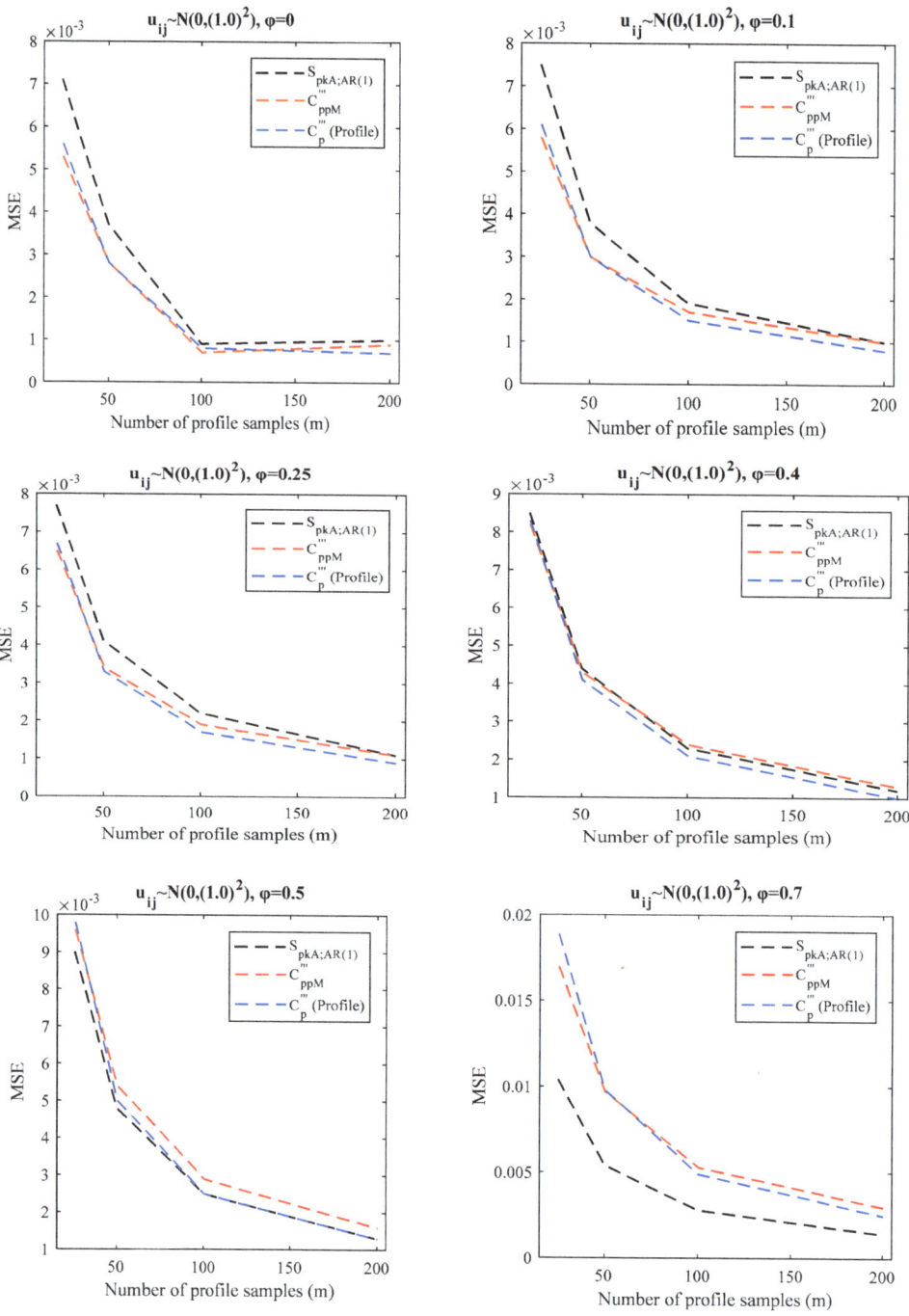

Figure 2. Comparison of various PCIs based on MSE criteria under different profile sample sizes m and parameters φ for SLP with BPAC when $u_{ij} \sim N(0, (1.0)^2)$.

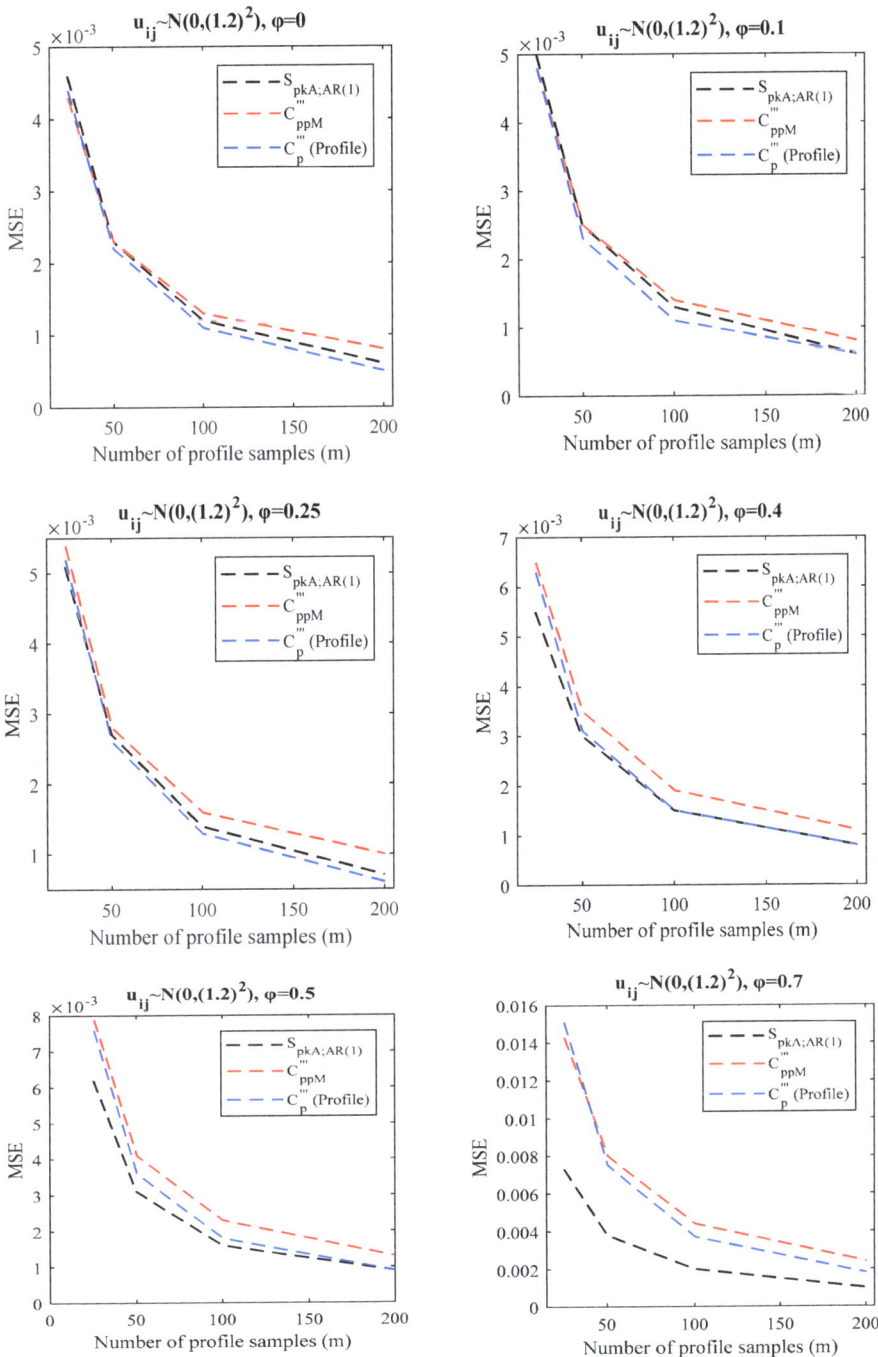

Figure 3. Comparison of various PCIs based on MSE criteria under different profile sample sizes m and parameters φ for SLP with BPAC when $u_{ij} \sim N\left(0, (1.2)^2\right)$.

4.2. Simulation Study for SLP in the Presence of WPAC

In this subsection, the following in-control SLP model in Equation (23) with within-profile autocorrelation ($\rho \geq 0$, $\varphi = 0$) and ten fixed X_i-values of 1, 2, 3, 4, 5, 6, 7, 8, 9, and 10 is considered similar to Wang and Tamirat [35]. The conditions in the simulated cases are the same as those in the Wang and Tamirat [35] simulation setting and include four profile samples ($m \in \{25, 50, 100, 200\}$), six correlation coefficients ($\rho \in \{0, 0.1, 0.25, 0.4, 0.5, 0.7\}$), and six error terms ($u_{ij} \sim N\left(0, (0.8)^2\right)$, $u_{ij} \sim N\left(0, (1.0)^2\right)$, $u_{ij} \sim N\left(0, (1.2)^2\right)$, $u_{ij} \sim N\left(0, (0.8)^2\right)$, $u_{ij} \sim N\left(0, (1.0)^2\right)$, $u_{ij} \sim N\left(0, (1.2)^2\right)$). The SLs and target values of the response variable at ten fixed X_i-values are shown in Table 1.

For each simulated case, we calculated the index $\hat{S}_{pkA;AR(1)}$ given by Equation (8) and description in Section 2.3.2, the index $\hat{C}_p'''(Profile)$ based on the method described in Section 3.2, and the index $\hat{C}_{ppM}'''$ using the same procedure as described in Section 3.2 for $\hat{C}_p'''(Profile)$. The simulations are repeated 10,000 times, and the average values of $\hat{S}_{pkA;AR(1)}$, $\hat{C}_p'''(Profile)$, and $\hat{C}_{ppM}'''$ are obtained and reported. For each simulated case, the true values of PCIs are calculated using the actual model parameters and listed (for more information, see Appendix B). Moreover, the bias and MSE for the PCIs are calculated. Tables 5–7 and Figures 4–6 illustrate the simulation findings.

Table 5. A comparison among $S_{pkA;AR(1)}$, C_{ppM}''', and $C_p'''(Profile)$ for SLP with WPAC when $u_{ij} \sim N\left(0, (0.8)^2\right)$.

ρ	m	$S_{pkA;AR(1)}$		C_{ppM}'''		$C_p'''(Profile)$	
		True Value	Estimated (Bias, MSE)	True Value	Estimated (Bias, MSE)	True Value	
0	25	1.2631	1.2071 (−0.0560, 0.0078)	1.2613	1.2563 (−0.0050, 0.0036)	1.3171	1.3184 (0.0013, 0.0043)
	50		1.2313 (−0.0318, 0.0037)		1.2562 (−0.0051, 0.0018)		1.3175 (0.0004, 0.0021)
	100		1.2467 (−0.0164, 0.0017)		1.2564 (−0.0049, 0.0009)		1.3174 (0.0003, 0.0010)
	200		1.2546 (−0.0085, 0.0008)		1.2560 (−0.0053, 0.0005)		1.3170 (−0.0001, 0.0005)
0.1	25	1.2579	1.2025 (−0.0554, 0.0078)	1.166	1.1601 (−0.0059, 0.0031)	1.2238	1.2243 (0.0005, 0.0038)
	50		1.2267 (−0.0312, 0.0037)		1.1601 (−0.0059, 0.0016)		1.2236 (0.0002, 0.0019)
	100		1.2413 (−0.0166, 0.0017)		1.1606 (−0.0054, 0.0008)		1.2238 (0.0000, 0.0009)
	200		1.2493 (−0.0086, 0.0008)		1.1611 (−0.0049, 0.0004)		1.2242 (0.0004, 0.0005)
0.25	25	1.2305	1.1787 (−0.0518, 0.0072)	1.0052	0.9992 (−0.0060, 0.0024)	1.0647	1.0650 (0.0003, 0.0031)
	50		1.2018 (−0.0287, 0.0034)		1.0000 (−0.0052, 0.0012)		1.0651 (0.0004, 0.0015)
	100		1.2151 (−0.0154, 0.0017)		0.9999 (−0.0053, 0.0006)		1.0646 (−0.0001, 0.0008)
	200		1.2229 (−0.0076, 0.0008)		1.0006 (−0.0046, 0.0003)		1.0652 (0.0005, 0.0004)

Table 5. Cont.

ρ	m	$S_{pkA;AR(1)}$		C'''_{ppM}		C'''_p (Profile)	
		True Value	Estimated (Bias, MSE)	True Value	Estimated (Bias, MSE)	True Value	Estimated (Bias, MSE)
0.4	25	1.1771	1.1320 (−0.0451, 0.0065)	0.8226	0.8168 (−0.0058, **0.0016**)	0.8818	0.8814 (−**0.0004**, 0.0023)
	50		1.1522 (−0.0249, 0.0032)		0.8172 (−0.0054, **0.0008**)		0.8812 (**0.0006**, 0.0011)
	100		1.1639 (−0.0132, 0.0015)		0.8179 (−0.0047, **0.0004**)		0.8817 (−**0.0001**, 0.0006)
	200		1.1703 (−0.0068, 0.0007)		0.8181 (−0.0045, **0.0002**)		0.8818 (−**0.0000**, 0.0003)
0.5	25	1.1247	1.0859 (−0.0388, 0.0057)	0.6892	0.6839 (−0.0053, **0.0011**)	0.7473	0.7468 (−**0.0005**, 0.0017)
	50		1.1031 (−0.0216, 0.0029)		0.6848 (−0.0044, **0.0006**)		0.7472 (−**0.0001**, 0.0008)
	100		1.1136 (−0.0111, 0.0013)		0.6851 (−0.0041, **0.0003**)		0.7471 (**0.0002**, 0.0004)
	200		1.1187 (−0.0060, 0.0007)		0.6852 (−0.0040, **0.0002**)		0.7472 (−**0.0001, 0.0002**)
0.7	25	0.9632	0.9383 (−0.0249, 0.0044)	0.395	0.3912 (−0.0038, **0.0004**)	0.4532	0.4517 (**0.0015**, 0.0007)
	50		0.9504 (−0.0128, 0.0022)		0.3921 (−0.0029, **0.0002**)		0.4525 (−**0.0007**, 0.0003)
	100		0.9566 (−0.0066, 0.0011)		0.3924 (−0.0026, **0.0001**)		0.4527 (−**0.0005**, 0.0002)
	200		0.9595 (−0.0037, 0.0006)		0.3927 (−0.0023, **0.0001**)		0.4530 (−**0.0002, 0.0001**)

Note: Bold values indicate the best-performing index in terms of Bias and MSE for each simulated case.

Table 6. A comparison among $S_{pkA;AR(1)}$, C'''_{ppM}, and $C'''_p(Profile)$ for SLP with WPAC when $u_{ij} \sim N(0, (1.0)^2)$.

ρ	m	$S_{pkA;AR(1)}$		C'''_{ppM}		$C'''_p(Profile)$	
		True Value	Estimated (Bias, MSE)	True Value	Estimated (Bias, MSE)	True Value	Estimated (Bias, MSE)
0	25	1.0553	1.0171 (−0.0382, **0.0040**)	1.0611	1.0559 (−0.0052, **0.0027**)	1.1053	1.1059 (**0.0006**, 0.0032)
	50		1.0345 (−0.0208, **0.0019**)		1.0563 (−0.0048, **0.0014**)		1.1058 (**0.0005**, 0.0016)
	100		1.0447 (−0.0106, **0.0009**)		1.0563 (−0.0048, **0.0007**)		1.1055 (**0.0002**, 0.0008)
	200		1.0493 (−0.0060, **0.0004**)		1.0563 (−0.0048, **0.0004**)		1.1053 (**0.0000**, 0.0004)
0.1	25	1.0511	1.0127 (−0.0384, **0.0041**)	0.9747	0.9681 (−0.0066, **0.0024**)	1.0181	1.0173 (**0.0008**, 0.0028)
	50		1.0306 (−0.0205 **0.0018**)		0.9697 (−0.0050, **0.0012**)		1.0183 (**0.0002**, 0.0014)
	100		1.0404 (−0.0107, **0.0009**)		0.9700 (−0.0047, **0.0006**)		1.0183 (**0.0002**, 0.0007)
	200		1.0460 (−0.0051, **0.0004**)		0.9701 (−0.0046, **0.0003**)		1.0182 (**0.0001**, 0.0004)

Table 6. Cont.

ρ	m	$S_{pkA;AR(1)}$		C'''_{ppM}		C'''_p (Profile)	
		True Value	Estimated (Bias, MSE)	True Value	Estimated (Bias, MSE)	True Value	Estimated (Bias, MSE)
0.25	25	1.0283	0.9931 (−0.0352, 0.0038)	0.833	0.8267 (−0.0063, **0.0018**)	0.8747	0.8737 (**−0.0010**, 0.0022)
	50		1.0094 (−0.0189, 0.0018)		0.8282 (−0.0048, **0.0009**)		0.8749 (**0.0002**, 0.0011)
	100		1.0186 (−0.0097, 0.0009)		0.8283 (−0.0047, **0.0005**)		0.8746 (**−0.0001, 0.0005**)
	200		1.0230 (−0.0053, 0.0004)		0.8287 (−0.0043, **0.0002**)		0.8747 (**0.0001**, 0.0003)
0.4	25	0.9839	0.9530 (−0.0309, 0.0036)	0.677	0.6720 (−0.0050, **0.0012**)	0.7169	0.7165 (**−0.0004**, 0.0015)
	50		0.9675 (−0.0164, 0.0017)		0.6719 (−0.0051, **0.0006**)		0.7160 (**−0.0009**, 0.0008)
	100		0.9754 (−0.0085, 0.0008)		0.6728 (−0.0042, **0.0003**)		0.7167 (**−0.0002**, 0.0004)
	200		0.9791 (−0.0048, 0.0004)		0.6733 (0.0037, **0.0002**)		0.7170 (**0.0001, 0.0002**)
0.5	25	0.9398	0.9132 (−0.0266, 0.0033)	0.5653	0.5601 (−0.0052, **0.0009**)	0.6044	0.6031 (**−0.0013**, 0.0011)
	50		0.9261 (−0.0137, 0.0016)		0.5615 (−0.0038, **0.0004**)		0.6042 (**−0.0002**, 0.0006)
	100		0.9325 (−0.0073, 0.0008)		0.5614 (−0.0039, **0.0002**)		0.6038 (**−0.0006**, 0.0003)
	200		0.9363 (−0.0035, 0.0004)		0.5622 (−0.0031, **0.0001**)		0.6045 (**0.0001, 0.0001**)
0.7	25	0.8023	0.7862 (−0.0161, 0.0028)	0.3227	0.3183 (−0.0044, **0.0003**)	0.3647	0.3621 (**−0.0026**, 0.0005)
	50		0.7937 (−0.0086, 0.0015)		0.3197 (−0.0031, **0.0002**)		0.3633 (**−0.0014, 0.0002**)
	100		0.7984 (−0.0039, 0.0007)		0.3202 (−0.0025, **0.0001**)		0.3640 (**−0.0007, 0.0001**)
	200		0.8003 (−0.0020, 0.0004)		0.3206 (−0.0021, **0.0000**)		0.3644 (**−0.0003**, 0.0001)

Note: Bold values indicate the best-performing index in terms of Bias and MSE for each simulated case.

Table 7. A comparison among $S_{pkA;AR(1)}$, C'''_{ppM}, and C'''_p (Profile) for SLP with WPAC when $u_{ij} \sim N\left(0, (1.2)^2\right)$.

ρ	m	$S_{pkA;AR(1)}$		C'''_{ppM}		C'''_p (Profile)	
		True Value	Estimated (Bias, MSE)	True Value	Estimated (Bias, MSE)	True Value	Estimated (Bias, MSE)
0	25	0.9109	0.8832 (−0.0277, 0.0024)	0.9139	0.9090 (−0.0049, **0.0022**)	0.9479	0.9485 (**0.0006**, 0.0025)
	50		0.8963 (−0.0146, 0.0011)		0.9096 (−0.0043, **0.0011**)		0.9485 (**−0.0006**, 0.0012)
	100		0.9034 (−0.0075, 0.0005)		0.9094 (−0.0045, **0.0006**)		0.9480 (**0.0001, 0.0006**)
	200		0.9072 (−0.0037, 0.0003)		0.9096 (−0.0043, **0.0003**)		0.9480 (**0.0001, 0.0003**)

Table 7. Cont.

ρ	m	$S_{pkA;AR(1)}$		C'''_{ppM}		C''_p (Profile)	
		True Value	Estimated (Bias, MSE)	True Value	Estimated (Bias, MSE)	True Value	Estimated (Bias, MSE)
0.1	25	0.9072	0.8802 (−0.0270, 0.0024)	0.8357	0.8302 (−0.0055, **0.0019**)	0.8682	0.8680 (−**0.0002**, 0.0022)
	50		0.8922 (−0.0150, 0.0011)		0.8306 (−0.0051, **0.0010**)		0.8678 (−**0.0004**, 0.0011)
	100		0.8997 (−0.0075, **0.0005**)		0.8308 (−0.0049, **0.0005**)		0.8677 (−**0.0005**, 0.0006)
	200		0.9034 (−0.0038, **0.0003**)		0.8315 (−0.0042, **0.0003**)		0.8683 (**0.0001**, 0.0003)
0.25	25	0.8873	0.8612 (−0.0261, 0.0023)	0.7099	0.7042 (−0.0057, **0.0014**)	0.7402	0.7394 (**0.0008**, 0.0017)
	50		0.8739 (−0.0134, 0.0011)		0.7056 (−0.0043, **0.0007**)		0.7402 (**0.0000**, 0.0008)
	100		0.8805 (−0.0068, **0.0005**)		0.7057 (−0.0042, **0.0004**)		0.7400 (−**0.0002**, 0.0004)
	200		0.8840 (−0.0033, **0.0003**)		0.7059 (−0.0040, **0.0002**)		0.7401 (−**0.0001**, 0.0002)
0.4	25	0.8482	0.8249 (−0.0233, 0.0023)	0.5742	0.5687 (−0.0055, **0.0009**)	0.6029	0.6014 (−**0.0015**, 0.0011)
	50		0.8366 (−0.0116, 0.0011)		0.5699 (−0.0043, **0.0005**)		0.6022 (−**0.0007**, 0.0006)
	100		0.8428 (−0.0054, **0.0005**)		0.5707 (−0.0035, **0.0002**)		0.6028 (−**0.0001**, 0.0003)
	200		0.8450 (−0.0032, **0.0003**)		0.5707 (−0.0035, **0.0001**)		0.6026 (−**0.0003**, 0.0001)
0.5	25	0.8094	0.7905 (−0.0189, 0.0022)	0.4784	0.4735 (−0.0049, **0.0006**)	0.5067	0.5053 (−**0.0014**, 0.0008)
	50		0.8000 (−0.0094, 0.0011)		0.4744 (−0.0040, **0.0003**)		0.5058 (−**0.0009**, 0.0004)
	100		0.8041 (−0.0053, **0.0005**)		0.4750 (−0.0034, **0.0002**)		0.5062 (−**0.0005**, 0.0002)
	200		0.8071 (−0.0023, 0.0003)		0.4754 (−0.0030, **0.0001**)		0.5065 (−**0.0002**, 0.0001)
0.7	25	0.6873	0.6768 (−0.0105, 0.0021)	0.2724	0.2677 (−0.0047, **0.0003**)	0.3049	0.3018 (−**0.0031**, 0.0003)
	50		0.6823 (−0.0050, 0.0011)		0.2696 (−0.0028, **0.0001**)		0.3037 (−**0.0012**, 0.0002)
	100		0.6849 (−0.0024, 0.0006)		0.2700 (−0.0024, **0.0001**)		0.3041 (−**0.0008**, 0.0001)
	200		0.6860 (−0.0013, 0.0003)		0.2704 (−0.0020, **0.0000**)		0.3044 (−**0.0005**, 0.0000)

Note: Bold values indicate the best-performing index in terms of Bias and MSE for each simulated case.

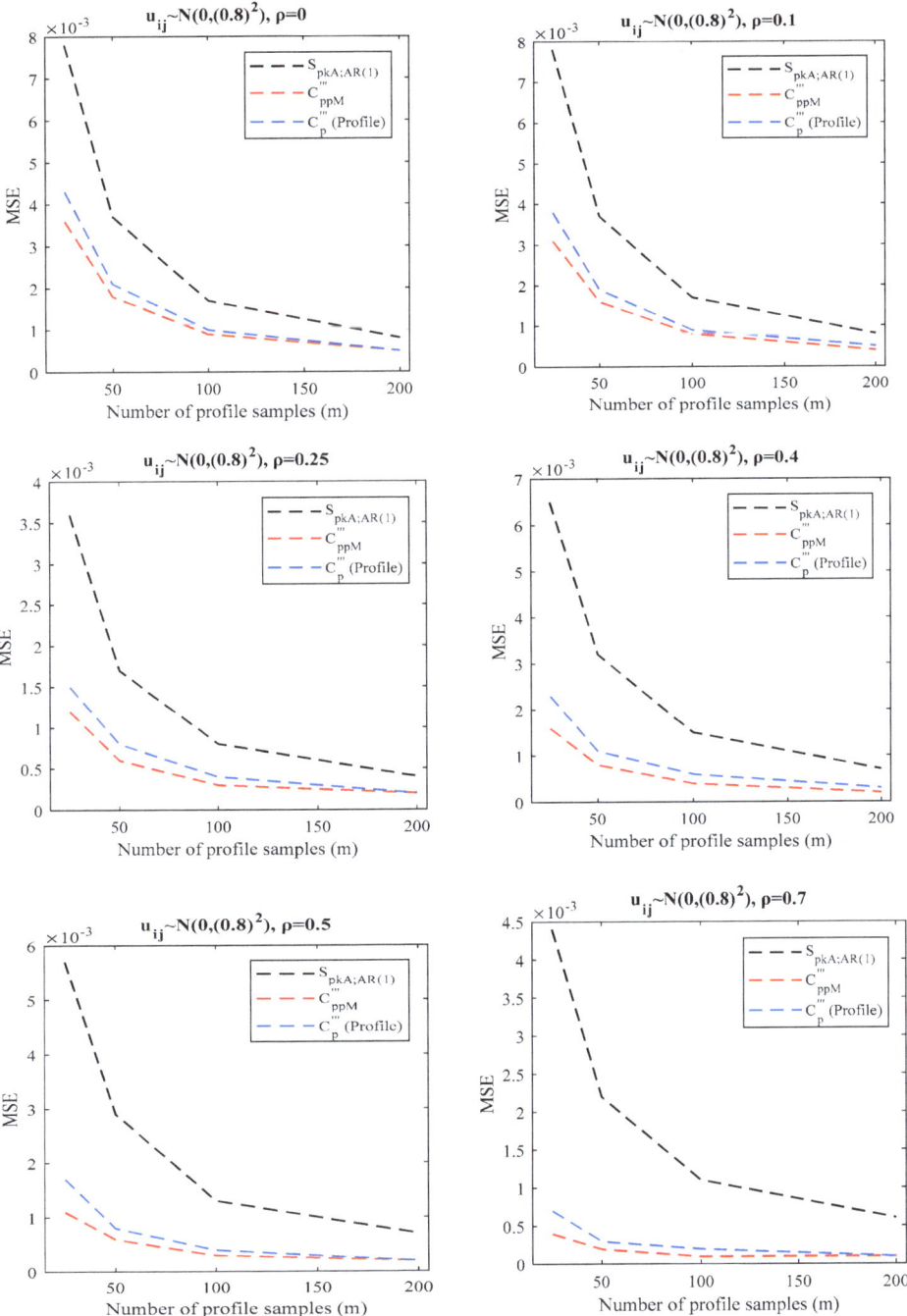

Figure 4. Comparison of various PCIs based on MSE criteria under different profile sample sizes m and parameters ρ for SLP with WPAC when $u_{ij} \sim N\left(0, (0.8)^2\right)$.

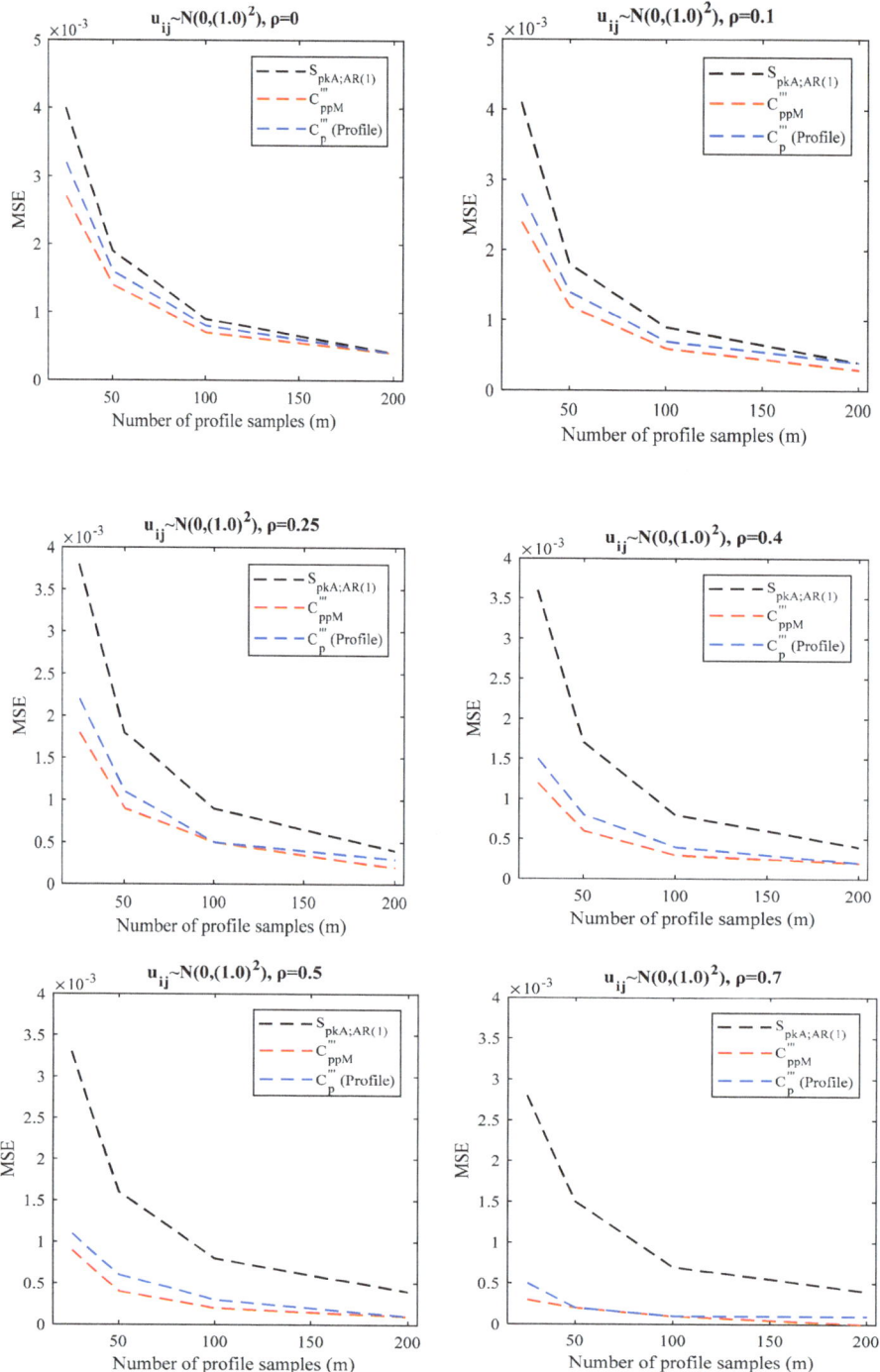

Figure 5. Comparison of various PCIs based on MSE criteria under different profile sample sizes m and parameters ρ for SLP with WPAC when $u_{ij} \sim N\left(0, (1.0)^2\right)$.

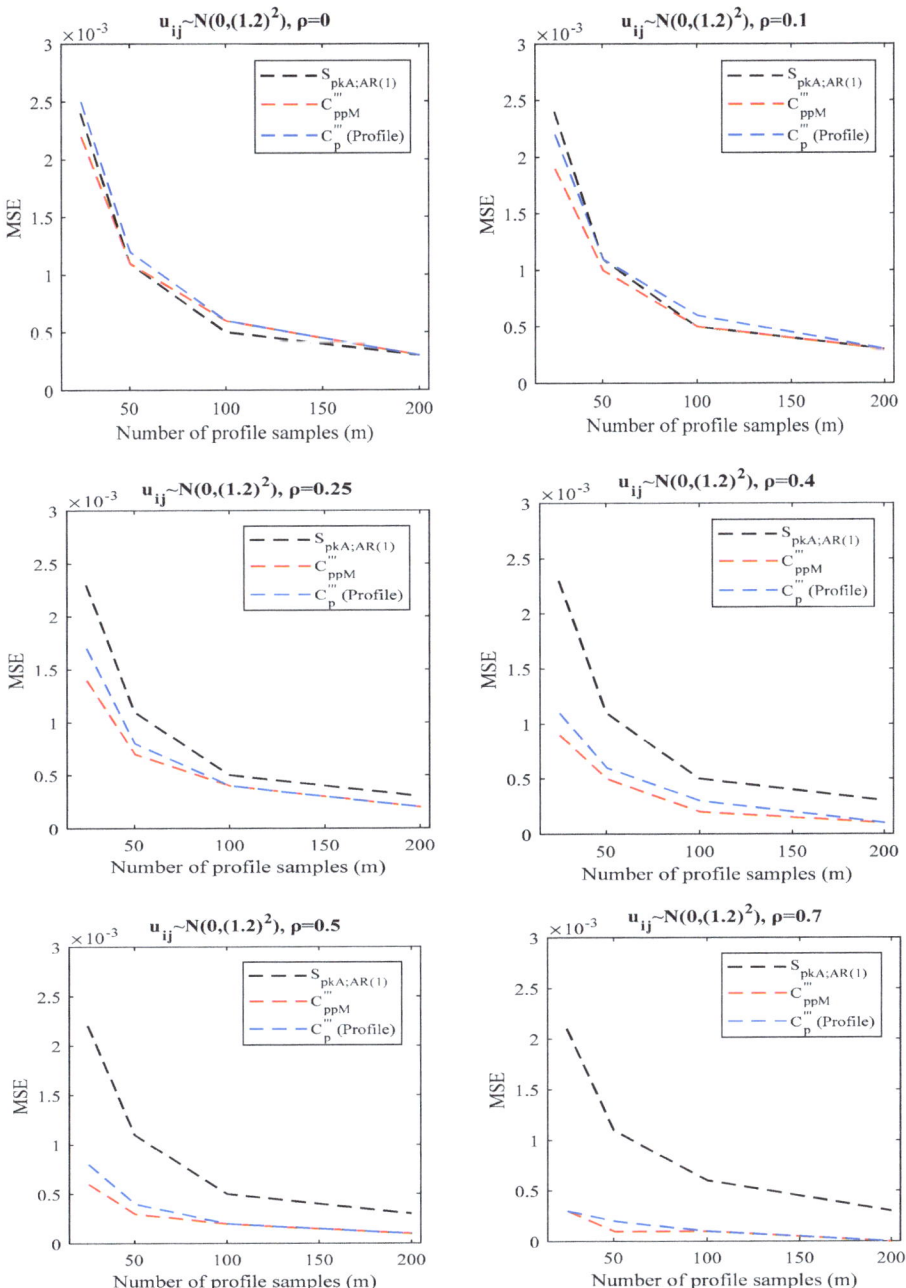

Figure 6. Comparison of various PCIs based on MSE criteria under different profile sample sizes m and parameters ρ for SLP with WPAC when $u_{ij} \sim N(0,(1.2)^2)$.

From Tables 5–7, it can be seen that the estimates of the index $C_p'''(Profile)$ have smaller bias and MSE values than $S_{pkA;AR(1)}$ in all simulated cases. Of course, these values are also improved by increasing the number of profile samples. Additionally, Tables 5–7 demonstrate the superiority of the index $C_p'''(Profile)$ over C_{ppM}''' in terms of smaller bias values. Furthermore, as can be seen in Figures 4–6, the differences in MSE values of the indices $C_p'''(Profile)$ and C_{ppM}''' are negligible, while the corresponding differences are significant for the indices $C_p'''(Profile)$ and $S_{pkA;AR(1)}$. The smallest bias and MSE values are highlighted in Tables 5–7. Similar results are obtained when the error terms follow $u_{ij} \sim N\left(0.1, (0.8)^2\right)$, $u_{ij} \sim N\left(0.1, (1.0)^2\right)$, and $u_{ij} \sim N\left(0.1, (1.2)^2\right)$. As a result, simulation studies show that the recommended index $C_p'''(Profile)$ has superior performance compared to the other two PCIs.

4.3. Simulation Study for SLP in the Presence of BAC

To investigate the performance of the proposed method in evaluating the capability of SLP with BAC effects, the in-control SLP in Equation (23) with BAC effects and four fixed X_i-values of 2, 4, 6, and 8 are considered. In this subsection, we consider the same simulated cases as before with the correlation coefficients ($\rho = \varphi \in \{0, 0.1, 0.25, 0.4, 0.5, 0.7\}$). The SLs and target values for the response variable at four fixed X_i-values are shown in Table 1.

For each simulated case, we calculated the index $\hat{C}_p'''(Profile)$ based on the method described in Section 3.3. The simulations are repeated 10,000 times, and the average values of $\hat{C}_p'''(Profile)$ are obtained and reported. For each simulated case, the true values of $C_p'''(Profile)$ are calculated using the actual model parameters and listed (for more information, see Appendix C). Moreover, the bias and MSE associated with the index $\hat{C}_p'''(Profile)$ are calculated. The results of the simulation study are presented in Table 8.

Table 8. The simulation results of $C_p'''(Profile)$ for SLP with BAC.

Cases u_{ij}		$u_{ij} \sim N(0, (0.8)^2)$		$u_{ij} \sim N(0, (1.0)^2)$		$u_{ij} \sim N(0, (1.2)^2)$	
ρ, φ	m	$C_p'''(Profile)$	$\hat{C}_p'''(Profile)$ (Bias, MSE)	$C_p'''(Profile)$	$\hat{C}_p'''(Profile)$ (Bias, MSE)	$C_p'''(Profile)$	$\hat{C}_p'''(Profile)$ (Bias, MSE)
$\rho = 0$ $\varphi = 0$	25	1.3493	1.3498 (0.0005, 0.0106)	1.1254	1.1250 (−0.0004, 0.0081)	0.9614	0.9603 (−0.0011, 0.0064)
	50		1.3510 (0.0017, 0.0051)		1.1252 (−0.0002, 0.0039)		0.9620 (0.0006, 0.0031)
	100		1.3499 (0.0006, 0.0026)		1.1255 (0.0001, 0.0019)		0.9606 (−0.0008, 0.0015)
	200		1.3496 (0.0003, 0.0013)		1.1254 (0.0000, 0.0010)		0.9615 (0.0001, 0.0008)
$\rho = 0.1$ $\varphi = 0.1$	25	1.1387	1.1378 (−0.0009, 0.0085)	0.9383	0.9370 (−0.0013, 0.0062)	0.7955	0.7937 (−0.0018, 0.0046)
	50		1.1398 (0.0011, 0.0043)		0.9386 (0.0003, 0.0030)		0.7945 (−0.0010, 0.0022)
	100		1.1388 (0.0001, 0.0021)		0.9383 (−0.0000, 0.0015)		0.7949 (−0.0006, 0.0011)
	200		1.1388 (0.0001, 0.0011)		0.9386 (0.0003, 0.0008)		0.7951 (−0.0004, 0.0006)
$\rho = 0.25$ $\varphi = 0.25$	25	0.8037	0.8016 (−0.0021, 0.0048)	0.6574	0.6545 (−0.0029, 0.0035)	0.5549	0.5507 (−0.0042, 0.0025)
	50		0.8029 (−0.0008, 0.0024)		0.6559 (−0.0015, 0.0016)		0.5532 (−0.0017, 0.0012)
	100		0.8032 (−0.0005, 0.0012)		0.6569 (−0.0005, 0.0008)		0.5546 (−0.0003, 0.0006)
	200		0.8031 (−0.0006, 0.0006)		0.6568 (−0.0006, 0.0004)		0.5545 (−0.0004, 0.0003)

Table 8. Cont.

Cases u_{ij}		$u_{ij} \sim N(0, (0.8)^2)$		$u_{ij} \sim N(0, (1.0)^2)$		$u_{ij} \sim N(0, (1.2)^2)$	
ρ, φ	m	C_p''' (Profile)	$\hat{C}_p'''$ (Profile) (Bias, MSE)	C_p''' (Profile)	$\hat{C}_p'''$ (Profile) (Bias, MSE)	C_p''' (Profile)	$\hat{C}_p'''$ (Profile) (Bias, MSE)
$\rho = 0.4$ $\varphi = 0.4$	25	0.5161	0.5118 (−0.0043, 0.0025)	0.4196	0.4145 (−0.0051, 0.0017)	0.3529	0.3466 (−0.0063, 0.0014)
	50		0.5146 (−0.0015, 0.0012)		0.4172 (−0.0024, 0.0008)		0.3499 (−0.0030, 0.0006)
	100		0.5147 (−0.0014, 0.0006)		0.4182 (−0.0014, 0.0004)		0.3514 (−0.0015, 0.0003)
	200		0.5158 (−0.0003, 0.0003)		0.4189 (−0.0007, 0.0002)		0.3522 (−0.0007, 0.0002)
$\rho = 0.5$ $\varphi = 0.5$	25	0.3552	0.3479 (−0.0073, 0.0016)	0.2877	0.2798 (−0.0079, 0.0012)	0.2414	0.2315 (−0.0099, 0.0011)
	50		0.3516 (−0.0036, 0.0008)		0.2836 (−0.0041, 0.0006)		0.2362 (−0.0052, 0.0005)
	100		0.3536 (−0.0016, 0.0004)		0.2855 (−0.0022, 0.0003)		0.2390 (−0.0024, 0.0002)
	200		0.3543 (−0.0009, 0.0002)		0.2868 (−0.0009, 0.0001)		0.2401 (−0.0013, 0.0001)
$\rho = 0.7$ $\varphi = 0.7$	25	0.1128	0.0946 (−0.0182, 0.0018)	0.0908	0.0685 (−0.0223, 0.0021)	0.0759	0.0501 (−0.0258, 0.0024)
	50		0.1033 (−0.0095, 0.0008)		0.0792 (−0.0116, 0.0009)		0.0627 (−0.0132, 0.0009)
	100		0.1076 (−0.0052, 0.0004)		0.0850 (−0.0058, 0.0004)		0.0691 (−0.0068, 0.0004)
	200		0.1103 (−0.0025, 0.0002)		0.0879 (−0.0029, 0.0002)		0.0724 (−0.0035, 0.0002)
$\rho = 0$ $\varphi = 0$	25	1.3737	1.3734 (−0.0003, 0.0109)	1.1409	1.1430 (0.0021, 0.0084)	0.9719	0.9690 (−0.0029, 0.0063)
	50		1.3734 (−0.0003, 0.0054)		1.1411 (0.0002, 0.0040)		0.9709 (−0.0010, 0.0031)
	100		1.3745 (−0.0008, 0.0026)		1.1405 (−0.0004, 0.0020)		0.9713 (−0.0006, 0.0015)
	200		1.3733 (−0.0004, 0.0013)		1.1409 (0.0000, 0.0010)		0.9714 (−0.0005, 0.0008)
$\rho = 0.1$ $\varphi = 0.1$	25	1.1675	1.1659 (−0.0020, 0.0088)	0.9559	0.9544 (−0.0015, 0.0065)	0.8072	0.8052 (−0.0020, 0.0047)
	50		1.1674 (−0.0005, 0.0043)		0.9562 (0.0003 0.0031)		0.8060 (−0.0012, 0.0022)
	100		1.1685 (0.0006, 0.0022)		0.9550 (−0.0009, 0.0015)		0.8065 (−0.0007, 0.0011)
	200		1.1677 (−0.0002, 0.0011)		0.9559 (0.0000, 0.0008)		0.8067 (−0.0005, 0.0006)
$\rho = 0.25$ $\varphi = 0.25$	25	0.8219	0.8195 (−0.0024, 0.0049)	0.6688	0.6653 (−0.0035, 0.0034)	0.5627	0.5595 (−0.0032, 0.0025)
	50		0.8209 (−0.0010, 0.0024)		0.6668 (−0.0020, 0.0017)		0.5608 (−0.0019, 0.0012)
	100		0.8212 (−0.0007 0.0012)		0.6679 (−0.0009, 0.0008)		0.5614 (−0.0013, 0.0006)
	200		0.8214 (−0.0005, 0.0006)		0.6683 (−0.0005, 0.0004)		0.5621 (−0.0006, 0.0003)

Table 8. Cont.

Cases u_{ij}		$u_{ij} \sim N(0,(0.8)^2)$		$u_{ij} \sim N(0,(1.0)^2)$		$u_{ij} \sim N(0,(1.2)^2)$	
ρ, φ	m	C_p''' (Profile)	$\hat{C}_p'''$ (Profile) (Bias, MSE)	C_p''' (Profile)	$\hat{C}_p'''$ (Profile) (Bias, MSE)	C_p''' (Profile)	$\hat{C}_p'''$ (Profile) (Bias, MSE)
$\rho = 0.4$ $\varphi = 0.4$	25	0.5261	0.5221 (−0.0040, 0.0024)	0.4263	0.4210 (−0.0053, 0.0017)	0.3578	0.3510 (−0.0068, 0.0013)
	50		0.5241 (−0.0020, 0.0012)		0.4238 (−0.0025, 0.0008)		0.3546 (−0.0032, 0.0006)
	100		0.5250 (−0.0011, 0.0006)		0.4247 (−0.0016, 0.0004)		0.3562 (−0.0016, 0.0003)
	200		0.5256 (−0.0005, 0.0003)		0.4256 (−0.0007, 0.0002)		0.3572 (−0.0006, 0.0001)
$\rho = 0.5$ $\varphi = 0.5$	25	0.3618	0.3549 (−0.0069, 0.0015)	0.2924	0.2843 (−0.0081, 0.0012)	0.2450	0.2351 (−0.0099, 0.0010)
	50		0.3584 (−0.0034, 0.0007)		0.2883 (−0.0041, 0.0006)		0.2400 (−0.0050, 0.0004)
	100		0.3597 (−0.0021, 0.0004)		0.2902 (−0.0022, 0.0003)		0.2426 (−0.0024, 0.0002)
	200		0.3608 (−0.0010, 0.0002)		0.2913 (−0.0011, 0.0001)		0.2439 (−0.0011, 0.0001)
$\rho = 0.7$ $\varphi = 0.7$	25	0.1160	0.0973 (−0.0187, 0.0018)	0.0933	0.0711 (−0.0222, 0.0020)	0.0779	0.0510 (−0.0269, 0.0024)
	50		0.1067 (−0.0093, 0.0007)		0.0823 (−0.0110, 0.0008)		0.0646 (−0.0133, 0.0009)
	100		0.1117 (−0.0043, 0.0003)		0.0872 (−0.0061, 0.0003)		0.0710 (−0.0069, 0.0004)
	200		0.1138 (−0.0022, 0.0002)		0.0905 (−0.0028, 0.0001)		0.0745 (−0.0034, 0.0002)

From Table 8, it is concluded that the index C_p''' (Profile) yields small bias and MSE values. The findings of the study reveal that the proposed method is quite efficient in evaluating the capability of SLP with BAC effects, and the performance of C_p''' (Profile) keeps increasing with an increase in the number of profile samples as well as the correlation coefficients ρ and φ values.

4.4. Findings of the Simulation Studies

In this section, we discuss how to select a suitable PCI to conduct capability analysis considering the four possible SLPs in practical applications using bias and MSE criteria.

- When there is no within and between profiles autocorrelation ($\rho = \varphi = 0$), the index C_p''' (Profile) that has superior performance than its competitors (see Tables 2–7, when $\rho = \varphi = 0$) is suggested.
- When there is a BPAC effect (($\rho = 0$, $\varphi > 0$)), one can use the transformations in Section 3.1 and Section 3.3 as two options for recommending PCI. To select the superior transformation, a comparison between the proposed indices in Section 3.1 and Section 3.3 is provided. We conducted the same simulation studies in Section 4.1 ($\rho = 0$, $\varphi = 0.1, 0.25, 0.4, 0.5, 0.7$) and calculated the proposed index in Section 3.3. Table 9 displays the outcomes of the simulations. Based on the simulation results in the last columns of Tables 2–4 and 9, as well as Figure 7, as φ increases, the difference between the two methods becomes more apparent, and the proposed index in Section 3.3 outperforms the proposed index in Section 3.1, especially for $\varphi \geq 0.25$ with regard to bias and MSE criteria. Additionally, Figure 7 clearly shows these findings based on MSE criteria. Similarly, it can be observed that the suggested index in

Section 3.3 generally has better bias and MSE than the index $S_{pkA;AR(1)}$ in Section 2.3.1 by comparing the first column of Tables 2–4 with the findings of Table 9. Thus, when there is a BPAC effect, we suggest using the proposed index $C_p'''(Profile)$ discussed in Section 3.3.

- When there is a WPAC effect ($\rho > 0$, $\varphi = 0$), the proposed index in Section 3.2 can be used because the transformation method in Equation (22) reduces to Equation (20), and according to simulation results in Section 4.2, the index $C_p'''(Profile)$ outperforms its competitors (see Tables 5–7 and Figures 4–6).
- When there is a BAC effect ($\rho > 0$, $\varphi > 0$), the proposed index $C_p'''(Profile)$ in Section 3.3 is recommended.

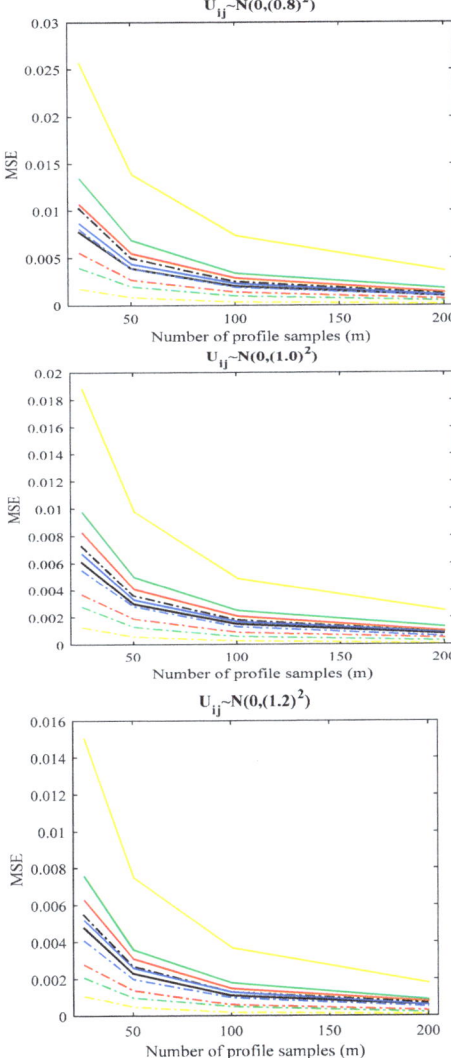

Figure 7. MSE of the proposed index in Section 3 (solid lines) and Section 3 (dotted lines) under different profile sample sizes m and $\varphi = 0.1$ (black lines), $\varphi = 0.25$ (blue lines), $\varphi = 0.4$ (red lines), $\varphi = 0.5$ (green lines), and $\varphi = 0.7$ (yellow lines) when there is a BPAC effect.

Table 9. The simulation results of $C_p'''(Profile)$ for SLP with BAC when there is a BPAC effect.

Cases u_{ij}			$u_{ij} \sim N(0,(1.2)^2)$		$u_{ij} \sim N(0,(1.0)^2)$		$u_{ij} \sim N(0,(0.8)^2)$	
m	ρ,φ	$C_p'''(Profile)$	$\hat{C}_p'''(Profile)$ (Bias, MSE)	$C_p'''(Profile)$	$\hat{C}_p'''(Profile)$ (Bias, MSE)	$C_p'''(Profile)$	$\hat{C}_p'''(Profile)$ (Bias, MSE)	
		25		1.2682 (0.0015, 0.0103)		1.0433 (−0.0010, 0.0073)		0.8835 (−0.0021, 0.0055)
$\rho=0$ $\varphi=0.1$		50	1.2667	1.2655 (−0.0012, 0.0050)	1.0443	1.0450 (−0.0007, 0.0036)	0.8856	0.8850 (−0.0006, 0.0027)
		100		1.2666 (−0.0001, 0.0025)		1.0439 (−0.0004, 0.0018)		0.8851 (−0.0005, 0.0013)
		200		1.2666 (−0.0001, 0.0012)		1.0442 (−0.0001, 0.0009)		0.8853 (−0.0003, 0.0007)
		25		1.0809 (−0.0018, 0.0081)		0.8829 (−0.0027, 0.0055)		0.7451 (−0.0024, 0.0041)
$\rho=0$ $\varphi=0.25$		50	1.0827	1.0814 (0.0013, 0.0039)	0.8856	0.8845 (−0.0011, 0.0028)	0.7475	0.7458 (−0.0017, 0.0020)
		100		1.0816 (−0.0011, 0.0019)		0.8847 (−0.0009, 0.0013)		0.7472 (−0.0003, 0.0010)
		200		1.0827 (0.0000, 0.0010)		0.8852 (−0.0004, 0.0006)		0.7472 (−0.0003, 0.0005)
		25		0.8808 (−0.0048, 0.0056)		0.7140 (−0.0052, 0.0037)		0.5985 (−0.0060, 0.0028)
$\rho=0$ $\varphi=0.4$		50	0.8856	0.8829 (−0.0027, 0.0027)	0.7192	0.7170 (−0.0022, 0.0019)	0.6045	0.6012 (−0.0033, 0.0014)
		100		0.8844 (−0.0012, 0.0014)		0.7178 (−0.0014, 0.0009)		0.6033 (−0.0012, 0.0006)
		200		0.8850 (−0.0006, 0.0007)		0.7184 (−0.0008, 0.0005)		0.6036 (−0.0009, 0.0003)
		25		0.7413 (−0.0062, 0.0040)		0.5973 (−0.0072, 0.0028)		0.5007 (−0.0061, 0.0021)
$\rho=0$ $\varphi=0.5$		50	0.7475	0.7446 (−0.0029, 0.0020)	0.6045	0.6011 (−0.0034, 0.0013)	0.5068	0.5032 (−0.0036, 0.0010)
		100		0.7455 (−0.0020, 0.0010)		0.6028 (−0.0017, 0.0006)		0.5047 (−0.0021, 0.0005)
		200		0.7465 (−0.0010, 0.0005)		0.6034 (−0.0011, 0.0003)		0.5058 (−0.0010, 0.0002)
		25		0.4447 (−0.0126, 0.0018)		0.3549 (−0.0124, 0.0013)		0.2936 (−0.0132, 0.0011)
$\rho=0$ $\varphi=0.7$		50	0.4573	0.4509 (−0.0064, 0.0009)	0.3673	0.3605 (−0.0068, 0.0006)	0.3068	0.3001 (−0.0067, 0.0005)
		100		0.4539 (−0.0034, 0.0004)		0.3640 (−0.0033, 0.0003)		0.3035 (−0.0033, 0.0002)
		200		0.4555 (−0.0018, 0.0002)		0.3657 (−0.0016, 0.0001)		0.3052 (−0.0016, 0.0001)

5. Bootstrap Confidence Intervals for $C_p'''(Profile)$ in the Presence of BAC

To enhance the reliability of the proposed index in Section 3.3, the nonparametric bootstrap method of Efron [44] is employed to determine confidence intervals for the proposed index as its underlying distribution is unknown. Both standard bootstrap (SB) and percentile bootstrap (PB) methods [45] will be used and evaluated in terms of relative coverage through a simulation study.

Let m profile samples from an in-control SLP process in the form (X_i, Y_{ij}); $i = 1, 2, \ldots, n$ and $j = 1, 2, \ldots, m$ with correlation coefficients $\rho \geq 0$, $\varphi \geq 0$ be collected. A bootstrap sample is denoted by $\left(X_i^*, Y_{ij}^*\right)$; $i = 1, 2, \ldots, n$ and $j = 1, 2, \ldots, m$ is a sample extracted by substitution using the original sample. Suppose the resampling process is repeated B times. We use the B samples to obtain B bootstrap estimates of the index $C_p'''(Profile)$ based on the proposed method in Section 3.3. The B bootstrap estimates of $C_p'''(Profile)$ are denoted by $\hat{C}_p'''(Profile)_k^*, k = 1, 2, \ldots, B$. These estimates are ordered from the smallest to the largest, denoted by $\hat{C}_p'''(Profile)_{(1)}^*, \hat{C}_p'''(Profile)_{(2)}^*, \ldots, \hat{C}_p'''(Profile)_{(B)}^*$. Subsequent sections detail the construction of two types of bootstrap confidence intervals.

5.1. Standard Bootstrap (SB) Confidence Interval

Let $\overline{C_p'''(Profile)}^*$ and S^* be the mean and standard deviation of $\hat{C}_p'''(Profile)_k^*$, $k = 1, 2, \ldots, B$, given by

$$\overline{C_p'''(Profile)}^* \frac{\sum_{k=1}^B \hat{C}_p'''(Profile)_k^*}{B}, \tag{24}$$

and

$$S^* = \sqrt{\frac{\sum_{k=1}^B \left(\hat{C}_p'''(Profile)_k^* - \overline{C_p'''(Profile)}^*\right)^2}{B-1}}. \tag{25}$$

Thus, $100(1-\alpha)\%$ SB confidence interval for $C_p'''(Profile)$ can be obtained by

$$\left(\overline{C_p'''(Profile)}^* - Z_{(1-\frac{\alpha}{2})}S^*, \overline{C_p'''(Profile)}^* + Z_{(1-\frac{\alpha}{2})}S^*\right). \tag{26}$$

where $Z_{(1-\frac{\alpha}{2})}$ is the $100(1-\frac{\alpha}{2})\%$ of the standard normal distribution.

5.2. Percentile Bootstrap (PB) Confidence Interval

According to ordered bootstrap estimates, $\hat{C}_p'''(Profile)_{(k)}^*$, $k = 1, 2, \ldots, B$, the $100(1-\alpha)\%$ PB confidence interval for $C_p'''(Profile)$ can be presented by

$$\left(\hat{C}_p'''(Profile)_{(B.(\frac{\alpha}{2}))}^*, \hat{C}_p'''(Profile)_{(B.(1-\frac{\alpha}{2}))}^*\right). \tag{27}$$

5.3. Simulation Study: Confidence Interval Evaluation

In this subsection, a series of simulations are carried out to assess the performance of the proposed confidence intervals for $C_p'''(Profile)$ in analyzing the capability of SLP with BAC. Using the same dataset as in Section 4.3, 10,000 simulation runs were performed for various error term and correlation coefficient combinations (see the first column of Table 10). For each simulation, 1000 bootstrap samples ($B = 1000$) were generated based on a different number of profile samples $m \in \{25, 50, 100, 200\}$, to construct 95% confidence intervals using both bootstrap methods. The performance of confidence intervals is evaluated by the relative coverage metric, which is the ratio of coverage percentage to the average length of the confidence interval. In Table 10, the average of estimated $C_p'''(Profile)$ over 10,000 runs as well as the average confidence interval ($\overline{CI}$) and relative coverage of $C_p'''(Profile)$ are reported. For better clarification, Pseudocode 2 represents the procedure for computing confidence intervals for $C_p'''(Profile)$ in the presence of BAC. The green lines are some comments about the codes.

Pseudocode 2. The procedure for computing the bootstrap confidence interval for $\hat{C}_p'''(Profile)$, $\overline{CI}$ and relative coverage using Monte Carlo simulation

Consider the in-control profile model, SLs' function, sigma = 1, n, m, ρ, and φ.
errorpre = normrnd (0,sigma, [1,n]); % Normal preerror with size n
while (we do not reach the number of generated samples)
    aij = normrnd (0,sigma,[1,n]);% Normal variable with size n
    % Generate data
    for i = 1:n
      if (i == 1)
        epsilon(i) = φ × errorpre(i) + aij(i);
      else
        epsilon(i) = ρ × epsilon (i−1) + φ × errorpre(i) − φ × ρ × errorpre (i−1) + aij(i);
      endif
    endfor
    $Y = A_0 + A_1 X +$ epsilon; % Equation (1)

    errorpre = epsilon; % Update the previous error
endwhile
% It is necessary to transfer the SLs as we apply the transformation of the in-control profile model.% The details for computations, were given in Section 3.3.
Apply the transformation on the SLs.
Compute $C_p'''(Profile)$ based on Equation (10) and the in-control parameters.
for rep = 1:1:10,000
    Generate m profiles by the in-control model.
    do a bootstrap loop 1000 times
      Generate a bootstrap with resampling profiles based on the m generated profile.
      % The parameter estimation is performed based on the computations given in Section 3.3.
      $(a_0, a_1, \hat{\sigma}^2)$ = Estimate the profile parameters, including intercept, slope, and standard deviation, based on the m profiles.
      Compute $\hat{\mu}_Y(X) = a_0 + a_1 X$ and $\hat{\sigma}^2$ based on the proposed method for SLP with BAC.
      $\hat{C}_p'''(Profile)_k^*$ = Estimate $\hat{C}_p'''(Profile)_k^*$ based on the Equation (10) by $\hat{\mu}_Y$ and $\hat{\sigma}^2$, $k = 1, 2, \ldots, B$.
      Store the values $\hat{C}_p'''(Profile)_k^*$.
      Sort $\hat{C}_p'''(Profile)_k^*$, $k = 1, 2, \ldots, B$ from the smallest to largest.
    endloop
    $\hat{C}_p'''(Profile)_{rep} = \frac{\sum_{k=1}^{1000} \hat{C}_p'''(Profile)_k^*}{1000}$.
    Compute the SB and PB confidence intervals.
    Record the length of intervals.
endfor
Compute the average of 10,000 $\hat{C}_p'''(Profile)_{rep}$ values.
Compute the average of confidence intervals ($\overline{CI}$).
Compute the average interval length, coverage percentage, and relative coverage for the SB and PB approaches.

Table 10. The simulation results of 95% bootstrap confidence intervals of C_p''' (Profile) for SLP with BAC.

Simulated Case	m	True Value C_p''' (Profile)	Estimated C_p''' (Profile)	Bootstrap Methods			
				SB		PB	
				CI	Relative Coverage	CI	Relative Coverage
$Y_{ij} = 3 + 2 X_i + \varepsilon_{ij}$ $i = 4,\ \varepsilon_{ij} \sim N\left(0, (1.0)^2\right)$ $\rho = \varphi = 0$	25	1.1254	1.126	(0.9559, 1.2961)	2.7597	(0.9676, 1.3074)	2.7697
	50		1.125	(1.0028, 1.2472)	3.8175	(1.0085, 1.2527)	3.841
	100		1.1265	(1.0404, 1.2125)	5.4931	(1.0430, 1.2148)	5.4885
	200		1.125	(1.0641, 1.1860)	7.7768	(1.0653, 1.1871)	7.7838

Table 10. Cont.

Simulated Case	m	True Value C_p''' (Profile)	Estimated C_p''' (Profile)	Bootstrap Methods			
				SB		PB	
				CI	Relative Coverage	CI	Relative Coverage
$Y_{ij} = 3 + 2X_i + \varepsilon_{ij}$ $i = 4,\ \varepsilon_{ij} \sim N(0, (1.0)^2)$ $\rho = \varphi = 0.1$	25	0.9383	0.9363	(0.7865, 1.0862)	3.1168	(0.7968, 1.0960)	3.1081
	50		0.9362	(0.8294, 1.0429)	4.4216	(0.8343, 1.0476)	4.4357
	100		0.9384	(0.8629, 1.0140)	6.2252	(0.8650, 1.0141)	6.2287
	200		0.9378	(0.8843, 0.9913)	8.8712	(0.8855, 0.9924)	8.9038
$Y_{ij} = 3 + 2X_i + \varepsilon_{ij}$ $i = 4,\ \varepsilon_{ij} \sim N(0, (1.0)^2)$ $\rho = \varphi = 0.25$	25	0.6574	0.6526	(0.5409, 0.7642)	4.1776	(0.5471, 0.7706)	4.1787
	50		0.6552	(0.5757, 0.7346)	5.9732	(0.5787, 0.7375)	5.9371
	100		0.6576	(0.6012, 0.7141)	8.4542	(0.6025, 0.7154)	8.3983
	200		0.6563	(0.6168, 0.6959)	11.958	(0.6174, 0.6966)	11.9074
$Y_{ij} = 3 + 2X_i + \varepsilon_{ij}$ $i = 4,\ \varepsilon_{ij} \sim N(0, (1.0)^2)$ $\rho = \varphi = 0.4$	25	0.4196	0.4098	(0.3282, 0.4913)	5.8561	(0.3291, 0.4930)	5.7814
	50		0.4154	(0.3582, 0.4725)	8.2756	(0.3586, 0.4732)	8.2351
	100		0.4176	(0.3776, 0.4577)	11.78	(0.3775, 0.4578)	11.6992
	200		0.4185	(0.3902, 0.4467)	16.7696	(0.3902, 0.4467)	16.7651
$Y_{ij} = 3 + 2X_i + \varepsilon_{ij}$ $i = 4,\ \varepsilon_{ij} \sim N(0, (1.0)^2)$ $\rho = \varphi = 0.5$	25	0.2877	0.2736	(0.2050, 0.3423)	6.947	(0.2011, 0.3396)	6.8325
	50		0.2798	(0.2319, 0.3277)	10.0152	(0.2296, 0.3258)	9.8405
	100		0.2846	(0.2514, 0.3177)	14.4352	(0.2501, 0.3166)	14.256
	200		0.2858	(0.2627, 0.3089)	20.6014	(0.2620, 0.3082)	20.4173
$Y_{ij} = 3 + 2X_i + \varepsilon_{ij}$ $i = 4,\ \varepsilon_{ij} \sim N(0, (0.8)^2)$ $\rho = \varphi = 0$	25	1.3493	1.3525	(1.1582, 1.5468)	2.4161	(1.1708, 1.5586)	2.4213
	50		1.3496	(1.2099, 1.4892)	3.3189	(1.2160, 1.4951)	3.3462
	100		1.3517	(1.2527, 1.4507)	4.8505	(1.2555, 1.4532)	4.8264
	200		1.3492	(1.2792, 1.4191)	6.7634	(1.2806, 1.4203)	6.7917
$Y_{ij} = 3 + 2X_i + \varepsilon_{ij}$ $i = 4,\ \varepsilon_{ij} \sim N(0, (0.8)^2)$ $\rho = \varphi = 0.1$	25	1.1387	1.1528	(0.9737, 1.3319)	2.5883	(0.9859, 1.3433)	2.5934
	50		1.1383	(1.0120, 1.2646)	3.7252	(1.0177, 1.2701)	3.7207
	100		1.1396	(1.0504, 1.2288)	5.2973	(1.0528, 1.2311)	5.2956
	200		1.1385	(1.0753, 1.2017)	7.5008	(1.0766, 1.2029)	7.5217

224

Table 10. Cont.

Simulated Case	m	True Value $C_p'''(Profile)$	Estimated $C_p'''(Profile)$	Bootstrap Methods			
				SB		PB	
				CI	Relative Coverage	CI	Relative Coverage
$Y_{ij} = 3 + 2X_i + \varepsilon_{ij}$ $i = 4, \varepsilon_{ij} \sim N(0, (1.2)^2)$ $\rho = \varphi = 0$	25	0.9614	0.9659	(0.8134, 1.1183)	**3.0412**	(0.8243, 1.1285)	3.0205
	50		0.9635	(0.8559, 1.0711)	**4.3634**	(0.8612, 1.0762)	4.3403
	100		0.9631	(0.8871, 1.0392)	**6.2731**	(0.8895, 1.0414)	6.2591
	200		0.9608	(0.9073, 1.0143)	**8.8543**	(0.9085, 1.0154)	8.8443
$Y_{ij} = 3 + 2X_i + \varepsilon_{ij}$ $i = 4, \varepsilon_{ij} \sim N(0, (1.2)^2)$ $\rho = \varphi = 0.1$	25	0.7955	0.7927	(0.6620, 0.9235)	3.5919	(0.6711, 0.9321)	**3.5941**
	50		0.7946	(0.7016, 0.8876)	5.0983	(0.7060, 0.8918)	**5.1159**
	100		0.7951	(0.7299, 0.8603)	**7.2434**	(0.7318, 0.8622)	7.207
	200		0.7948	(0.7487, 0.8409)	10.2791	(0.7497, 0.8419)	**10.3151**

Note: Bold values indicate the best-performing bootstrap method in terms of relative coverage for each simulated case.

Table 10 presents the relative coverage values for the confidence intervals of the index $C_p'''(Profile)$ for SLP with BAC based on the SB and PB methods under various simulation cases. The results consistently demonstrate that the SB method outperforms the PB method in terms of related coverage in almost all cases, indicating higher coverage percentages with shorter interval lengths. The higher values of relative coverage are highlighted in Table 10. Furthermore, a larger number of profile samples m and higher correlation coefficients improve relative coverage for both methods. While the SB method generally exhibits superior performance, it is essential to note that the relative performance of the two methods might vary under different conditions. Future research could explore the sensitivity of the SB and PB methods to different autocorrelation structures, sample sizes, and process characteristics.

6. Illustrative Example

To bridge the gap between simulation and real-world applications, a general framework in the following steps to a proper implantation in real-world scenarios in different fields, such as manufacturing, metrology, and environmental monitoring, is provided.

- **Data Collection:** Real-world data should be collected in accordance with appropriate sampling plans to ensure representativeness and statistical power.
- **Data Preprocessing:** Data cleaning and preprocessing steps are essential to address issues such as missing values, outliers, and trends.
- **Parameter Estimation:** The parameters of the autocorrelation models can be estimated using standard statistical methods, such as maximum likelihood or least squares.
- **Phase I Profile Monitoring:** The process stability should be assessed and the in-control profile parameters estimated.
- **Process Capability Assessment:** The proposed process capability index can be calculated based on the estimated parameters and the observed data.
- **Interpretation:** The results should be interpreted in the context of the specific application and used to inform decision making.

To streamline the illustrative example, this paper concentrates solely on the final steps of the proposed methodology. Consequently, the processes of data collection, preprocessing, and parameter estimation are excluded from the current analysis. Instead, the primary focus is on the evaluation of process capability through the proposed index and the subsequent interpretation of results within the specific application context. Hence, we considered a calibration system used in the chemical industry to illustrate the use of the proposed methodology and verify its applicability in real-life situations. The chemical experimental instruments usually consist of gas sensors to detect toxic and unnatural processes. They are usually kept away from the laboratory to ensure accurate measurements and prevent potential safety hazards. When moving these devices to a new sensing environment, some challenges arise in the new conditions, such as (i) the fact that the calibration parameters may need to be adjusted or recalibrated to ensure accurate readings. (ii) Gas sensors can exhibit cross-sensitivity, meaning they may respond not only to the target gas but also to other gases present in the environment. (iii) Different sensing environments may present varying temperatures, humidity levels, and atmospheric pressures. These conditions can impact the performance and reliability of gas sensors. (iv) Moving gas sensing devices to a new environment involves proper physical installation. Factors such as location, mounting, and accessibility need to be considered to ensure representative and effective gas detection. To handle this situation, a profile model can be applied to address calibration issues and ascertain their proper performance over time. The calibration of metal oxide (MOX) as a gas sensor has been discussed and monitored in other related studies [18,46,47]. According to these studies, a functional relationship is considered between the resistance (R) of the sensor as the response variable and the concentrations of carbon monoxide as the explanatory variable. Four fixed levels of carbon monoxide (CO) concentration as 25, 100, 125, and 150 ppm are considered. Based on 3287 in-control SLPs, the calculated reference profile is $Y_{ij} = 71.741 + 0.0176 X_i + \varepsilon_{ij}$, where $\varepsilon_{ij} \sim N(0, 0.142)$. Nadi et al. [18] investigated the effect of adding additive material to increase the efficiency of the process. Although it can be useful for this purpose, some changes in the reference relationship may occur in such a way that the new formula is obtained between the resistance and carbon concentration, as given by

$$Y_{ij} = 71.741 + 0.0176 X_i + \varepsilon_{ij},$$
$$\varepsilon_{ij} = 0.289 \varepsilon_{(i-1)j} + a_{ij},$$
$$a_{ij} = 0.565 \varepsilon_{i(j-1)} + u_{ij}, \qquad (28)$$
$$u_{ij} \sim N(0, 0.0742),$$
$$i = 1, 2, 3, 4. j = 1, 2, \ldots.$$

To illustrate how to measure the calibration of MOX capability, we generate datasets of 200 profiles based on the in-control model in Equation (28). Since the SLs for this process do not exist, we set the SLs for the resistance (R) of the MOX at different concentrations of carbon monoxide based on the natural tolerance limits of the process. Therefore, the functional SLs and target value are calculated as $LSL_Y(X) = 70.9238 + 0.0176X$, $USL_Y(X) = 72.5582 + 0.0176X$, and $T_Y(X) = 71.7410 + 0.0176X$, respectively, where $X \in [25, 150]$. The methods proposed in Section 3 can be used to assess process capability for the calibration process. According to Equation (22), $LSL_{Y'}(X') = 21.9357 + 0.0077X'$, $USL_{Y'}(X') = 22.4412 + 0.0077X'$, and $T_{Y'}(X') = 22.1844 + 0.0077X'$, and $\mu_{Y'}(X') = 22.2558 + 0.0070X'$ are obtained, where $X' \in [92.7750, 113.8750]$, and $\hat{\sigma}^2 = 0.0774$. These transformed functional limits for the calibration of MOX data are depicted in Figure 8.

The estimated value for the index $C_p'''(Profile)$ is 0.3001, with the following 95% confidence intervals:

$$SB : (0.2867, 0.3265), \ PB : (0.2872, 0.3270) \qquad (29)$$

According to Figure 8, it is shown that although the process mean almost coincides with the target line, the process variation is undesirable. Since the value of the capability index is less than 1 with 95% confidence intervals, we conclude that the calibration process is incapable. Therefore, corrective actions to reduce process variance are required.

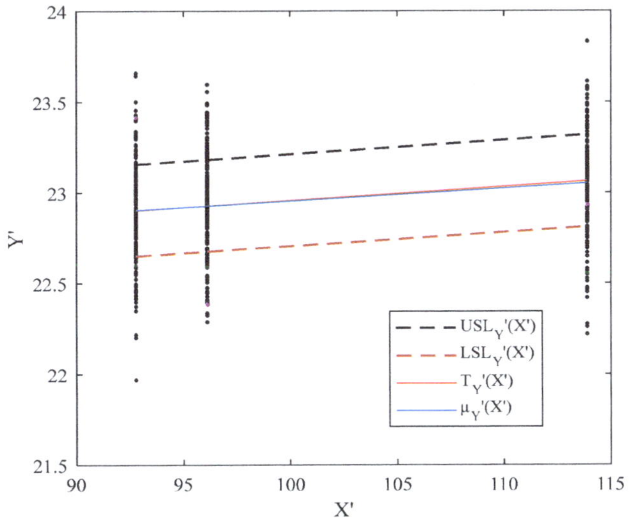

Figure 8. The transformed functional limits of the calibration of metal oxide (MOX) data.

7. Conclusions and Future Works

In this article, we studied the capability analysis of SLP with different possible autocorrelation effects in real applications. The SLP with a general error model, which is adaptable enough to model four possibilities for the SLP in practical applications, was taken into consideration. The new functional capability index $C_p'''(Profile)$ for the SLP in the absence of autocorrelation effects was introduced. In order to evaluate the capability of the SLP with BPAC, WPAC, and BAC, where autocorrelation structures follow an AR(1) model, three distinct approaches were presented. The basic idea was first applying the transformation to eliminate BPAC, WPAC, and/or BAC, and then estimating PCIs for SLP using independent statistics. The simulation studies in the presence of each of the two autocorrelation effects (BPAC and WPAC) were conducted to investigate and compare the performance of the proposed PCIs with those of the existing ones proposed by Wang and Tamirat [34] and Wang and Tamirat [35] in terms of bias and MSE criteria. Moreover, we compared these PCIs with another existing PCI for SLP (index C_{ppM}''') to draw conclusions about the proposed method. In brief, the results of the simulation study showed that:

- In the case of BPAC, the new index $C_p'''(Profile)$ had a bias that was noticeably lower than the bias of the indices $S_{pkA;AR(1)}$ and C_{ppM}''' for all the values of correlation coefficients φ, and its MSE was also lower than the MSE of $S_{pkA;AR(1)}$ especially for $\varphi < 0.7$, and approximately equal to that of C_{ppM}'''.
- In the case of WPAC, the new index $C_p'''(Profile)$ outperformed the index $S_{pkA;AR(1)}$ due to its smaller bias and MSE in all simulated cases. In addition, the index $C_p'''(Profile)$ performed better than C_{ppM}''' due to its close MSE to the index C_{ppM}''' and extremely small bias.
- In two cases, BPAC and WPAC, all PCIs performed well with a larger sample size.

The results of the simulation study showed that the recommended index $C_p'''(Profile)$ had superior performance over its competitors. Additionally, the effectiveness of the suggested index for SLP with BAC was examined in various simulation instances; bootstrap confidence intervals were obtained using the two methods SB and PB, and their performance was assessed using simulation studies. The results indicated that the new index can accurately reflect the actual performance of SLP with BAC. Additionally, when only the BPAC effect was present, the proposed index for SLP with BAC performed better than the proposed index for SLP with the BPAC effect, particularly for $\varphi \geq 0.25$. Additionally, the SB method bootstrap confidence intervals performed better than the PB method for almost all simulated cases in terms of the relative coverage metric. Finally, an illustrative example in a calibration system from the chemical industry demonstrated the usefulness of the proposed PCI for SLP with BAC in practice.

Future studies may include analyzing the proposed methods in the case of random explanatory variable X or when the assumption of normality of error terms in the SLP model is violated. In addition, introducing robust PCIs for SLP as well as other types of profiles like multivariate profiles, multiple linear profiles, and logistic regression profiles in the presence of both within- and between-profile autocorrelations can be considered as future works. Analyzing the proposed methods in the case of within- and between-profile autocorrelation with other autocorrelation patterns, such as autoregressive moving average (ARMA) and vector ARMA (VARMA), also requires further study.

Author Contributions: All authors contributed to the study's conception and design. All authors have read and agreed to the published version of the manuscript.

Funding: The authors declare that no funds, grants, or other support were received during the preparation of this manuscript.

Data Availability Statement: The source code and data used to produce the results and analyses presented in this manuscript are available and will be sent, if peer reviewers will be asked to assess.

Conflicts of Interest: The authors declare that they have no conflicts of interest.

Appendix A

Calculating the true values of $S_{pkA;AR(1)}$, C'''_{ppM}, and $C'''_p(Profile)$ for SLP with BPAC when $u_{ij} \sim N\left(0, (0.8)^2\right)$, $\varphi = 0.1$, $X_i \in \{2, 4, 6, 8\}$, $LSL_i \in \{2.5, 6.85, 11.25, 16.25\}$, $USL_i \in \{10, 14.35, 18.75, 23.75\}$, $T_i \in \{6.25, 10.60, 15, 20\}$.

- **Calculating the true values of $S_{pkA;AR(1)}$**: By considering LSL_i, USL_i, T_i, $\mu_i = 3 + 2X_i$, and $\sigma^2 = \frac{\sigma^2}{1-(\varphi)^2} = \frac{(0.8)^2}{1-(0.1)^2} = 0.6465$, Equation (6) is used to first determine the values of S_{pk_i} for 4 levels of the explanatory variable, which are calculated as 1.3008, 1.4401, 1.5547, and 1.2015, respectively. The values of P_i are then calculated using Equation (7) and are equal to 0.9999, 1, 1, and 0.9997. Therefore, $P = \frac{\sum_{i=1}^{4} P_i}{4}$ value is calculated, which is equal to 0.9999. Using Equation (8), the index $S_{pkA;AR(1)}$ is finally calculated to be 1.2916.

- **Calculating the true value of C'''_{ppM}**: By considering LSL_i, USL_i, T_i, $\mu_i = 3 + 2X_i$, and $\sigma^2 = (0.8)^2$, we must first calculate the values of C'''_{ppi} for each of the 4 levels of the explanatory variable using Equation (4). These values are 1.0943, 1.3816, 1.5625, and 0.9067, respectively. Finally, the $C'''_{ppM} = \frac{\sum_{i=1}^{4} C'''_{ppi}}{4}$ is calculated to be 1.2363.

- **Calculating the true value of $C'''_p(Profile)$**: By considering LSL_i, USL_i, T_i, $\mu_Y(X) = 3 + 2X$, and $\sigma^2 = (0.8)^2$, firstly, the regression lines are fitted to the values of LSL_i, USL_i, and T_i, to yield the functional SLs and target line as $LSL_Y(X) = -2.2 + 2.2825X$, $USL_Y(X) = 5.3 + 2.2825X$, and $T_Y(X) = 1.55 + 2.2825X$, where $X \in [2, 8]$. Then, to calculate $C'''_p(Profile)$, the location of $\mu_Y(X)$ relative to $T_Y(X)$ is determined based on the method proposed in Pakzad and Basiri [39]. Figure A1 shows that $\mu_Y(X)$ and $T_Y(X)$ cross at $x_m = 5.1327$ and $C'''_p(Profile)$ is calculated as 1.3160 using Equation (16).

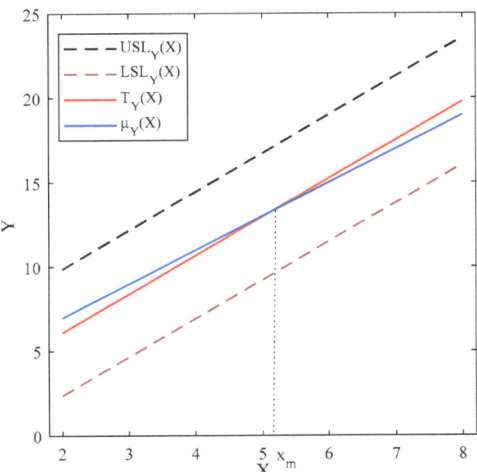

Figure A1. The location of $\mu_Y(X)$ relative to $T_Y(X)$ for SLP with BPAC.

Appendix B

Calculating the true values of $S_{pkA;AR(1)}$, C'''_{ppM}, and $C'''_p(Profile)$ for SLP with WPAC when $u_{ij} \sim N\left(0, (0.8)^2\right)$, $\rho = 0.1$, $X_i \in \{1, 2, \ldots, 10\}$, LSL_i, USL_i, T_i where results are shown in Table 1.

- **Calculating the true values of $S_{pkA;AR(1)}$:** By considering LSL_i, USL_i, T_i, $\mu_i = 3 + 2X_i$, and $\sigma^2 = \frac{\sigma^2}{1-(\rho)^2} = \frac{(0.8)^2}{1-(0.1)^2} = 0.6465$, first Equation (6) is used to calculate the values of S_{pk_i} for 10 levels of the explanatory variable as 1.1344, 1.3008, 1.3566, 1.4401, 1.5517, 1.5547, 1.3964, 1.2015, 1.1738, and 1.4975, respectively. The values of P_i are then calculated using Equation (7) as 0.9993, 0.9999, 1, 1, 1, 1, 1, 0.9997, 0.9996, and 1. $P = \frac{\sum_{i=1}^{10} P_i}{10}$ is thus calculated to be 0.9998. Using Equation (8), the index $S_{pkA;AR(1)}$ is finally calculated to be 1.2579.

- **Calculating the true value of C'''_{ppM}:** By considering LSL_i, USL_i, T_i, $\mu_i = 3 + 2X_i$, and $\sigma^2 = (0.8)^2$, first the transformation method of Soleimani et al. [13] is applied. Therefore, we have $X'_i \in \{1.9, 2.8, 3.7, 4.6, 5.5, 6.4, 7.3, 8.2, 9, 1\}$, $LSL'_i \in \{2.492, 4.39, 6.386, 8.515, 10.33, 12.635, 14.874, 16.695, 17.668\}$, $USL'_i \in \{9.242, 11.14, 13.136, 15.265, 17.08, 19.385, 21.624, 23.445, 24.418\}$, $T'_i \in \{5.867, 7.765, 9.761, 11.89, 13.705, 16.01, 18.249, 20.07, 21.043\}$, and $\mu'_i = 2.7 + 2X'_i$. The values of C'''_{ppi} for 9 levels of X'_i are then determined using Equation (4) as 1.0640, 1.1396, 1.2817, 1.4061, 1.4064, 1.1587, 0.8347, 0.8208, and 1.3818, respectively. The $C'''_{ppM} = \frac{\sum_{i=1}^{9} C'''_{ppi}}{9}$ is finally calculated as 1.1660.

- **Calculating the true value of $C'''_p(Profile)$:** By considering LSL_i, USL_i, T_i, $\mu_Y(X) = 3 + 2X$, and $\sigma^2 = (0.8)^2$, as well as using the transformation method proposed by Soleimani et al. [13], we obtain X'_i, LSL'_i, USL'_i, T'_i, and $\mu_{Y'}(X') = 2.7 + 2X'$. Then, the regression lines are fitted to the values of LSL'_i, USL'_i, and T'_i, to obtain the functional transformed SLs and target line as $LSL_{Y'}(X') = -1.6486 + 2.1984X'$, $USL_{Y'}(X') = 5.1014 + 2.1984X'$, and $T_{Y'}(X') = 1.7264 + 2.1984X'$, where $X' \in [6.5, 20.9]$. Finally, to calculate $C'''_p(Profile)$, the location of $\mu_{Y'}(X')$ relative to $T_{Y'}(X')$ is determined based on the method proposed in Pakzad and Basiri [39]. As we can see in Figure A2, $\mu_{Y'}(X')$ and $T_{Y'}(X')$, intersect at $x'_m = 4.9064$ and $C'''_p(Profile)$ derived as 1.2238 based on Equation (16).

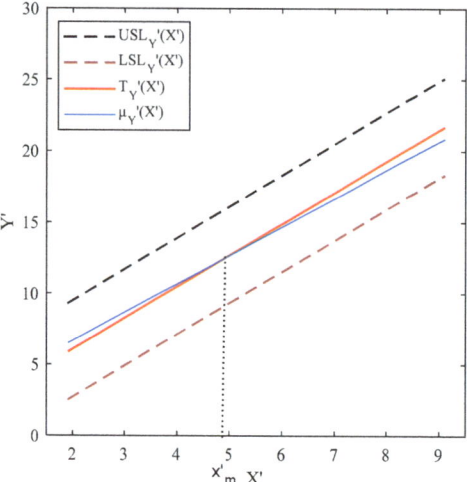

Figure A2. The location of $\mu_{Y'}(X')$ relative to $T_{Y'}(X')$ for SLP with WPAC.

Appendix C

Calculating the true value of $C_p'''(Profile)$ for SLP with BAC when $u_{ij} \sim N(0, (0.8)^2)$, $\rho = \varphi = 0.1$, $X_i \in \{2, 4, 6, 8\}$, $LSL_i \in \{2.5, 6.85, 11.25, 16.25\}$, $USL_i \in \{10, 14.35, 18.75, 23.75\}$, $T_i \in \{6.25, 10.60, 15, 20\}$.

- **Calculating the true value of $C_p'''(Profile)$**: By considering LSL_i, USL_i, T_i, $\mu_Y(X) = 3 + 2X$, and $\sigma^2 = (0.8)^2$, the regression lines are first fitted to the values of LSL_i, USL_i, and T_i, resulting in the functional SLs and target line as $LSL_Y(X) = -1.7820 + 2.0543X$, $USL_Y(X) = 4.2930 + 2.0543X$, and $T_Y(X) = 1.2555 + 2.0543X$, where $X \in [2, 8]$. Then the transformation by Soleimani et al. [13] is used. According to Equation (22), $LSL_{Y'}(X') = -1.6486 + 2.1984X'$, $USL_{Y'}(X') = 5.1014 + 2.1984X'$, and $T_{Y'}(X') = 1.7264 + 2.1984X'$, and $\mu_{Y'}(X') = 2.43 + 1.8X'$, where $X' \in [3.8, 7.4]$. Then, to calculate $C_p'''(Profile)$, the location of $\mu_{Y'}(X')$ relative to $T_{Y'}(X')$ is determined based on the method proposed in Pakzad and Basiri [39]. Figure A3 demonstrates that $\mu_{Y'}(X')$ and $T_{Y'}(X')$ intersect at $x'_m = 4.6195$ and $C_p'''(Profile)$ obtained as 1.1387 based on Equation (16).

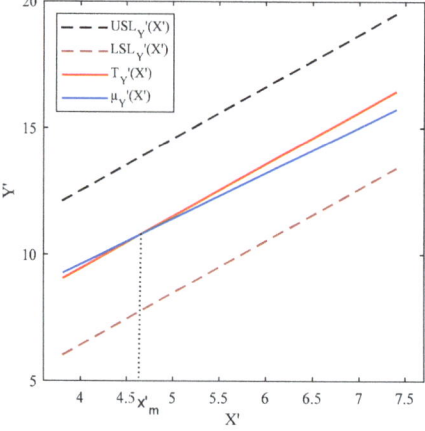

Figure A3. The location of $\mu_{Y'}(X')$ relative to $T_{Y'}(X')$ for SLP with BAC.

References

1. Gupta, S.; Montgomery, D.C.; Woodall, W.H. Performance evaluation of two methods for online monitoring of linear calibration profiles. *Int. J. Prod. Res.* **2006**, *44*, 1927–1942. [CrossRef]
2. Kang, L.; Albin, S.L. On-Line Monitoring When the Process Yields a Linear Profile. *J. Qual. Technol.* **2000**, *32*, 418–426. [CrossRef]
3. Kim, K.; Mahmoud, M.A.; Woodall, W.H. On the Monitoring of Linear Profiles. *J. Qual. Technol.* **2003**, *35*, 317–328. [CrossRef]
4. Li, Z.; Wang, Z. An exponentially weighted moving average scheme with variable sampling intervals for monitoring linear profiles. *Comput. Ind. Eng.* **2010**, *59*, 630–637. [CrossRef]
5. Xu, L.; Wang, S.; Peng, Y.; Morgan, J.P.; Reynolds, M.R.; Woodall, W.H. The Monitoring of Linear Profiles with a GLR Control Chart. *J. Qual. Technol.* **2012**, *44*, 348–362. [CrossRef]
6. Chen, S.; Yu, J.; Wang, S. Monitoring of complex profiles based on deep stacked denoising autoencoders. *Comput. Ind. Eng.* **2020**, *143*, 106402. [CrossRef]
7. Nassar, S.H.; Abdel-Salam, A.-S.G. Robust profile monitoring for phase II analysis via residuals. *Qual. Reliab. Eng. Int.* **2022**, *38*, 432–446. [CrossRef]
8. Maleki, M.R.; Amiri, A.; Castagliola, P. An overview on recent profile monitoring papers (2008–2018) based on conceptual classification scheme. *Comput. Ind. Eng.* **2018**, *126*, 705–728. [CrossRef]
9. Noorossana, R.; Amiri, A.; Soleimani, P. On the Monitoring of Autocorrelated Linear Profiles. *Commun. Stat. Theory Methods* **2008**, *37*, 425–442. [CrossRef]
10. Wang, Y.-H.T.; Huang, W.-H. Phase II monitoring and diagnosis of autocorrelated simple linear profiles. *Comput. Ind. Eng.* **2017**, *112*, 57–70. [CrossRef]
11. Khedmati, M.; Niaki, S.T.A. Phase II monitoring of general linear profiles in the presence of between-profile autocorrelation. *Qual. Reliab. Eng. Int.* **2016**, *32*, 443–452. [CrossRef]
12. Wang, Y.-H.T.; Lai, Y. Monitoring of autocorrelated general linear profiles. *J. Stat. Comput. Simul.* **2019**, *89*, 519–535. [CrossRef]
13. Soleimani, P.; Noorossana, R.; Amiri, A. Simple linear profiles monitoring in the presence of within profile autocorrelation. *Comput. Ind. Eng.* **2009**, *57*, 1015–1021. [CrossRef]
14. Jensen, W.A.; Birch, J.B.; Woodall, W.H. Monitoring Correlation Within Linear Profiles Using Mixed Models. *J. Qual. Technol.* **2008**, *40*, 167–183. [CrossRef]
15. Zhang, W.; Niu, Z.; He, Z.; He, S. Online monitoring of profiles via function-on-scalar model with an application to industrial busbar. *Qual. Reliab. Eng. Int.* **2022**, *38*, 3816–3828. [CrossRef]
16. Niaki, S.T.A.; Khedmati, M.; Soleymanian, M.E. Statistical Monitoring of Autocorrelated Simple Linear Profiles Based on Principal Components Analysis. *Commun. Stat. Theory Methods* **2015**, *44*, 4454–4475. [CrossRef]
17. Yeganeh, A.; Johannssen, A.; Chukhrova, N.; Abbasi, S.A.; Pourpanah, F. Employing machine learning techniques in monitoring autocorrelated profiles. *Neural Comput. Appl.* **2023**, *35*, 16321–16340. [CrossRef]
18. Nadi, A.A.; Yeganeh, A.; Shadman, A. Monitoring simple linear profiles in the presence of within-and between-profile autocorrelation. *Qual. Reliab. Eng. Int.* **2023**, *39*, 752–775. [CrossRef]
19. Alevizakos, V. Process Capability and Performance Indices for Discrete Data. *Mathematics* **2023**, *11*, 3457. [CrossRef]
20. Perakis, M.; Xekalaki, E. On the relationship between process capability indices and the proportion of conformance. *Qual. Technol. Quant. Manag.* **2016**, *13*, 207–220. [CrossRef]
21. Afshari, R.; Nadi, A.A.; Johannssen, A.; Chukhrova, N.; Tran, K.P. The effects of measurement errors on estimating and assessing the multivariate process capability with imprecise characteristic. *Comput. Ind. Eng.* **2022**, *172*, 108563. [CrossRef]
22. Wang, S.; Chiang, J.-Y.; Tsai, T.-R.; Qin, Y. Robust process capability indices and statistical inference based on model selection. *Comput. Ind. Eng.* **2021**, *156*. [CrossRef]
23. Alevizakos, V.; Koukouvinos, C. Evaluation of process capability in gamma regression profiles. *Commun. Stat. Simul. Comput.* **2022**, *51*, 5174–5189. [CrossRef]
24. Guevara G, R.D.; Alejandra Lopez, T. Process capability vector for multivariate nonlinear profiles. *J. Stat. Comput. Simul.* **2022**, *92*, 1292–1321. [CrossRef]
25. Wang, F.K.; Guo, Y.C. Measuring process yield for nonlinear profiles. *Qual. Reliab. Eng. Int.* **2014**, *30*, 1333–1339. [CrossRef]
26. Bera, K.; Anis, M.Z. Process incapability index for autocorrelated data in the presence of measurement errors. *Commun. Stat. Theory Methods* **2023**, *53*, 5439–5459. [CrossRef]
27. Cohen, A.; Amin, R. The effects of normal mixtures and autocorrelation on the fraction non-conforming. *Commun. Stat. Simul. Comput.* **2017**, *46*, 8105–8117. [CrossRef]
28. Lundkvist, P.; Vannman, K.; Kulahci, M. A comparison of decision method for Cpk when data are autocorrelated. *Qual. Control. Appl. Stat.* **2013**, *58*, 451–452.
29. Hosseinifard, S.Z.; Abbasi, B. Evaluation of process capability indices of linear profiles. *Int. J. Qual. Reliab. Manag.* **2012**, *29*, 162–176. [CrossRef]
30. Hosseinifard, S.Z.; Abbasi, B. Process Capability Analysis in Non Normal Linear Regression Profiles. *Commun. Stat. Simul. Comput.* **2012**, *41*, 1761–1784. [CrossRef]
31. Ebadi, M.; Shahriari, H. A process capability index for simple linear profile. *Int. J. Adv. Manuf. Technol.* **2013**, *64*, 857–865. [CrossRef]
32. Wang, F.-K. A Process Yield for Simple Linear Profiles. *Qual. Eng.* **2014**, *26*, 311–318. [CrossRef]

33. Wang, F. Measuring the Process Yield for Simple Linear Profiles with one-Sided Specification. *Qual. Reliab. Eng. Int.* **2014**, *30*, 1145–1151. [CrossRef]
34. Wang, F.-K.; Tamirat, Y. Process yield analysis for autocorrelation between linear profiles. *Comput. Ind. Eng.* **2014**, *71*, 50–56. [CrossRef]
35. Wang, F.; Tamirat, Y. Process Yield Analysis for Linear Within-Profile Autocorrelation. *Qual. Reliab. Eng. Int.* **2015**, *31*, 1053–1061. [CrossRef]
36. Mehri, S.; Ahmadi, M.M.; Shahriari, H.; Aghaie, A. Robust process capability indices for multiple linear profiles. *Qual. Reliab. Eng. Int.* **2021**, *37*, 3568–3579. [CrossRef]
37. Ganji, Z.A.; Gildeh, B.S. A new process capability index for simple linear profile. *Commun. Stat. Theory Methods* **2023**, *52*, 3879–3894. [CrossRef]
38. Pakzad, A.; Razavi, H.; Gildeh, B.S. Developing loss-based functional process capability indices for simple linear profile. *J. Stat. Comput. Simul.* **2022**, *92*, 115–144. [CrossRef]
39. Pakzad, A.; Basiri, E. A new incapability index for simple linear profile with asymmetric tolerances. *Qual. Eng.* **2023**, *35*, 324–340. [CrossRef]
40. Ganji, Z.A.; Gildeh, B.S. A class of process capability indices for asymmetric tolerances. *Qual. Eng.* **2016**, *28*, 441–454. [CrossRef]
41. Boyles, R.A. Brocess capability with asymmetric tolerances. *Commun. Stat. Simul. Comput.* **1994**, *23*, 615–635. [CrossRef]
42. Bothe, D.R. *Measuring Process Capability: Techniques and Calculations for Quality and Manufacturing Engineers*; McGraw-Hill Companies: New York, NY, USA, 1997.
43. Keshteli, R.N.; Kazemzadeh, R.B.; Amiri, A.; Noorossana, R. Developing functional process capability indices for simple linear profile. *Sci. Iran.* **2014**, *21*, 1096–1104.
44. Efron, B. *The Jackknife, the Bootstrap and other Resampling Plans*; SIAM: Philadelphia, PA, USA, 1982.
45. Efron, B.; Tibshirani, R. Bootstrap Methods for Standard Errors, Confidence Intervals, and Other Measures of Statistical Accuracy. *Stat. Sci.* **1986**, *1*, 54–75. [CrossRef]
46. Abbas, T.; Mahmood, T.; Riaz, M.; Abid, M. Improved linear profiling methods under classical and Bayesian setups: An application to chemical gas sensors. *Chemom. Intell. Lab. Syst.* **2020**, *196*, 103908. [CrossRef]
47. Mahmood, T.; Riaz, M.; Omar, M.H.; Xie, M. Alternative methods for the simultaneous monitoring of simple linear profile parameters. *Int. J. Adv. Manuf. Technol.* **2018**, *97*, 2851–2871. [CrossRef]

Disclaimer/Publisher's Note: The statements, opinions and data contained in all publications are solely those of the individual author(s) and contributor(s) and not of MDPI and/or the editor(s). MDPI and/or the editor(s) disclaim responsibility for any injury to people or property resulting from any ideas, methods, instructions or products referred to in the content.

MDPI AG
Grosspeteranlage 5
4052 Basel
Switzerland
Tel.: +41 61 683 77 34

Mathematics Editorial Office
E-mail: mathematics@mdpi.com
www.mdpi.com/journal/mathematics

Disclaimer/Publisher's Note: The title and front matter of this reprint are at the discretion of the Guest Editors. The publisher is not responsible for their content or any associated concerns. The statements, opinions and data contained in all individual articles are solely those of the individual Editors and contributors and not of MDPI. MDPI disclaims responsibility for any injury to people or property resulting from any ideas, methods, instructions or products referred to in the content.